All About Energy

Printed in Mexico

ISBN 978-0-15-362218-2
ISBN 0-15-362218-0

1 2 3 4 5 6 7 8 9 10   050   16 15 14 13 12 11 10 09 08 07

Visit *The Learning Site!*
www.harcourtschool.com

Energy to Move

Cars move. People move. Think of all the ways you can move. You can move your arm. You can move your foot or your leg. You can move your whole body. You can move other people. Have you ever picked up a baby or rocked one to sleep? You move things, too. You pick up books. You throw balls. All this movement has one thing in common. It all takes energy! **Energy** is the ability to make something move or change.

This car moves very fast. It burns gasoline to get the energy to move.

Eating the right foods gives your body the energy to play soccer.

Cars get energy by burning gasoline. How do you get energy? You eat food. You eat breakfast before you leave for school. Your body changes the food you eat into energy. Your brain uses energy to pay attention and learn new things in math, science, and reading. Your muscles need energy so you can run and play and not get too tired. Without food, your body would not be able to produce energy for you to think, work, and play.

 MAIN IDEA AND DETAILS What is energy?

Different Kinds of Energy

People get their energy to move and grow from food. What about plants? Plants need energy, too. It takes energy for a seed to sprout. It takes energy for a tree or a carrot to grow. Plants make their own food by using the energy from sunlight.

Sunlight is a kind of energy called light energy. The main source of light energy on Earth is light from the sun. The sun also supplies heat. Most of the energy on Earth comes from the sun's heat and light.

These flowers need the sun to grow.

A siren on a fire truck is sound energy traveling away from it.

"R-r-r-i-i-i-n-n-g!" goes the telephone. "Honk" goes the car. You hear sounds all the time. You make sounds, too. But did you know that sound is energy? You have learned that energy is the ability to make something move or change. Sound energy is made when matter vibrates, or moves back and forth very fast. Sound energy travels away from the vibrating matter. When you hear something, it is because sound energy has reached your ears.

Light, heat, and sound are all different kinds of energy.

 COMPARE AND CONTRAST How are heat and light energy the same as and different from sound energy?

Kinetic and Potential Energy

Energy can be grouped in two ways. **Kinetic energy** is the energy of motion. A rock rolling down a hill has kinetic energy. Without kinetic energy, that rock would stay at the top of the hill. It would not roll or move. Anything that is moving has kinetic energy. When you zoom down the street on a skateboard, you have kinetic energy. When a bird swoops in the sky, it has kinetic energy.

Think back to that rock at the top of the hill. Even before it started to move, it had energy. It had potential energy. **Potential energy** is energy of position. You can also think of potential energy as the energy that is stored in something. There is energy present, but it is not being used. Nothing is moving. Nothing is changing.

This turtle may be slow, but when it moves, it has kinetic energy.

This rock has potential energy now. If it starts to roll down the hill, the potential energy will change to kinetic energy.

A rock has potential energy and kinetic energy, but it does not have them at the same time. The rock at the top of the hill is not moving. Its energy is potential. The rock rolling down the hill is moving. Its energy is kinetic. When the rock began rolling, its potential energy became kinetic energy. When it stopped, that kinetic energy became potential energy.

 COMPARE AND CONTRAST When does an object have kinetic energy, and when does it have potential energy?

How We Use Energy

We could not live without energy. Plants need the energy from sunlight to grow. We need energy from food, and we get it from eating plants and animals. Food gives our muscles the energy they need to make our bodies move.

Machines use energy to move, too. Many machines get energy from combustion. **Combustion** is another word for burning. When we burn fuels, such as wood, coal, gas, and oil, heat energy is given off. Engines in cars, buses, trains, airplanes, and boats burn gasoline. The burning gasoline gives off energy. The energy makes the cars and other vehicles move.

Burning gasoline powers a car.

Stoves can be powered by electricity or by gas. Which kind do you have at your home?

Heat released by combustion boils the water and cooks the eggs.

There are many other uses for the heat that is released by combustion. Thermal energy is used to cook food and boil water. It warms the air in our homes and other buildings during cold weather. Have you ever sat in front of a fireplace to get warm? If so, you have felt how burning wood gives off heat.

Heat is also produced when electricity is used. That is why lightbulbs become hot. Hair dryers use electricity to produce air warm enough to dry your hair.

 MAIN IDEA AND DETAILS Why do you get warm when you stand in front of a fireplace?

We Need Energy!

You use energy in many ways. It takes energy for you to grow bigger and taller. It takes energy to walk and run and play. It even takes energy to sleep! When you are asleep, you are still breathing. Your heart is still beating. You use energy every moment of the day.

In a way, communities, towns, and cities are like people. Communities need energy every moment of the day. Every minute, even in the middle of the night, someone in the community is using energy. Someone is cooking; that takes energy. Someone has turned a light on; that takes energy. Others are driving their cars or using their computers or watching TV. All of these things take energy.

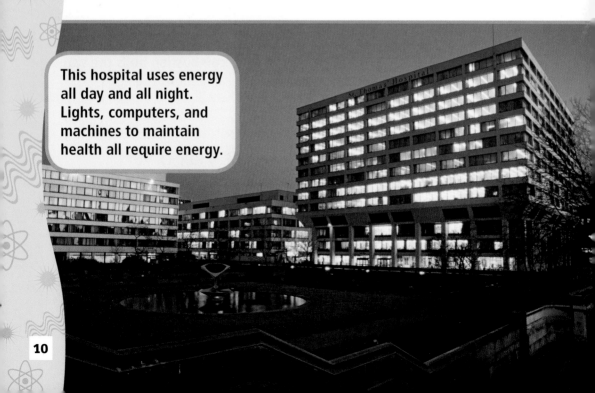

This hospital uses energy all day and all night. Lights, computers, and machines to maintain health all require energy.

Like other fossil fuels, oil is found under the surface of Earth.

Coal, oil, and natural gas are energy resources. We use coal, oil, and natural gas to release energy. A **resource** is something in nature that people use.

Coal, oil, and gas are called fossil fuels. **Fossil fuels** come from the remains of plants and animals that lived long ago. It took many years for fossil fuels to form.

 MAIN IDEA AND DETAILS What do people and communities have in common?

Renewable or Nonrenewable?

A resource that we cannot replace during a human lifetime is called a **nonrenewable resource.** Fossil fuels are examples of nonrenewable resources. Every time we use fossil fuels, the supply gets smaller.

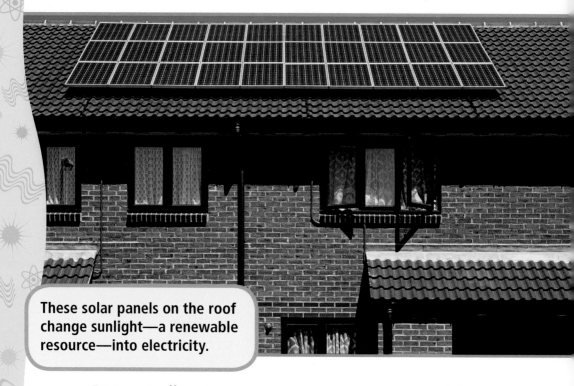

These solar panels on the roof change sunlight—a renewable resource—into electricity.

But not all energy sources are nonrenewable. A **renewable resource** is one that can be replaced. Energy from the wind is a renewable resource. Windmills can use wind to make electricity. Energy from the sun is also a renewable resource. The sun's energy can be used to make electricity. This electricity can be used to heat and cool buildings, to run computers, and to give us light.

Windmills use wind to make energy.

Think of all the ways you use energy. Think of all the ways your community uses energy. Think of how your community is growing. Each new building needs energy. Each new car needs energy. Each new person needs energy for living and working. Our nonrenewable resources could disappear. Scientists are always looking for more renewable resources.

 MAIN IDEA AND DETAILS Name some energy resources that are renewable. Name some that are nonrenewable.

Save Our Energy

Since we do not want our resources to run out, we need to try to save them. The easiest way to save energy is not to waste it. After you finish watching TV, turn it off. If you are the last person to leave a room, turn off the light. You can walk up the stairs instead of taking the elevator.

It takes energy to make all the things you use every day. So, if you reuse something, you save energy.

 MAIN IDEA AND DETAILS What is one way a person can help save energy?

No one is watching this TV. Energy is being wasted.

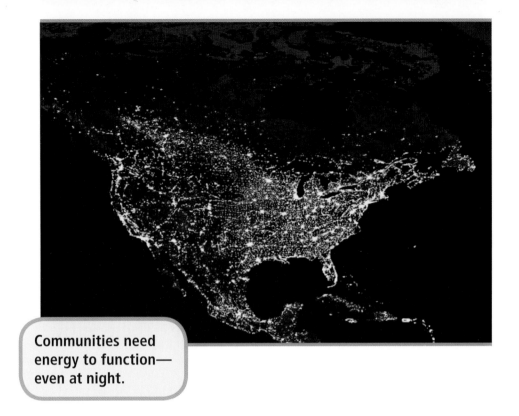

Communities need energy to function—even at night.

Summary

Energy is the ability to make something move or change. When something moves, it has kinetic energy. When something is not moving, it has potential energy. Potential energy is energy of position. People need energy to live. Communities need energy to function. Some energy resources, such as coal, oil, and gas, are fossil fuels. Fossil fuels are nonrenewable, which means that we cannot replace them. Other energy sources, such as wind and sunlight, are renewable. We can save energy by using less of it.

Glossary

combustion (kuhm•BUS•chuhn) Another word for burning (8, 9)

energy (EN•er•jee) The ability to make something move or change (2, 3, 4, 5, 6, 7, 8, 10, 11, 12, 13, 14, 15)

fossil fuels (FAHS•uhl FYOO•uhlz) Resources that come from the remains of plants and animals that lived long ago (11, 12, 15)

kinetic energy (kih•NET•ik EN•er•jee) The energy of motion (6, 7, 15)

nonrenewable resource (nahn•rih•NOO•uh•buhl REE•sawrs) A resource that cannot be replaced during a human lifetime (12, 13, 15)

potential energy (poh•TEN•shuhl EN•er•jee) Energy due to position (6, 7, 15)

renewable resource (rih•NOO•uh•buhl REE•sawrs) A resource that can be replaced quickly (12, 13, 15)

resource (REE•sawrs) A material that is found in nature and that is used by living things (11, 12, 13, 14, 15)

Credit Risk Management and Basel II

Credit Risk Management and Basel II

An Implementation Guide

By Mohan Bhatia

Published by Risk Books, a Division of Incisive Financial Publishing Ltd

Haymarket House
28–29 Haymarket
London SW1Y 4RX
Tel: +44 (0)20 7484 9700
Fax: +44 (0)20 7484 9800
E-mail: books@incisivemedia.com
Sites: www.riskbooks.com
       www.incisivemedia.com

ISBN 1 904339 43 3

British Library Cataloguing in Publication Data
A catalogue record for this book is available from the British Library

Publisher: Laurie Donaldson
Assistant Editor: Steve Fairman
Designer: Rebecca Bramwell

Typeset by Mizpah Publishing Services Private Limited, Chennai, India

Printed and bound in Spain by Espacegrafic, Pamplona, Navarra

Contents

About the Author

Mohan Bhatia has been working in the risk management area for the past decade. Allied to this he has worked in the areas of credit risk and operational risk with international financial institutions across the globe. Mohan has provided consultancy in constructing credit risk solutions, designing policies and processes for credit and operational risks and building and validating models for credit risk.

At i-flex consulting, Mohan occupies the role of senior principal consultant and is spearheading the implementation team for Reveleus analytics for credit and operational risks. He has also been involved with business strategy, development of capabilities and offerings in the risk management arena, various HR and KM initiatives and client management.

Mohan has a post graduate degree in software technology and has designed applications on service oriented architecture. In the process, he developed expertise in the web services security, and technology risk management. He has earned a CISM from ISACA.

1

Introduction to Credit Risk Management

There has been a substantial surge in interest in credit risk management in the past decade or so. This has primarily been because credit risk management is undergoing a transition. The quest is for methods of credit risk measurement to make the transition from qualitative-based measurements to quantitative-based measurements; to develop abilities to participate in the credit risk transfer mechanisms being developed by the financial services industry; to improve methods to correctly price credit risk for facing competition; to address the challenge of unbundling of financing, servicing and risk taking that happens in the financial services industry. In addition, supervisors are catalysts in the development of greater sensitivities towards credit risk by linking regulatory capital to the credit risk quantum. An additional factor is – transition to quantitative risk measurement will not be easier due to the paucity of relevant data.

While credit risk has been in existence since financial institutions came into being, it was traditionally managed through collaterals, covenants and selections of obligors. As we will see in the credit scoring chapter, only since the 1960s has credit risk been measured quantitatively. The financial industry was exposed to market risk in 1973, when the US dollar started floating. The financial-services industry quickly learned to manage market risk, and risk measurement was in place by the mid-1980s in most of the Tier I and Tier II banks, however banks are only now starting to implement credit risk management systems. The probable reason is that, unlike market risks, credit risk is difficult to measure and manage. This book aims to help financial institutions to make the transition to the next level of credit risk management.

COMPARISON WITH MARKET RISK MEASURES

Let us first consider, what is risk? Risk is a state in the future, which may not be to our liking. And what don't we like in the financial world? We don't like or want "loss". In a nutshell, risk measurement is a process of identifying and measuring the probability and the quantum of future losses. We know, however, that it is almost impossible to predict the future and future losses (for that matter, even future profits cannot be measured).

Experts have developed indirect methods of measuring future losses by making certain assumptions. The most important assumption is that the past will continue into the future and past performance is a good indicator of future performance. If the past is a good indicator of future, a similar distribution pattern would continue. Therefore, for measuring future distributions, the distribution pattern observed in the past can be a guide. However, the past itself is represented by various states and large numbers of data. Hence, all the statistical techniques required to understand the large dataset are used. Techniques on measuring central tendencies, distribution patterns, testing of hypotheses, classifications, best fit, statistical-significance tests, building a series of techniques and so forth are used to understand and predict the data, data patterns and distribution. The quantum of data is the key to effective use of statistical techniques.

Variance, or standard deviation or some form of standard-deviation measure, is used to measure risk or loss. In order to compute risk, risk factors are identified. Risk factors are the relevant attributes that can help in loss prediction (eg, FX price, interest-rate curve, obligor financial ratios, equity prices, credit spread). Risk factors work as the dataset on which various statistical techniques are applied.

In market risk, there is only one variable and that is price. The price variable has good frequency with observable data. So, a frequency distribution can be plotted and patterns observed, and the variance and other central tendencies estimated. The profit and loss of the seller are a mirror image of loss and profit of the pur-

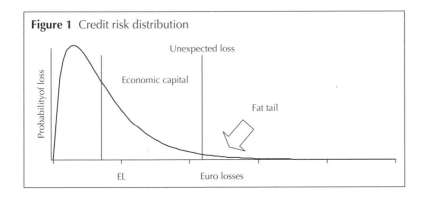

Figure 1 Credit risk distribution

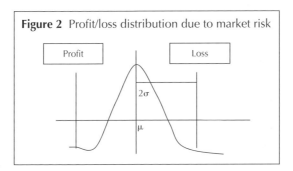

Figure 2 Profit/loss distribution due to market risk

chaser, ie, the prices are normally distributed. If the market is highly liquid, every exposure/position (whether purchased from bank A or bank B) has exactly the same market risk (risk of loss due to changes in market prices). Furthermore each dollar within an exposure of say US$1 million has the same market risk. Normally distribution patterns are parametric – the central-tendency parameters are sufficient to represent the data. Correlation is ignored in basic models.

Now compare this with credit risks – the first problem is the data. For most banks, there are hundreds of corporate customers, thousands of SME customers and millions of retail customers. Taking a 1% default rate per annum, one corporate customer, 10 SME customers and a thousand retail customers are likely to fail in a year. These customers may not fail suddenly but credit quality typically deteriorates over a period of time. These credit-quality changes are more difficult to measure. As against this, the FX price is available every four seconds. For a 10-hour trading window, around 9,200 prices/data points are available daily or around 2 million data points every year.

In the FX market, the distribution is symmetrical (see Figure 2). There is equal chance for both profit and loss. For credit risk, there is no probability of upside or profit. There is only downside or a chance of loss. At the most, the investment in the defaultable assets is repaid. This makes credit distribution skewed (see Figure 1) with no profit potential – a long or fat loss tail around expected loss. Loss beyond expected loss is called *unexpected loss*. Loss at a certain confidence level is cushioned with their own capital. This is called *economic capital*.

Concentration risk: For credit risk, each dollar within the same portfolio, sub-portfolio or exposure is not exposed to the same risk. The concentration risk has a major impact on the credit risk measure.

Correlation: Correlation plays a major role in credit risk as compared with market risk. Market risk (volatility in the price) is of the order of 20% per annum while credit risk (volatility in the credit loss) is of the order of 1%. An addition of 1% due to correlation doubles the credit risk but increases the market risk to 21%. We have further explained correlation impact in the chapter "Credit Risk Portfolio Models".

CREDIT RISK MANAGEMENT IS NOT THE SAME AS CREDIT ADMINISTRATION

Banks have been dealing with credit risk for sometime now and have also been using quantitative credit-analysis models for decades. The question is how their existing processes and models differ from credit risk management best practices. Broadly, here are the key differences.

❑ Banks have been using quantitative tools mainly for selecting obligors fit for lending.
❑ Before the publication and implementation of the Basel Market Risk Amendment 1996, most banks did not have credit risk measurement in place for most of their products. Quantitative models were developed by banks to measure and control specific counterparty risk after implementation of the Amendment.

❑ Credit analysis for all products was focused only on ordinal measurement of the credit risk. The whole emphasis was on identifying whether or not credit proposals were good for lending, differentiating between the accepted proposals (cardinal measure of risk). External credit-rating agencies were also involved in assessment, but their methodology till the late 1980s was ordinal and not cardinal. The ordinal measure aims to sequence the credit proposal in the order of credit risk and a certain cut-off point is applied to reject the unfit or high-risk borrower, while in cardinal measures risk is measured for each proposal.

❑ In the absence of quantitative cardinal measures of risk, banks were not able to price their assets correctly.

❑ The entire focus of credit processes or credit admin in a bank was to:

 ❑ control or limit the exposure;
 ❑ ensure and enforce adherence to covenants; and
 ❑ secure, perfect and enforce collaterals.

For credit risk management, in addition to the credit admin processes, the following credit risk measurement processes are implemented by banks to improve their decision making.

❑ Different asset classes are exposed to different amounts of credit risk primarily due to different borrower and risk characteristics, the size of exposure and different systemic factors.

❑ The accepted credit is differentiated and priced.

❑ The portfolio is managed actively. Assets acquisition, financing and servicing activities are decoupled and managed as separate activities.

❑ Risk mitigation is quantified and managed through risk transfers and credit insurance, in addition to the traditional collateral and guarantees.

❑ Credit risk is granularised and measured through risk components.

❑ Correlation is managed beyond exposure limits.

❑ Regulatory capital is sensitised to the risk quantum.

UNBUNDLING CREDIT LOSS

Credit risk is the credit loss emanating from a borrower or counterparty failing to meet their obligation in accordance with the agreed terms, or a diminution in the value of the assets due to a diminution in the credit quality of the borrower, counterparty, guarantor or the assets (collateral) supporting the credit exposure. Therefore, credit *risk* is the possibility of credit *loss*.

We have seen in previous paragraphs that the available data points are few and far between, and are not suitable for statistical treatment. If we can divide the loss into various components, we will be able to multiply the data points and may be able to generate sufficient data points for enabling us to measure credit risk.

COMPONENTS OF CREDIT LOSS

Credit loss occurs for several reasons.

❏ Obligor default – an obligor defaults only if they are unable or unwilling to pay in full. On default, a portion of exposure is lost. Therefore, the default rate is an indicator or component of loss.

❏ The entire exposure may not be lost because collateral or recovery or other risk-mitigation mechanisms may be available. After default, attempts are made to recover the exposure or sell off the exposure to asset-reconstruction companies. Defaulted exposures fetch discounted prices. Expenses are also incurred for recovery. This is called loss given default.

❏ Empirical studies have proved that before default, the obligor generally draws the unutilised credit to the extent possible, resulting in an increase in the exposure at the time of default. In addition, for certain types of assets like derivative the replacement value of the assets includes the potential exposures in addition to current exposures. Both these factors result in an increase in the exposure of the assets which default.

❏ Maturity of exposure also contributes to risk. Longer maturity means exposure to the credit risk for a longer time – higher average credit risk. The duration concept applicable to fixed income securities is also applicable. The changes in credit quality change the credit risk premium portion of the interest. Longer duration securities lose more value for the same change in credit quality (and interest rates).

❏ Correlation is the most important contributor to credit risk. Obligors are correlated with each other as supplier, customer and competitors. Economic, business and industry cycles impact obligors simultaneously. This results in an increase in credit losses.

However, no causal relationship can be established. Correlation is extremely difficult to measure, correlation also changes continuously.

❏ Concentrations in a bank's credit portfolio are key drivers of correlation, unexpected loss, portfolio risk and credit risk capital. Risk concentrations are caused by concentrations of exposure to individual names as well as to a single sector (geographic region or industry) or to several highly correlated sectors. Single-name risk concentrations are relatively straightforward to measure and manage. Measuring and managing sector concentrations is difficult and requires quantitative techniques to identify sector concentration. Empirically, it is observed that obligors within the same sector are exposed similarly to the economic and business cycle, resulting in correlation.

Risk components

On unbundling credit loss, Basel has identified a number of credit risk components.

❏ *Exposure* is the amount lent to the borrower and is the simple and most direct measure of credit risk. Special measurement techniques are needed to measure non-funded and derivative instruments.

❏ *Default* is a discrete state of the borrower or counterparty. This state occurs at the frequency of *probability of default* (PD).

❏ *Exposure at default* is the economic value of the claim at the time of default.

❏ *Loss given default* (LGD) represents the fractional loss due to inability to recover the claim. This is the fraction of exposure which is not recovered.

❏ *Maturity* is the maturity period of the exposure.

MEASURING CREDIT LOSSES

Like all other risks, credit risk is measured by measuring losses. However, unlike other losses, credit losses have their own complexities and peculiarities. Some of them are listed below.

Definition of default: This defines at what point in time default is recognised. Although Basel provides a standard definition, the

Table 1 Two-state vs multi-state

Multi-state default models	Two-state default models
Changes in the credit quality are measured	Only defaults during the holding period are measured
The short-term horizon is appropriate for mark-to-market models	The longer-term horizon is appropriate for default-mode models
Distance to default, credit spreads and changes in the transition matrix measure the changes in the credit quality	Credit ratings and historical default rates measure the default rates

definition varies from bank to bank and country to country. Definition also differs among the banking, accounting and capital-market regulators. We will deal with default definition in detail in the chapter "Probability of Default".

Default mode and mark-to-market (MTM) mode: Default mode recognises the loss as a sudden event when the default occurs. Under the MTM regime (see Table 1), changes in the values/credit grades are recognised on a mark-to-market basis. Traditionally, for loans, an MTM paradigm is difficult. However, the development of risk-transfer techniques such as securitisation and credit derivatives are now enabling MTM even for loans.

Banks have two types of system: a multi-state system, which has many grades in non-default mode and a two-state system, which has only one non-default grade.

The choice of number of non-default grades depends upon the types of product, business and risk management policies of the bank and granularity of the default-rate measurement. With the implementation of the Basel Accord, strict and transparent loan-loss provisioning, linking of loss provisions with the capital definition, economic value of the assets, implementation of fair-value provisioning and so forth, banks are moving towards the mark-to-market mode.

Default mode (DM): This is also called two-state (default and non-default) methods (see Table 1) to measure only the default. The MTM method measures not only the probability of default but also the change in the default probability or change in rating grade. This is a multi-state or mark-to-market model. In MTM, a credit loss can arise in response to deterioration in an asset's credit quality, short of default.

The mark-to-market models are similar to market-risk models, since they attempt to estimate losses because of changes in the asset value. The basic difference between two-state and multi-state models is the method for evaluating the losses to the lender. The two-state models consider only the losses that result from defaulted loans; the multi-state models, in addition, also consider deterioration in the creditworthiness of the borrower. The deterioration in the creditworthiness causes the decrease in the value of the loan. Thus, the multi-state loan model is a pseudo-mark-to-market procedure.

Diminution in the credit quality

PD is the likelihood of a borrower's defaulting within a certain period (one year) in the future. For most of the obligors, the state does not change to default in one go, but there are many intermediate stages. PD is measured in terms of credit ratings and there are many grades for these ratings. Diminution in the PD or credit quality takes place when an obligor's credit rating is downgraded. Rating grades can be categorised broadly into three types:

❑ *Investment grades*: These are higher or better grades with a very low default probability. The obligor can move lower within the investment grades or into speculative grades or default grades. Generally, the obligor will not move directly to the default grade from this grade.
❑ *Speculative grades*: These are the non-default grades with high-default probabilities. The obligor can move higher to investment grades, lower within the speculative grades or to default grade.
❑ *Default grade*: On the occurrence of a default event, the obligor or the exposure is assigned to default grade. This is an absorbing grade. Once assigned to default grade, the obligor or exposure cannot move up to speculative or investment grade. Generally, speculative grades will move to default grade.

Downgrade or migration in the credit quality: Risk that changes in the possibility of a future default by an obligor will adversely affect the present value of the contract with the obligor today. This is reflected through change in the credit spreads. A variety of ratings agencies provide these assessments to the public, giving the investor a perceived level of confidence in the issuer's ability to make good on the repayment schedules to which it is committed.

Credit losses

Expected loss (EL) represents the anticipated average loss rate that a bank should expect over time from a sub-portfolio of defaultable assets with common risk characteristics. This anticipated average loss rate should be reflected directly in loan pricing. EL is not the risk but the cost of providing for credit, which must be recovered as a part of the pricing.

$$\text{EL (Expected Loss)} = \text{Exposure} \times \text{LGD} \times \text{Probability of default}$$

Unexpected loss (UL) represents the volatility of actual loss rates that occur around EL. It is the presence of UL that creates the requirement for a capital "cushion" to ensure the viability of the bank during the year when losses are unexpectedly high. Credit risk is not the risk of expected losses from the credit but the risk of unexpected losses from credit. The entire focus in credit risk is to predict and manage unexpected loss. We will explained further unexpected loss in the chapter "Credit Risk Portfolio Management".

Types of credit risk

Credit risk involves nonpayment of an obligation. In the previous paragraphs, we have seen definitions of credit risk on loans and investments. Other types of transaction are also exposed to credit risk. For example, settlement risk arises from the lag between the value date and settlement date of securities transactions or from the exchange of principals in different currencies during a short window of say one day.

Presettlement risk is the risk of loss due to non-performance of the obligation during the life of the transaction. Transactions include loan, bond and derivative contracts. Presettlement risk exists over a long period – running into years. The risk starts from the date of the contract to its settlement. Due to changes in financial variables, the value of the contract keeps changing during the tenure of the transaction. Whenever, a contract is in-the-money (the sum of current and potential exposure is positive), the bank is exposed to pre-settlement risk.

Credit loss occurs not only due to the inability of the borrower/counterparty to perform its obligation, but also due to restrictions imposed by the country on foreign-exchange remittance or repayment on foreign-denominated assets to a foreign lender. The borrower may be able and willing to pay in local currency, but due to government restrictions, the debt may remain unpaid. This is called *country risk* or *sovereign risk*. We explain sovereign risk in detail in the chapter on "Credit Rating".

Counterparty risk: A counterparty is a borrower with whom a bank generally deals in traded assets and who is sanctioned to use a credit line where exposure keeps fluctuating. Counterparty risk is a risk of exposure to settlement and pre-settlement risk.

DEVELOPMENTS IN MEASUREMENT OF CREDIT RISK

Credit risk for financial institutions has substantially increased in the past three decades or so due to an increase in the volatility in the underlying physical economy, integration of financial markets, convergence of financial products and institutions and increased competition among financial institutions. To manage the increased risks, institutions and markets are developing instruments and mechanisms to manage and measure risks.

Figure 3 shows how regulators have gradually increased the focus and granularity of credit risk measurement.

To ensure the safety and soundness of the banking system over the last two decades, regulators are attempted to introduce the credit risk measurement process into banks through regulatory requirements. The Basel Accord II is an attempt to fine-tune this process by empowering banks and providing handholding and baseline for the credit risk management process so that banks can move to the next level of risk management.

Regulators are interested in having credit risk management techniques adopted by banks not only for regulatory capital computation, but also to enable regulated institutions to face competition appropriately and reduce systemic risks and bank failures. At individual institutions, certain levels of refinement in risk management

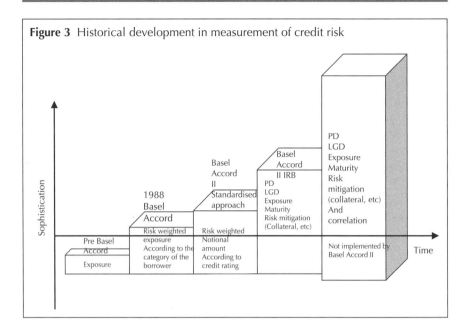

Figure 3 Historical development in measurement of credit risk

and measurement techniques will help them handle competition and measure performance better.

With the adoption of the Basel II Accord, due to regulatory requirements and increasing competition, banks are under increasing pressure to improve their credit risk management systems. Under the new process, banks are required to re-engineer and upgrade their credit risk management policies, processes, measurement techniques, systems and organisational structures.

Establishing credit risk management in a bank is a long, drawn-out process and requires years of developing best practices, data collection, and developing, testing and calibrating models, establishing the required organisational structures and developing products and implementing systems.

The changing role of banks – transferring credit risk

Banks have traditionally been selling (see Figure 4) non-tradable and tradable defaultable assets. The decision making on the selling of loans and investment is done at origination. There may be many origination desks for products and groups of products – eg, separate retail, SME, corporate and Treasury origination desks.

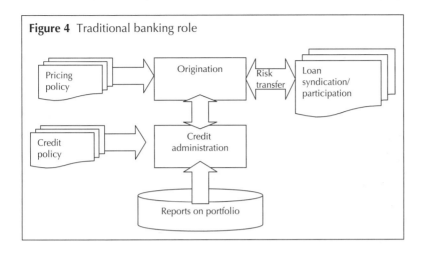

Figure 4 Traditional banking role

Origination desks are empowered to price assets according to pricing policy. Each of the origination desks is assisted by a credit administration (CA) department. The job of credit administration is to assist origination in the underwriting of credit, approval and management of obligor and group limits, enforcement and management of covenants and receipt and management of collaterals.

It also provides support for implementing various portfolio-concentration limits. CA may be supporting a product or group of products. In the absence of well-developed systems for credit underwriting, limits and collateral, inter- and intra-department coordination among origination and CA departments is a challenge. Traditionally, risk is managed through covenants, collateral and loan syndication. Securitisation, credit-derivative and credit risk management are not integrated by front-office-driven decisions.

Under the credit risk management regime (see Figure 5), the entire credit management is reorganised; underwriting gains independence from origination with better control of credit risk management, and risk-transfer mechanisms are managed by the CRM.

For better management and decision making (see Figure 6), funding is separated from lending through Fund Transfer Pricing (FTP), Credit Assets Liability Committee (ALCO) and Risk-Adjusted Return On Capital (RAROC). CRM now need to work closely with the Treasury and finance departments. All this helps

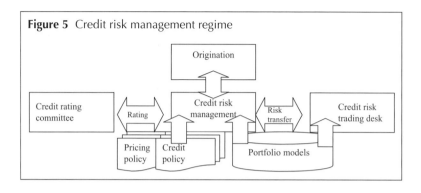

Figure 5 Credit risk management regime

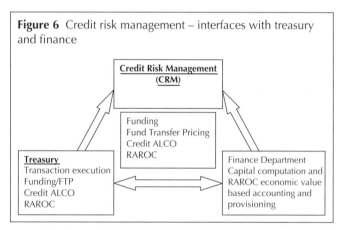

Figure 6 Credit risk management – interfaces with treasury and finance

to remove the conflict of interest and provides mechanisms for active credit-portfolio management.

We will explain credit risk management in the chapter "Credit Risk Fortification". Figure 6 shows interfaces of credit risk management with treasury and finance department.

UNBUNDLING CREDIT RISK

Credit losses are the result of inability and unwillingness to repay credit (see Table 2). Every economic entity that includes individuals and organisations works in macroeconomic and business surroundings. The ability to repay is dependent on the ability to generate cashflow. The ability to generate cashflow is influenced by idiosyncratic factors and macroeconomic and business cycles. Willingness to repay is influenced by the existing claims on the cashflow and the charge on the assets of the obligor.

Table 2 Inability and Unwillingness to pay

Inability to repay	Cashflow/income of the borrower	Value of the assets available to lender
Unwillingness to repay	Leverage and other claims on the cashflow	Seniority of the charge on the assets

Linkages with macroeconomic factors

Repayment abilities are influenced by macroeconomic and business factors. However, it is extremely difficult to identify the factors and then measure their impact (many of them have the opposite effect). So, the other way is to map the factor correlation to the loss correlation.

Note that factors are modelled on correlation or loss volatility. Multifactor models are complex – most of the models currently being used (including Basel Accord and CreditMetrics) are single-factor models.

Idiosyncratic factors

Idiosyncratic factors are firm-specific factors. These are measured by various accounting ratios, management quality (ability to churn assets to generate cashflow) and equity prices (called *structural models*). Cashflow to repay comes either from the cashflow generated from churning or selling the assets, or fresh borrowings. These are also called *cashflow* or *asset-based* models. These factors are linked with credit losses/risks. Two types of model (structural and reduced-form) have been developed to link various idiosyncratic factors to measure credit risk drivers.

Figure 7 indicates interaction of various factors under two paradigms – cashflow-based lending and assets-based lending.

Willingness to repay

Willingness to repay is a factor that varies over exposures. Furthermore, this factor can vary within the same exposure. Let's say that, if within each portfolio, every exposure is the size of one euro, each euro can be considered to be exposed to the same risk. Willingness to pay stems from the seniority of the charge on the cashflow or assets. Willingness to pay after default is measured by LGD. Willingness to pay is managed or enhanced by applying credit risk mitigation techniques (collateral, guarantees and nettings, insurance).

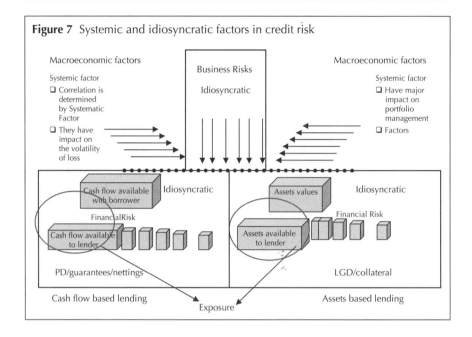

Figure 7 Systemic and idiosyncratic factors in credit risk

Calibration issues

None of the factors (ability or willingness to pay) can be directly linked to the amount of credit loss. Unlike market risks, calibration plays a major role in risk measurement in credit risks and this is the most difficult aspect of risk measurement. The Accord has not been able to prescribe a framework to validate the risk models (the job of models is to measure risk by measuring various factors).

Understanding correlation

Credit risk is variability in the credit loss (unexpected loss). Loss variability emanates from:

❑ the average default rate not being a true measure of actual default rate;
❑ change in the default rate due to correlation; and
❑ variability in the exposure size.

Let us deal with each of these.

(1) *The average default rate not being a true measure of actual default rate*: Default rates are mapped to a risk grade. Obligor and

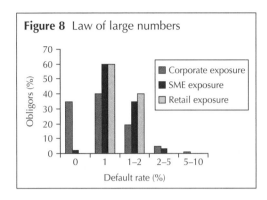

Figure 8 Law of large numbers

facility are classified into a risk grade. Within a risk grade, the default rate is not a single number but a distribution. The average default rate measure may not be a true measure of the distribution for the following reasons:

❑ Due to sample size, there are outliers whose PD is many times higher.

❑ Laws of large numbers implies that the average of a random sample from a large population is likely to be close to the mean of the whole population. We know that in most of the banks the number of corporate exposures is of the order of hundreds, the number of SME exposures is of the order of thousands and the number of retail exposures is of the order of millions. Due to large numbers, retail exposures concentrate around the mean default rate while outliers are found for corporate exposure.

Figure 8 on the Law of large numbers shows that corporate and SME default rate exists at 2–5%. Retail exposures do not have both these default rates. Some of the corporate have exposure rates of even 5–10%.

(2) *Change in the default rate due to correlation*: Correlation has an impact on the probability of one obligor's defaulting when other obligor has already defaulted.

❑ Default correlation does not imply that one credit's default directly *causes* the change in another credit's default probability.

❑ Negative correlation is not common in credit risk.

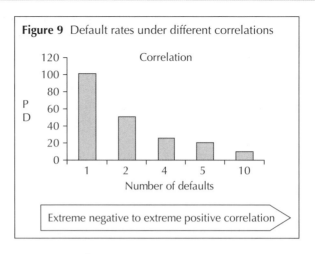

Figure 9 Default rates under different correlations

Suppose we have a portfolio of 10 bonds and an average default rate of 10% (see Figure 9). There are various possible outcomes depending upon the correlation.

Empirically, every year default rates are different. There are two reasons for change: volatility due to correlation and changes in the base default rate. Volatility due to correlation means the bank had obligors in its portfolio that were impacted adversely by the same economic, business and industry cycles.

Correlation changes the conditional default rate substantially. For a pair of obligors, the conditional default rate is the default rate of the remaining obligor given that one obligor has already defaulted.

Conditional default rate $P_{A/B} = (p_{AB}/p_B)$

We know $P_{AB} = P_A P_B + \rho_{AB}\sqrt{P_A(1-P_B)P_B(1-P_B)}$

$$P_{A/B} = P_A + \rho_{AB}\sqrt{P_A(1-P_B)(1-P_B)/P_B}$$

We know default probabilities are always very small 0
Therefore $P_A = 0$ and $P_B = 0$ and $P_{A/B} = \rho_{AB}$
Conditional probabilities are equivalent to the correlation coefficient, which is in the range of 10–20%, as compared the default rate, which is in the range of 1%.

The joint probability between the two obligors is approximately equal to $\rho_{AB}p$, which is smaller than the independent probabilities and is an irrelevant measure of the credit risk.

Variability in the exposure size: The primary problem with credit risk is that, if an obligor defaults (with a default probability of, say, 1%), the entire exposure is defaulted and not the 1% of its default. Let us consider two obligors (A & B) with similar risk characteristics, that is, a PD rate of say 1%. Obligor A has exposure of EUR €100,000 and obligor B has exposure of EUR €1 million. Both of them have a PD rate of 1%. If obligor B defaults, the bank loses EUR €1 million. If obligor A defaults, the bank loses EUR €100,000. This is a common phenomenon for corporate and SME assets class. The exposure size is managed by two approaches.

❑ *Granularity criteria*: The granularity is reflected in the tail shape of the curve. Loss volatility is higher with fat tails. Experts disagree about granularity adjustment due to different opinions on the shape of the tail. Granularity is also described in terms of the absolute exposure size to a single obligor. For the retail segment, the Basel Accord has prescribed the granularity criteria for an exposure to be classified as retail. It has set a limit of 0.20% of the portfolio size for exposure to a single counterparty.

❑ *Loss given default and collateral*: Basel has recognised reduction in the effective exposure through the use of collaterals. LGD is likely to vary according to the exposure size. One of the factors for assets class segmentation is the size of exposure. Basel prescribes different measurements for LGD of different assets classes.

Factors for measuring correlation: Correlations are not directly observable. Correlation is measured indirectly through asset correlation, equity correlation, rating migration or economic-factor correlation – or any other proxy or factor correlation.

We saw earlier that credit risk cannot be measured directly. Risk is divided into risk components such as PD, LGD, EAD, maturity and correlation. Since risk components are also not observed directly, borrowers' characteristics, called risk factors relevant for risk measurement, are identified and mapped to the risk

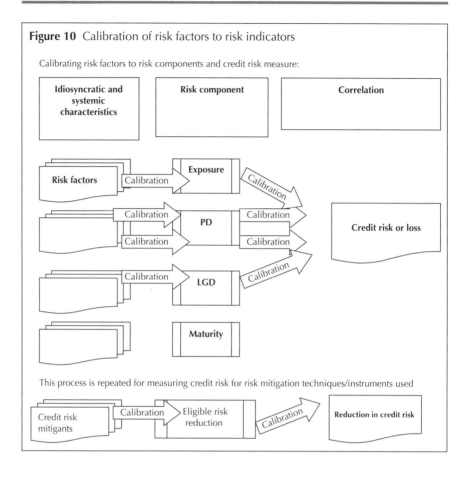

Figure 10 Calibration of risk factors to risk indicators

component to establish a relationship between risk factors and risk components. It is very difficult to get accurate risk-component data, so an intermediate step of mapping risk factors to reference data is followed. We explain the calibration process in the chapters "Validation of Credit Risk Process" and "Probability of Default". We also explain the process of computing credit risk in the chapter "Credit Risk Portfolio Management".

Data for calibration

Default is a rare phenomenon and sample size has an impact on the accuracy of the models or prediction. Therefore, internal bank data may be insufficient for a given risk segment and therefore needs to be supplemented with external data.

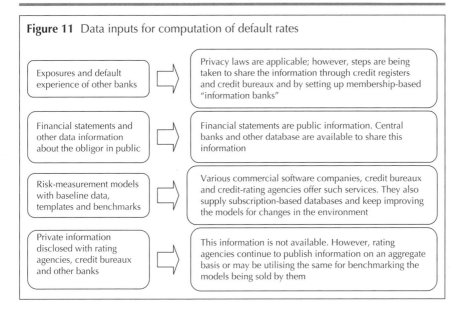

Figure 11 Data inputs for computation of default rates

Exposures and default experience of other banks	Privacy laws are applicable; however, steps are being taken to share the information through credit registers and credit bureaux and by setting up membership-based "information banks"
Financial statements and other data information about the obligor in public	Financial statements are public information. Central banks and other database are available to share this information
Risk-measurement models with baseline data, templates and benchmarks	Various commercial software companies, credit bureaux and credit-rating agencies offer such services. They also supply subscription-based databases and keep improving the models for changes in the environment
Private information disclosed with rating agencies, credit bureaux and other banks	This information is not available. However, rating agencies continue to publish information on an aggregate basis or may be utilising the same for benchmarking the models being sold by them

Broadly, the data inputs to the default rate can be divided into four types, as shown in Figure 11.

Most default estimation models attempt to estimate an obligor's probability of default to assign a quantitative risk score. Generally, they do not take into account a facility's structural elements, such as collateral, that can moderate the impact of a borrower's default.

We discuss data requirements for each approach of PD estimation in relevant chapters.

All obligors (individuals, small business, corporates, financial institutions and sovereigns) are economic entities and operate in a business/economic environment. Each of the economic entities is linked with other economic entities as a supplier, consumer or competitor. At any point in time, every economic or business environment is passing through economic/business cycles of varying intensity. Not all economic entities are impacted in the same way by these cycles. Through systemic factors, economic/business cycles determine the default correlation and therefore default rates. Each obligor has idiosyncratic characteristics; these impact on ability and willingness to pay. The entire quest is to identify such factors and collect historical data for them.

BEST PRACTICES

The Basel Committee has identified the following four best practices for the implementation of credit risk management.

❏ Establishing an appropriate credit risk environment.
❏ Operating under a sound credit-granting process. (The Accord has identified the risk factors to be considered when assessing the credit risk.)
❏ Maintaining an appropriate credit administration, measurement and monitoring process.
❏ Ensuring adequate controls over credit risk.

This book aims to assist banks in the implementation of these best practices.

Best practices from Basel

Economic capital is unexpected loss due to credit risk at a certain confidence level. Unlike the case with market risk, where banks have been computing economic capital, banks are not doing so for credit risk. For the time being, banks are aiming at computing regulatory capital. To compute regulatory capital and to measure the basic aspects of credit risk, banks are establishing the systems and processes to compute risk components – PD, LGD and EAD. Basel has published more than 100 reports and working papers to help banks to compute PD and LGD for a substantial portion of their assets. The reports and working papers broadly cover the following areas.

❏ A framework and best practices in the areas of credit risk management. This covers reports on principles for management of credit risk; internal rating-based methods (IRB); surveys of credit risk measurement practices; models used by the banks; risk-mitigation techniques used by banks; sound practices for loan-loss provisioning and reports on expected loss and unexpected loss; and sound credit risk assessment and valuation of loans.
❏ Reports on implementing risk components measurement systems and practices in the bank. These include reports and working papers on IRB and LGD, and studies on validation of IRB.

❑ Basel Accord as a framework to compute regulatory capital. In addition, BIS has conducted five Quantitative Impact Studies (QIS). The spreadsheet published as part of the QIS are useful templates to compute the regulatory capital.

❑ A framework and best practices for risk transfer is provided in the securitisation framework, report on credit risk transfer and report on fair-value accounting.

ORGANISATION OF THE BOOK

Chapter 1, "Introduction to Credit Risk Management" identifies and explains the risk components, basic building blocks and challenges in implementing credit risk management. The subsequent chapters elaborate on the techniques to measure each of the risk components.

Chapter 2, "Probability of Default", elaborates default definition as given by Basel, ISDA and Rating agencies. It establishes a Framework for measuring probability of default. It also establishes ex-ante PD estimation systems and models. Transition Matrices are defined and explained as a tool to measure PD.

Chapter 3, "Credit Scoring", explains credit scoring as a tool to measure PDs. The types of credit scoring model, their inputs, issues, limitation and challenges are discussed. Various techniques used by scoring models along statistical/mathematical equations and their logic are explained.

Chapter 4 "Market Information Based Models", deals with this model in a novel way. We identify the underlying processes and building blocks for structural and reduced-form models. We explain a spectrum of assumptions and methods for modelling incomplete information. We identify the assets-value process, risk-free interest rate process, recovery process and default process as basic building blocks. We developed mathematical equations for each of the processes along the lines of assumptions. Using the building blocks, we explain well known models such as Merton; KMV; Jarrow and Turnbull; Jarrow, Lando and Turnbull; the Duffie–Singleton Model.

Chapter 5, "Credit Rating", explains the assessment of business risk, financial strengths, credit structures and sovereign credit rating. The chapter defines a generic rating process for external and internal rating.

Chapter 6, "Credit Risk Mitigation", identifies risk mitigation and transfer techniques. The chapter deals with collaterals, credit derivatives and securitisation as tools for credit risk mitigation.

Chapter 7, "Loss Given Default", explains the loss given default and recovery rates. It also identifies risk factors determining LGD. The chapter explains the treatment of LGD under the Basel II Accord.

Chapter 8, "Credit Risk Fortification", identifies, credit risk policies at high level, explains various types of credit limit, analyses the organisational structure recommended for managing credit risk, identifies the accounting policies and issues to support credit risk measurement, and explains active credit risk portfolio management.

Chapter 9, explains "Credit Risk Portfolio Models" in a new way. The chapter identifies generic approximation and substitution assumptions made by various portfolio models. It identifies the risk factors and components required to measure portfolio risk. The chapter provides a methodology to generate risk components and loss distribution and applies the loss measure to manage credit risk.

Chapter 10, "Validating the Risk Measurement Process" explains the validation of risk management processes. A generic risk measurement process and the relationship of risk factors, risk measurement models and risk components are explored. The chapter identifies quantitative and qualitative validation techniques. The importance of documentation in the risk management process is also explained. Model risks and model-validation techniques are looked at. Stress testing for credit risk is explained and the best practices of the international financial institutions to address the challenge of data paucity and to validate credit scoring and credit rating are also evaluated.

Chapter 11, "Software and Data" provides an overview of the software and data for credit risk management. The chapter identifies the commercial systems, their functionality and technology, credit rating systems, collateral management systems, LGD systems, regulatory capital computation systems and portfolio-management systems. The chapter provides a methodology to enable IT systems for Basel II.

REFERENCES

Basel Accord, 2004, published by Bank for International Settlement.

Basel Market Amendment, 1996, published by Bank for International Settlement.

Quantitative Impact Studies (QIS) 3, 4, 5 published in 2002, 2004 and 2005 by Bank for International Settlement.

Probability of Default

DEFINING DEFAULT

Default risk is the uncertainty regarding a borrower's ability to service its debts or obligations. Default can be quantified by measuring probability of default (PD). PD reflects the probabilistic assessment of the likelihood that an obligor or counterparty will default on its contractual obligations within a certain period of time; ie, default on the repayment of principal and interest within a certain time period (year/quarter).

PD assumes a significant importance in commercial transactions since the lenders want compensation for the uncertainty around the likelihood of default and will demand a spread over the risk-free interest rates. There is a considerable variation in default probabilities across firms: eg, a AAA-rated firm has approximately two in 10,000 per annum default probability, an A rated firm as five times higher the default probability, ie, 10 in 10,000 and a CCC rated firm has a 200 times higher default probability, ie, 400 in 10,000 default probability. Hence, lenders demand a substantial spread for CCC rated firm *vis a vis* an A rated or AAA-rated firm.

In general, default is not an abrupt process. There is always a normal and continuous deterioration of a firm's financial position and asset quality that ultimately leads to default. A default situation is also difficult to detect because the firm may be technically insolvent, ie, liabilities are greater than short-term liquid assets but still may not be in default. This is because the firm may be able to raise resources to meet its liabilities. This would always be dependent on the organisations reputation, strength and financial market structure.

This leads us to an important question of what constitutes default. The default definition can seriously impact the PD and credit loss estimations. Let us examine some of the definitions and their relevance.

Standard & Poor's (S&P's) defines default as "The first occurrence of a payment default on any financial obligation, rated or unrated, other than a financial obligation subject to a bona-fide commercial dispute; an exception occurs when an interest payment missed on the due date is made within the grace period".

Basel definition of default

Under paragraph 452 default is defined in the Basel Accord as follows: "A default is considered to have occurred with regard to a particular obligor when either or both of the two following events have taken place.

❏ The bank considers that the obligor is *unlikely to pay* its credit obligations to the banking group in full, without recourse by the bank to actions such as realising security (if held).
❏ The obligor is *past due* more than 90 days on any material credit obligation to the banking group. Overdrafts will be considered as being past due once the customer has breached an advised limit or been advised of a limit smaller than current outstanding."

Paragraph 453 has defined unlikely to pay as follows:

❏ The bank puts the credit obligation on non-accrued status.
❏ The bank makes a charge-off or account-specific provision, resulting from a significant perceived decline in credit quality subsequent to the bank taking on the exposure.
❏ The bank sells the credit obligation at a material credit-related economic loss.
❏ The bank consents to a distressed restructuring of the credit obligation where this is likely to result in a diminished financial obligation, caused by the material forgiveness, or postponement, of principal, interest or fees (where relevant).
❏ The bank has filed for the obligor's bankruptcy or a similar order in respect of the obligor's credit obligation to the banking group.

❑ The obligor has sought or has been placed in bankruptcy or some similar protection where this would avoid or delay repayment of the credit obligation to the banking group.

The International Swaps and Derivatives Association definition of default

The International Swaps and Derivatives Association (ISDA) defines default as one of the following events.

❑ *Bankruptcy under the laws*: The efficiency and effectiveness of the legal environment and the bankruptcy laws vary from country to country. In the US, the bankruptcy is defined as:

  ❑ dissolution of the obligor other than merger;
  ❑ insolvency or inability to pay its debt;
  ❑ assignment of claims;
  ❑ institution of bankruptcy proceedings;
  ❑ appointment of receivership;
  ❑ attachment of substantially all assets by a third party.

❑ *Failure to pay*: This means failure of the creditor to make a payment due above a certain amount. This is usually triggered after an agreed grace period.

❑ *Obligation/cross default*: This means the occurrence of a default (other than the failure to make a payment) on any other similar obligation.

❑ *Obligation/cross acceleration*: The occurrence of a default (other than the failure to make a payment) on any other similar obligation that results in that obligation becoming due immediately.

❑ *Repudiation/moratorium*: This means that the counterparty rejects or challenges the validity of the obligation.

❑ *Restructuring*: This covers a waiver, deferral or rescheduling of the obligation with the effect that the terms are less favourable than before.

❑ *Downgrade*: The credit rating is lower than a previous rating, or is withdrawn.

❑ *Currency inconvertibility*: The imposition of exchange controls or other currency restrictions.

❑ *Government action*: This impairs the validity of the obligation. The occurrence of war or other armed conflict that impairs the functioning of the government or banking activities.

❑ *Covenant*: Covenant triggered default events.

The ISDA definition is designed to minimise the legal risks by precisely wording the definition of the credit event and capturing the default event early with an expanded scope for cross default.

Cross default

Cross default means default at a facility level that also considers default on similar or different facilities extended to the obligor. For example, a corporate may default on one of the exposures, then all other exposures will also be taken as defaulted. Depending upon the contractual agreements, the default may be even a "market" default, ie, the borrower defaults on repayment of bonds issued to the public. Even if the bank does not have direct exposure in such bonds, the borrower will be considered to be in default if the bank has exposure on the same borrower and such facilities are not in default. Such a contractual definition does not only cover the obligor but also the group and other members of the group. This is called the propagation of the default or cross default.

The primary determinant factor for cross default is the capabilities of the obligor to raise funds at a short notice.

❑ For individuals, a default on one facility is generally not propagated to the other facilities.
❑ Corporates have higher capabilities to raise funds. Therefore, default at facility level may be propagated up to obligor level but not up to group level. Small and medium sized enterprises (SMEs) fall somewhere in between these two regimes.

The discussion above can be presented in a diagram, as seen in Figure 1.

❑ Institutions like banks have access to the money markets, and therefore they have very high capability to raise funds at very short notice. So if a bank defaults on an obligation, it means it has exhausted all means to raise funds and is therefore probably bankrupt at the group level.

Impact of cross default on credit risk estimation and modelling
A credit default is a discrete state: either it happens or it does not happen. A wider definition of default captures default as early as possible but increases the cost of credit administration and also increases the loss volatility substantially.

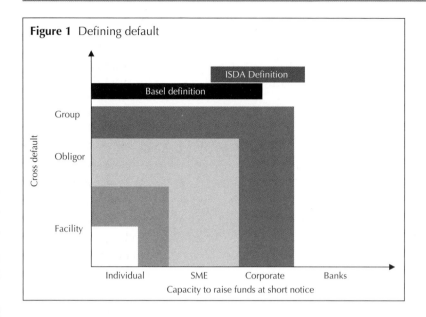

Figure 1 Defining default

Covenant triggered default events

According to contractual terms, the breach of a covenant triggers a "technical" default; eg, the failure to comply with financial ratio covenants during the tenure of the loan. Technical default generally requires a waiver from a lender before continuing operations. Such events usually initiate negotiation. Some covenants trigger cross default and cross acceleration. Cross acceleration specifies that all repayments are due immediately, although this might result in the start of negotiations. Defining default for securitised transactions is challenging. One of the important reasons for this challenge is the practice of capitalising unpaid interest without triggering default. In general, securitisation structures and funds are in default when a breach of the covenant occurs. The loss for a securitisation is more difficult to define than a loan. Default event under securitisation triggers early amortisation of a pool of securitised assets or liquidation of assets, so it is very difficult to say if the loss is due to credit loss or due to early amortisation or liquidation of assets.

The entire discussion can be distilled into two concepts:

❑ obligor or issuer default;
❑ transaction, facility or issue default and the ways to propagate transaction default to an obligor.

> **Definition of reference default**
>
> A reference default definition is needed if banks are interested in using external default data. If there is a mismatch between the internal default definition and the default definition used for external data collection, necessary adjustments must be made in the calculations to account for the difference in default definition.
>
> Efforts are required by banks, industry associations, regulators and accounting bodies to converge the definition around the Basel definition.

The treatment for the transaction default depends on the type of risk segment. For retail exposures, the definition of default can be applied at the particular facility level rather than at the level of the obligor. As such, default by a borrower on one obligation does not require a bank to treat all other obligations to the banking group as defaulted.

The distinction between facility/issue default and obligor/issuer default is important if different facilities/issues fall under different jurisdictions or issues/facilities are legally encircled and made independent (eg, using a special purpose vehicle) or is excluded or included in/from cross default.

MEASURING PROBABILITY OF DEFAULT

There are two approaches to measure default risk.

❑ *Stand-alone risk*: Default probability, the probability that the counterparty or borrower will fail to service its obligations.
❑ *Portfolio risk*: Default correlations, the degree to which the default risks of the borrowers and counterparties in the portfolio are related.

In this chapter we will focus on learning about stand-alone default risk measurement. Portfolio risk has been discussed in the chapter on credit portfolio models.

Default rates or default risk are not directly observable from the data. Default risk measurement starts with the identification and measurement of risk drivers or risk indicators that best represent the quantum of default risk. The risk drivers/indicators are such that any changes in these factors are an indication of changes in the PD. The next step is to model the risk factors to compute the default risk. The model may be quantitative or qualitative or both. The main problem is that for a model to work it has to be calibrated so that it is sensitive enough to signify changes in the PD. However, such calibration is difficult, primarily due to the following reasons.

❏ *Unexpected losses due to correlation*: Borrowers default together since they are connected with each other as a supplier/customer, competitor or partner or are subjected to a similar impact of the macroeconomic factors.
❏ Errors in PD measurement or estimation are also due to the following causes:

  ❏ incorrect selection of risk drivers/indicators;
  ❏ drivers or indicators do not have the required sensitivity to the risk changes or have a lagging effect (eg, the rating provided by the external rating agencies lag behind the actual changes in credit quality);
  ❏ mapping or calibration error.

This leads to the following two approaches to measure PD.

❏ *Ex ante*: This is to measure the PD at the start of the period. This is the expected loss (EL).
❏ *Ex post*: This is to measure the default at the end of the period. This is the realised loss.

It is easier to measure the *ex post* PD. This simply represents the number of defaults compared to total number of exposures at the beginning of the period. The most difficult part is the *ex ante* measurement of PD. The whole exercise of PD measurement is measuring PD *ex ante*.

Credit rating: does it measure probability of default?

A rating, in its most generalised definition, is a summary indicator of the credit quality or default rate. In general, it is an assessment of a client's probability to fail to meet its obligations in accordance with the agreed terms. Ratings present a much broader approach to estimating default probabilities than the purely market-data-oriented measurements. We have further dealt with this in the following section in this chapter and in the separate chapter on "Credit Rating".

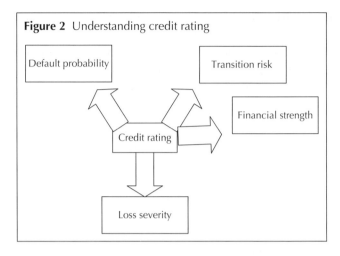

Figure 2 Understanding credit rating

Whether ratings really measure PD

Traditionally, the credit rating provided by rating agencies does not only measure the default rates but also other risk measures (see Figure 2) such as transition risk, loss severity and the financial strength of the obligor, in addition to the default probability. In general, credit ratings are evaluated in terms of historical default rates, while other attributes are also important.

The rating agencies have been taking steps to achieve greater comparability across markets by an increased emphasis on the EL rate – the product of the expected default rate and loss severity – as the primary measure of credit quality. However, the relative importance of each of the risk factors measured varies across the industry. In response to the needs of investors who are highly sensitive to default and transition risk, rating agencies will continue to overweight these aspects of credit risk in certain sectors.

Transition risk refers to the uncertainty with respect to the levels and timing of credit events. High-transition risk credits have relatively high probabilities of large rating movements. Examples of risk segments subject to high-transition risk are banks, sovereigns, investor-owned utilities and local government authorities. Issuers in these segments derive significant credit strength from external sources of support – bank regulators, multilateral institutions, rate commissions and state governments. Furthermore, confidence-sensitive issuers with high levels of short-term funding, such as securities firms, also face high transition risk. Transition risk is low for retail portfolios and securities products.

Financial strength refers to the intrinsic creditworthiness, which abstracts from potential (and uncertain) external support elements (such as a rescue by a third party). With the higher financial strength, ie, guarantees by governments, the recovery and credit loss risk is generally very low. However, investors of such assets are definitely exposed to substantial price risk due to low liquidity. The relative importance of each of the factors is not the same across the risk segments. The relative emphasis on the different aspects of credit quality is varied to meet the expectations of investors in these different markets.

Reasons why PD and credit rating are used interchangeably
For a given sector, bonds with the same rating tend to be quite comparable, both with respect to the overall credit quality and to the specific credit quality attributes given above. At today's level of risk component measurement, expected severity of loss in the event of default, ratings volatility and transition risks is assumed to be uniform across issuers or assumed to be constant. Therefore, ratings correlate closely with the likelihood of default (assuming other factors to be constant). Each factor has a different significance for different industries and whether the rating is an investment rating or a speculative rating. The investment-grade bonds are generally held by investors who, in general, are averse to default risk, irrespective of severity. The distinction and risk segmentation of borrowers is being eroded due to deregulation, technological change and increased competition. Since default rates are the only observable attributes, to showcase consistency across rating categories, sectors, time horizons and geographies the weight given to the EL rate in assessing credit quality has to increase.

Time horizon to measure the default rate
The probability for a company to get into financial distress is dependent on the length of time under consideration. Default probabilities are estimated for each sub-portfolio of asset class over a certain time horizon, usually a year.

The time slices over which to measure the default rate: In general, the default is measured for one year (see Table 1). However, the criticism

Table 1 Time horizon in PD

Factors impacting the choice of time horizon
❑ time required to take loss mitigation action;
❑ publication of default data;
❑ new information from the obligor;
❑ time horizon for the model, which is being calibrated and requires data reconciliation.

Types of time horizons
❑ Multiple time horizons, driven by the maturity of facility.
❑ Multiple time horizons, leading to a different set of PDs. The time horizon is driven by the business cycle.
❑ Averaging PDs over a complete business cycle to estimate a one-year default. This method is generally adopted by rating agencies.
❑ One-time horizon/one-year horizon for all the exposures.

against this approach is that a one-year time horizon does not reflect actual risks.

Default rates are measured in terms of yearly rates. Yearly rates may be for one year or cumulative for many years. The many years horizon may be a hold to maturity. For example, CreditRisk+ provides two approaches to measure the default rate:

❑ constant time horizon;
❑ hold-to-maturity time horizon (see Figure 3).

Under the constant time horizon, one year is the most frequently used interval. Alternatively, a hold-to-maturity time horizon allows the full term structure of default rates over the lifetime of the exposures to be recognised. The hold-to-maturity time horizon is actually a cumulative yearly default rate for the maturity period. These two rates can be interchanged.

Approaches to validity of rating period:

❑ *The point-in-time (PIT) rating:* In the PIT rating method, risks are evaluated based on the current condition of a firm regardless of the phase of the business cycle at the time of evaluation. This estimation of default rates is valid for a short-term constant time horizon.
❑ *Through-the-cycle system (TTC rating):* In the TTC rating method, risk default rates are estimated for a borrower's conditions at the worst point in the economic or industry cycle. Although, each bank considers the stress condition for the stress testing, banks generally

Figure 3 Cumulative default rates for rating grades

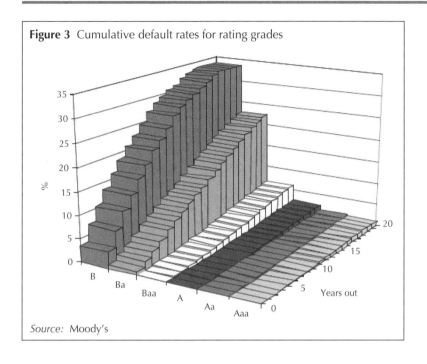

Source: Moody's

assess borrowers on the current (PIT) conditions and not on stress conditions. While rating, banks generally consider a one-year horizon. For short-term exposure, this may be the correct approach. However, for exposure beyond one year a TTC approach is recommended. Moody's and S&P's rate through the cycle. They analyse the borrower's current condition at least partly to obtain an anchor point for determining the severity of the downside scenario. The borrower's projected condition in the event of the downside scenario is the primary determinant of the rating. This is further explained in detail in the chapter on "Credit Rating".

Review cycle for PD measures: The time horizon for external ratings always aims to provide stability to the ranking, ie, it is not changed unless there are strong reasons for doing so. It is reviewed and changed whenever there is an event warranting changes in the rating (it is event based). Internal ratings are generally reviewed once a year or it is also event based.

Every rating agency aims to provide a smooth path for default, ie, the debt should not default suddenly, especially for investment grade.

Each credit grade is divided into three notches. For example credit grade A has three notches (A+, A, A−). Smooth path means, there should not be more than a 2–3 notches jump at a time. Rating agencies aim that an investment grade bond should not default within three years and should move to speculation grade before default.

Rating agencies generally aim at achieving ordinal accuracy (better rating grade has less default rate in comparative and not in absolute terms) and do not aim at maintaining constant default rates (the default rate difference between the two grades may not be constant) for a rating grade. Otherwise, ratings have to be changed *en masse* in response to changes in cyclical conditions. Rather, ratings are changed "one-at-a-time" as needed, in order to maintain the current rank ordering of credit risk. Cardinal ranking measures the risk in absolute terms. Under cardinal rating system risk grades are mapped to an idealised or historical average default rate with a given variability in realised loss rates. The variability is measured by mean absolute error or root mean squared error statistics. A validity period of a particular rating is the time during which these parameters are within a range, and the agencies attempt to achieve stability in relative ranking of the credit risk at each point in time. The power of ratings with respect to expected credit losses as realised over a longer horizon is that low ratings should, on average, prove to be more risky than credits that have high ratings. Over time, this is measured as changes in the pool of ratings assigned to multiple credits, and possibly even the same credits, observed at different points in time. To get meaningful relative rankings, the meaning of ratings should be consistent across the time horizon.

Rating agencies do not differentiate risks on the basis of the maturity of the obligation. A long-term loan and a short-term loan may respond similarly to defaults within the chosen time horizon. However, a downgrade from, say, Baa to Ba will tend to have a larger relative price effect on a 10-year loan compared with a one-year loan. Banks adopting best practices recognise the importance of more formal maturity adjustments and more institutions are moving in this direction.

Measuring PD ex *post*

This is the realised PD. The *ex post* PD plays an important role in model validation as well as in model calibration. Basel has prescribed at least five years historical data to estimate PD. PD measures are closely linked with the rating system and different rating systems will give different PDs for the same obligors. Basel recommends that the bank must have been using the rating system for at least three years prior to qualification.

Default rate is measured for issuer and issues. The default rate is either assumed to be constant over a given time horizon or it can be assumed to be variable over the time horizon due to the macroeconomic cycle. The number of observations has an impact on the precision with which the PD can be measured. Therefore, higher grades require a larger sample. Let us compute default rates for an issuer under the assumption of constant rates. The same approach can be extended to issues. There are two approaches to this problem.

Computing default rate for an issuer for one year time horizon
Let p be the PD, n be the number of issuers in a year, d be the number of default in year and p be the default rate ($= d/n =$ the default rate in a year).

Multi-year observations are necessary in order to reduce the standard deviation, which is likely to be about 1.5 times the default probability for a diversified portfolio of investment-grade credits and approximately equal to the default probability for a diversified portfolio of speculative-grade credits.

Computing default rate for an issuer for a time horizon of T (year)
For computation of default rates two variables are needed. The number of issuers who were enjoying the credit during the year and number of defaults during the year. Since the number of issuers keep changing, the generally average number of issuers are considered.

$$n = \sum_{t=1}^{t=T} Number\ of\ Issuer\ during\ each\ period$$

$$d = \sum_{t=1}^{t=T} Number\ of\ defaults\ during\ each\ period$$

$$P = \frac{d}{n}$$

where P is the default rate, which is equal to the weighted default rate over T years.

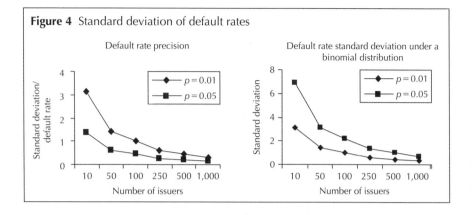

Figure 4 Standard deviation of default rates

In large samples, the empirical default rates will be approximately normally distributed with their means equal to p and their standard deviations given by the following equation (see Figure 4):

$$\sigma = \sqrt{\frac{p(1-p)}{n}}$$

The standard deviation of the historical default rate is an extremely useful measure of empirical precision: the true underlying default probability is highly unlikely to be more than two or three standard deviations (if the standard deviation is measured correctly) from the underlying default rate.

The historical default rate is actually a less precise estimator when the underlying default rate is low. As a result, relatively more observations are necessary to evaluate consistency for investment-grade ratings than for speculative-grade ratings. Now let us relax some conditions of PD such as it being constant over a time period of T years. Let us assume that PD varies over time, either due to fluctuations in the macroeconomic environment or conditions within a specific bond market sector.

To incorporate the concept of time-varying underlying default probabilities, we assume that each year's default rate is normally distributed around a long-term underlying probability, p, with an annual shock of mean zero and standard deviation, σ:

$$p_t = p + \xi_t$$

where

$$\xi_t \sim N(0, \sigma)$$

and

$$P \sim N\left(p, \sqrt{\frac{p(1-p)}{n}} + \sigma^2 \right)$$

Ex post PDs should converge to ex ante *PDs*

The *ex post* PDs should be sufficiently close to the *ex ante* PD associated with that grade. However, this equality is not expected each year. Indeed, the very reason that a bank holds capital is that unexpected events do occur. However, over relatively long periods of time, the *ex post* PD and *ex ante* PD should converge. The question that is always asked is "over what time period should we expect this equivalence to hold?" Another question is "how are we going to measure whether a given difference is within tolerable limits?" This is measured through the validation tests explained in Chapter 10. *Ex post* the realised default rates for low-quality grades should higher than the default rates for high quality grades. This is one of the most important pieces of evidence to show that PD grades have a meaning.

Test of significance for differences between
historical default rates

Statisticians generally address this question by estimating the probability of observing the difference in default rates, under the "null hypothesis" assumption, ie, that the true underlying default probabilities are the same and equal to the weighted average of their historical default rates. The default rates are different because of the different level of parameters (economic and business cycles) to which they are exposed to at different levels of correlation. The difference in the historical default rate is really due to change in the default rates for a smaller sample size, larger default volatilities, high level of parameters and smaller correlation of default rates with shock parameters.

Whether ex post *PD measured across the banks
should be the same*

Apparently, it may appear that the default rate across the banks should be similar as, in general, banks use similar rating criteria. As a result, one might expect that that the average *ex post* default frequency over a five-year period for a given ratings grade should be pretty much the same across all banks. However, this is not the case – average default rates for some grades differ widely across banks. They vary widely because each bank has a different historical sample data, and their models have different powers.

Under the Basel Accord, risk weights are determined both by the credit rating and the asset class (sovereigns, banks and corporates). Sovereigns have lower risk weights when compared with banks and corporates. Corporates from a certain sector have lower risk weights. The assignment of different risk weights for different sectors is an indication that similar ratings do not reflect the same default rate across the industry.

Credit rating agencies assign similar letter grade scales for a wide variety of issuers – sovereigns, municipal governments, industrial firms and financial institutions located in many countries. Historical experience indicates that the historical default rates vary widely across industries such as utilities, corporates, municipals and structured finance. Ratings across time periods also vary significantly. Therefore, the same letter grade may not represent the same default rate across the industry sector and geography For example (see Figure 5), US banking firms have experienced significantly more defaults than US industrial firms over the period from 1983 to 1998. These results are at odds with the proposal that, for some rating levels, bank obligations carry a lower capital requirement than an otherwise identical liability of an industrial firm.

Rating agencies have started to address the problem of PD differences across the sectors and are harmonising ratings across sectors. However, only a very limited amount of literature is available to the public on sectoral differences in the measurement of credit risk. Few papers address default rates directly and we know of no previous work that undertakes a systematic statistical analysis of sectoral differences in default rates.

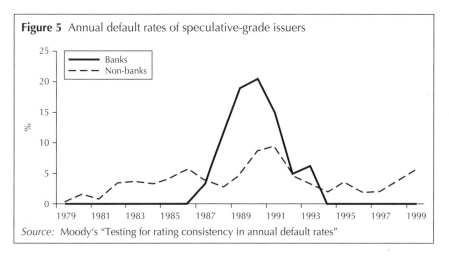

Figure 5 Annual default rates of speculative-grade issuers

Source: Moody's "Testing for rating consistency in annual default rates"

Split ratings: will different rating agencies provide the same ratings?
Credit ratings are a subjective assessment of default rate. Being subjective, ratings will differ across the rating agencies due to differences
in opinion, methodology, rating scale, etc. Some of the empirical findings on the rating differences are summarised in Table 2.

Thus, different agencies measuring default rate at different levels for the same obligor at the same time will give rise to differences
in minimum capital. To minimise the differences in capital measures, Basel mandates the mapping of assessment categories of the
rating agencies to the risk weights. This process is likely to take
care of consistent differences in the risk measures of the rating
agencies. In addition, the Accord has prescribed risk weight rules
for multiple assessments.

Maturity impact on the ratings or default rates
Maturity or duration is not considered by ratings agencies. A single
rating is applied to all similar bond issues of an individual issuer
despite their different durations. A bond rating is never upgraded
with time (other factors remaining same). Rating categories are not
associated with a single EL rate. Instead, each rating category is
associated with a schedule of EL that varies by time horizon.

Loan ratings versus bond ratings
Empirical studies at Moody's using 1996–2001 data have found
that the historical credit losses on similarly rated North American

Table 2 A summary of the empirical findings on rating differences across rating agencies

Ederington (1986)	Compares Moody's and S&P's ratings to the newly issued industrial bonds between 1975 and 1980, and finds that the two ratings are different in approximately 13% of the cases. Ederington concludes that this difference is due to chance and not due to differences in the rating standard
Beattie and Searle (1992)	Used Financial Times credit ratings data and finds overall disagreement rate of 56% Beattie and Searle conclude that this is due to different methodologies
Cantor and Packer (1995)	Uses Moody's and S&P's credit ratings and concludes that for lower credit ratings, the difference in opinion is higher for sovereigns than for the corporates. Cantor and Packer conclude that this may be due to higher uncertainties in the measurement of credit risk
Cantor and Packer (1996)	Using corporate issuer ratings published by five US agencies Cantor and Packer infer that firms request an additional rating with an expectation of improvements on their existing rating and concluded that the difference is due to different non-identical rating scales
Morgan (2002)	Difference in opinion is different for different risk segments. Using a sample of almost 8,000 US bonds issued between 1983 and 1993, with Moody's and S&P's ratings, Morgan concludes that raters disagree more on banks and insurance companies than for the other industries and this opinion is lopsided – one agency is more conservative than the other
Santos (2003)	Concludes that disagreement is more for the industries, countries and time horizon under recession. This is due to different opinions on the quantum of impact of the recession
Several studies	Smaller credit rating agencies, whose assessments will also be used in Basel II, tend to assign more favourable credit ratings than those issued by Moody's, S&P's and Fitch

loans and bonds of issuers with both outstanding loans and bonds are different. Loss rates of bonds are three times that of the loss rates of loans; their default rate is 1.25 times higher and LGD 2.25 times higher. Seniority and collateral offered plays an important role in this difference. These findings suggest that loan ratings need to be higher relative to bond ratings.

Risk-neutral default probabilities
Credit spread is an important source for measuring default probabilities observed from the bond prices. The market provides a risk-free yield curve for government debt and risky yield curves for

Table 3 Credit spreads (in basis points (bps)) over risk-free interest rates for different ratings and maturities

Rating	Maturity			
	1–3 years	**3–5 years**	**5–7 years**	**7–10 years**
AAA	49.50	63.86	70.47	73.95
AA	58.97	71.22	82.36	88.57
A	88.82	102.91	110.71	117.52
BBB	168.99	170.89	185.34	179.63
BB	421.20	364.55	345.37	322.32
B	760.84	691.81	571.94	512.43

Source: adopted from Amato and Remolona, 2003 "Credit Spread Puzzle"

Table 4 Transition matrix – under MTM and default paradigm

Rating at beginning of the year (i)	Mark-to-market rating at the end of the year (j)							Default
	AAA	**AA**	**A**	**BBB**	**BB**	**B**	**CCC**	
AAA	0.8352	0.1335	0.0176	0.0092	0.0032	0.0001	0.0002	0.0001
AA	0.0058	0.9030	0.0612	0.0194	0.0086	0.0015	0.0004	0.0001
A	0.0002	0.0274	0.9265	0.0300	0.0109	0.0018	0.0007	0.0007
BBB	0.0275	0.0340	0.0404	0.8312	0.0575	0.0065	0.0013	0.0016
BB	0.0016	0.0356	0.0474	0.6502	0.0343	0.1732	0.0467	0.0110
B	0.0016	0.0068	0.1287	0.0451	0.0376	0.5590	0.1714	0.0498
CCC	0.0004	0.0011	0.0612	0.0030	0.0045	0.4116	0.3256	0.1926

different ratings for other debt. Bonds are generally grouped on the basis of rating. Credit spread is the difference between the risky yield and risk-free yield for a given rating. Obviously, the credit spread varies for credit rating. Table 3 shows the credit spreads (in basis points (bps)) over risk-free interest rates for different ratings and maturities.

Risk-neutral default probabilities are the default probabilities adjusted for the risk premium. These probabilities are estimated from the credit spreads for the risky bonds.

Risk neutral means that investors are indifferent between the expected value of the risky cashflow and risk-free cashflow. There are two ways to value risky debt: discounted contractual cashflow

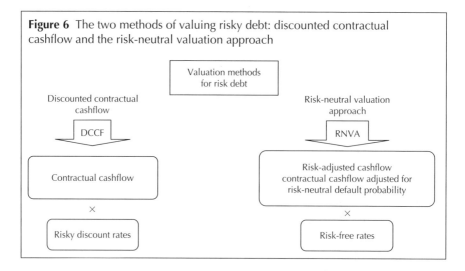

Figure 6 The two methods of valuing risky debt: discounted contractual cashflow and the risk-neutral valuation approach

(DCCF) and the risk-neutral valuation approach (RNVA), as summarised in Figure 6.

The primary difference is whether to adjust the cashflow according to the risk or adjust discount rates for risk. The credit spreads observed are adjustment in the discount rates. Over a given period, the cashflow changes, for the asset which moves into the default category, the recovery value is measured.

Adjustment in the discounting factors is called the DCCF approach. Adjustment in the cashflow is called the risk-neutral PD approach.

Limitations for using the DCCF approach
There are a few limitations in using the DCCF approach.

❑ DCCF does not reflect the recovery rates. Senior and subordinated loans to a single firm would have the same cashflow regardless of differences in the expected recovery.
❑ Loans to two identically rated firms receive the same discount rates, even if the two firms are not equally sensitive to the business cycle or to other systematic factors.

Under no-arbitrage conditions, the values of both the DCCF and risk-neutral cashflow approaches should be the same.

The risk-free rates, the contractual cashflow and the risky discount rates are all observable. The default probabilities are not

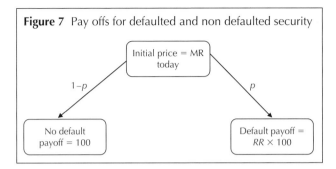

Figure 7 Pay offs for defaulted and non defaulted security

Initial price = MR today

1−p

p

No default payoff = 100

Default payoff = RR × 100

determined from the assumptions that values from the two methods are the same. The default probability to make this value equal is called the risk-neutral default probability.

Assume for simplicity that US$100 is repayable on a bond after one year. Under the DCCF approach and for a given market rate of MR for this bond,

$$PV = Price = MP = \frac{100}{1+R_r}$$

Now let us compute the risk-free cashflow. Let p be the PD. An investor pays today's market price. And this market price will be equal to the future cash flow discounted at "risky" rate. There are two payoffs, as summarised in Figure 7.

Therefore,

$$PV = Price = MP = \frac{100}{1+R_r} = \left[\frac{100}{(1+R_r)} \times (1-p) + \frac{RR \times 100}{(1+R_f)} \times p\right]$$

$$\therefore 1+R_f = (1-R_r)(1-p\,(1-RR))$$

$$\therefore Risk\text{-}neutral\ default\ probability = p = \frac{1}{1-RR}\left(1-\frac{1+R_f}{1+R_r}\right)$$

Volatility in risk-neutral probability of default
As time passes, three types of changes may take place.

❑ A firm's rating changes.
❑ The average spread required for a rating grade, ie, the risky rate demanded, may change.

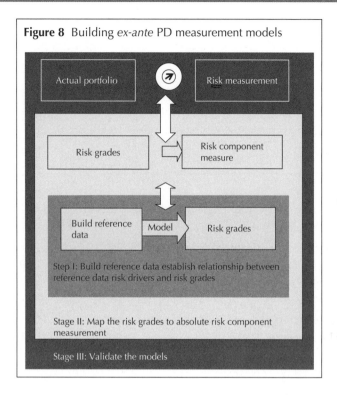

Figure 8 Building *ex-ante* PD measurement models

❑ The gap between the firm's idiosyncratic spread and the average spread of credit exposures with the same ratings changes. However, this gap can be assumed to have been diversified away.

ESTIMATING *EX ANTE* PROBABILITY OF DEFAULT

Establishing PD estimation models for *ex ante* PD measurement is a three-step process (see Figure 8).

Step 1: Build reference data

Banks have traditionally been using the ordinal measures of credit risk. Ordinal measures of credit risk broadly classify the obligors into risk grades. However, no absolute risk number or PD is attached to risk grades. The Basel Accord aims to attach risk measures to each of the risk grade and recommends techniques to measure PD. This is called cardinal risk measurement.

As discussed in Chapter 1 on Credit Risk Management the paucity of default data is a major area of concern and a source of model inaccuracies. Historical default data over a sufficient time

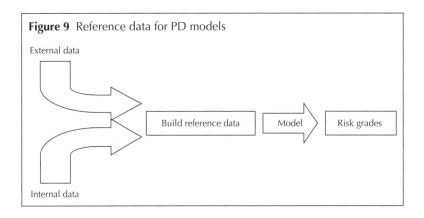

Figure 9 Reference data for PD models

External data

Build reference data

Model

Risk grades

Internal data

period is a primary requirement to start credit risk measurement. An external supplement (see Figure 9) is required since the amount of internal data may be too small for statistical inference.

It is very difficult to get exactly matching external data. Data generally differs in the composition of the firms, industrial sectors and size of business, geographical areas and definition of default. The external database needs to be modified for such discrepancies so that a reasonably conservative estimate of PD is made. Both internal and external data consists of risk drivers and default and credit quality diminution history. There are two types of risk factors: systemic and idiosyncratic.

PD is measured using idiosyncratic risk factors. Systemic factors are either considered constant (by averaging out the historical losses over a larger period or by adjusting the historical losses for the systemic factors). The most important step is to identify idiosyncratic risk factors. In the chapters on "Credit Scoring", "Credit Rating" and "Market Information Based Models" we have identified idiosyncratic risk factors.

Historical PD is a historical mean value for a risk grade. The following list is a non-exhaustive list of assumptions made to estimate historical default:

❑ for a risk grade, default probability equals its observed historical average default rate;
❑ default probability is assumed to be constant over time, and not influenced by the present position in the business cycle or long-term changes in the general economic situation;

❏ default rates are serially independent.

The Accord recommends historical data for five years and simple annual average for historical PD.

The definition of default has a major impact on the historical mean value. External data is adjusted for the changes in the default definition. External data can be obtained from an external agency, a credit register, etc.

Build models

Based on the quantitative techniques used, default models can be broadly divided into six types: default models; statistical, structural and reduced form models; macroeconomic factor models; and hybrid models.

Statistical models have been explained in the chapter on Credit Scoring. Structural and reduced form models have been explained elsewhere in the chapter Market Information based Models. The statistical approach, which is the most frequently use in the literature, maps a reduced set of risk factors and other information to a risk scale, dependent variable or the PD. The mapping acts as a statistical distillation of the historical data and can be used to discriminate between good (unlikely to default) and bad (likely to default). Credit ratings are based on the hybrid form models and the same are explained in the chapter on Credit Ratings.

Different risk segments/assets classes assess different risk factors at different frequencies. The frequency is determined by data availability and cost of assessment. The underlying exposure should justify the cost. Therefore, the applicability of a model varies according to asset class, exposure size and type of counterparty. Figure 10 shows a way of classifying models according to the market and exposure they are aiming at.

Step II: Map the risk grades to absolute risk component measurement

According to paragraph 462 of the Accord:

> *Banks may associate or map their internal grades to the scale used by an external credit assessment institution or similar institution and then attribute the default rate observed for the external institution's grades to the bank's grades. Mappings must be based on a comparison of internal rating*

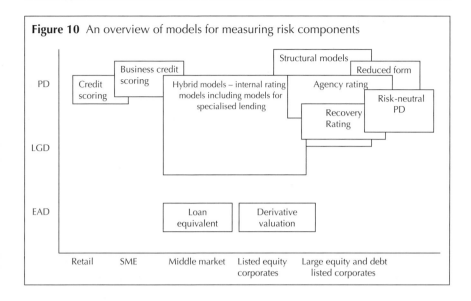

Figure 10 An overview of models for measuring risk components

criteria to the criteria used by the external institution and on a comparison of the internal and external ratings of any common borrowers. Biases or inconsistencies in the mapping approach or underlying data must be avoided. The external institution's criteria underlying the data used for quantification must be oriented to the risk of the borrower and not reflect transaction characteristics.

Step III: Validate risk models

We have explained model validation in the chapter on Validating Risk Measurement Process.

TRANSITION MATRIX

Default probability is measured using risk factors. The change in the default probability or the volatility in PD is measured through a transition matrix (TM). While PD measurement helps in measuring risk at the instrument level, PD volatility helps in measuring risk at the portfolio level. The likelihood of a customer migrating from its current risk-rating category to any other category within the time horizon is frequently expressed in terms of a rating TM. The use of TMs has been gaining prominence in modelling techniques for risk and valuation. A TM can be represented by a matrix which has k rows and k columns, where within each row each entry is the probability of moving from a given rating to the rating at the end of the year.

A transition matrix P is represented by

$$P = \begin{bmatrix} P_{11} & P_{12} & \cdots & P_{1k} \\ P_{21} & P_{22} & \cdots & P_{2k} \\ \cdots & \cdots & \cdots & \cdots \\ P_{k1} & P_{k2} & \cdots & P_{kk} \end{bmatrix}$$

where

$$\sum_{j=1}^{k} P_{ij} = 100$$

The following matrix is an example TM:

Under default mode, only rating migrations into the default state would be relevant. Within the Mark-to-Market paradigm, the other columns of the TM also play a critical role.

Time horizon for a transition matrix

A TM measures change in the PD. The time horizon over which the PD change is to be measured is a very important factor for use and modelling of the TM. The time horizon for which "change" data is available may not be the same as the time horizon for which data is required in a credit risk model. Therefore, building models to convert the TM of one time horizon into another is a central theme of TM modelling.

Building a transition matrix

TM modelling is generally used to try to solve the following two problems:

❑ building a TM;
❑ converting a TM from one time horizon to another.

Credit rating and credit scores are used as a PD proxy to build TMs. The following assumptions are made to use credit rating and credit scores as proxy.

❑ Rating and scoring standards are applied diligently and consistently across the risk segments, industries and countries. The same rating category means the same default rate across the industries, countries and time.

❑ Obligors within a rating grade or score band exhibit similar behaviour for rating migration, in addition to similar behaviour for default rates.

❑ Sampling error is minimised – the number of firms multiplied by the number of years of observation is large enough to yield the required level of granularity. For example, data of 2000 firm years may be able to provide a granularity of 0.05% (2000 × 0.05% will yield 1) but may not provide granularity of 0.02%. Also, cumulative default likelihood should not violate proper rank order, ie, the total number of defaults for AA over a period of 10 years should not exceed the total number of defaults for BBB.

There are several sources and methods to build transition metrics.

External credit rating data-based transition metrics
One of the important sources for constructing transition metrics are the publicly available ratings. TMs are either built using rating data or are built and supplied by rating agencies using their historical data. The rating agencies generally build TMs for 1, 2, 3 or 5 years and longer periods (say 10 and 20 years). Table 5 shows a S&P published TM. The longer period TM is called a cumulative default table. One of the important concerns for a TM is to consider the impact of the economic cycle. In general, they use data of a longer period to even out the impact of economic cycles on PD changes.

Table 5 S&P's global average one-year transition rates, 1981–2005 (%)

From/to	AAA	AA	A	BBB	BB	B	CCC/C	D	N.R.
AAA	88.20	7.67	0.49	0.09	0.06	0.00	0.00	0.00	3.49
AA	0.58	87.16	7.63	0.58	0.06	0.11	0.02	0.01	3.85
A	0.05	1.90	87.24	5.59	0.42	0.15	0.03	0.04	4.58
BBB	0.02	0.16	3.85	84.13	4.27	0.76	0.17	0.27	6.37
BB	0.03	0.04	0.25	5.26	75.74	7.36	0.90	1.12	9.29
B	0.00	0.05	0.19	0.31	5.52	72.67	4.21	5.38	11.67
CCC/C	0.00	0.00	0.28	0.41	1.24	10.92	47.06	27.02	13.06

Source: S&P's "Annual 2005 Global Corporate Default Study and Rating Transitions"

Table 6 Internal transition matrix with degree of concentration around diagonal

From/To	AAA	AA	A	BBB	BB	B	CCC/C	D
AAA	0.92	0.07	0.01	0.00	0.00	0.00	0.00	0.00
AA	0.01	0.91	0.08	0.01	0.00	0.00	0.00	0.00
A	0.00	0.02	0.91	0.07	0.00	0.00	0.00	0.00
BBB	0.00	0.00	0.05	0.89	0.06	0.01	0.00	0.00
BB	0.00	0.00	0.00	0.05	0.85	0.08	0.01	0.01
B	0.00	0.00	0.00	0.00	0.06	0.84	0.04	0.05
CCC/C	0.00	0.00	0.01	0.02	0.02	0.14	0.54	0.27

Rating agency supplied data

Another source is to build an internal TM based on the rating agency supplied data. Table 6 is an example of a TM built internally based on S&P's supplied data over the time period from 1980 to 2002.

The largest values in the TM are along the prime diagonal, reflecting the fact that the most likely rating for an issuer at the end of a given year is the rating with which the issuer began the year.

Internal rating models and credit scoring

The third source is to use internal rating models and credit scoring. Banks have established internal rating systems and credit scoring and have been working on building TMs for their internal rating systems. An example of Z-score-based TM for one year is given in Table 7.

It is clear from the two tables that, while default rates are comparable, the volatility in the TM for the Z-score rating is about twice the volatility in the S&P's TM.

Transition matrix for one year

TMs show the default rates and migration rates for a rating grade over a period of time. These rates can be assumed to be constant in a stable environment. A TM may be a good fit for the entire debt but not a good fit for a newly issued debt. A TM does not represent the migration and default rates for newly issued debt. This is due to a seasoning effect, where sub investment grade usually have a low default rate in the initial period.

Table 7 An example of a Z-score-based TM for one year

	Z-score ratings							Default
	Z < -1.40	$-0.84 <$ Z < 1.44	$1.45 <$ Z < 2.28	$2.29 <$ Z < 3.17	$3.19 <$ Z < 4.47	$4.50 <$ Z < 5.92	Z < 5.99	
$Z < -1.40$	0.85	.14	0.01	0.00	0.00	0.00	0.00	0.00
$-0.84 < Z$ < 1.44	0.07	0.75	0.16	0.01	0.00	0.00	0.00	0.00
$1.45 < Z$ < 2.28	0.00	0.12	0.65	0.19	0.03	0.01	0.00	0.00
$2.29 < Z$ < 3.17	0.00	0.01	0.16	0.55	0.26	0.02	0.00	0.00
$3.19 < Z$ < 4.47	0.00	0.00	0.01	0.12	0.67	0.20	0.00	0.01
$4.50 < Z$ < 5.92	0.00	0.00	0.00	0.00	0.09	0.80	0.09	0.01
$Z < 5.99$	0.00	0.00	0.00	0.00	0.00	0.15	0.75	0.09

Source: Altman and Rijken, 2002, "Benchmarking the rating agencies: In search of true credit rating transition matrix"

Default over a period of time is represented by a cumulative average default table (see Table 8).

The paucity of data is the most important problem to consider when building a TM as it makes it difficult to obtain the required level of granularity in the risk measurement. This is especially the case for recent years. There will only exist five tables of cumulative default rates for 20 years if the data available is from 1981 to 2005. Furthermore, within each year, 19 years of data are overlapped and therefore the tables are not independent. Omitting or adding a few customers can drastically alter the reported default rate. The major issue with the cumulative default rate is the error size, which is substantial when compared to default rate for low-default-rate credit ratings.

Markov chain model

A TM model is an example of the Markov process. A Markov process is a state space model that allows the next progression to be determined only by the current state and not by the information of previous states. Therefore, the default process is modelled using a Markov process.

Table 8 S&P's cumulative average default rates, 1981–2005 (%)

Rating	Time horizon (year)											
	1	2	3	4	5	6	7	8	9	10	11	12
AAA	0.00	0.00	0.03	0.06	0.10	0.17	0.24	0.36	0.40	0.44	0.44	0.44
AA	0.01	0.04	0.09	0.19	0.29	0.40	0.52	0.62	0.71	0.81	0.91	1.01
A	0.04	0.12	0.23	0.38	0.59	0.81	1.06	1.29	1.55	1.83	2.06	2.26
BBB	0.27	0.76	1.32	2.06	2.83	3.56	4.15	4.76	5.27	5.82	6.37	6.80
BB	1.12	3.33	5.96	8.45	10.65	12.77	14.45	15.90	17.26	18.29	19.25	19.97
B	5.38	11.80	17.14	21.24	24.16	26.45	28.37	29.91	31.15	32.38	33.48	34.44
CCC/C	27.02	35.63	40.93	44.39	47.56	48.78	49.98	50.64	52.17	53.05	53.79	54.57
Investment grade	0.11	0.31	0.54	0.85	1.18	1.51	1.81	2.10	2.37	2.65	2.91	3.12
Speculative grade	4.65	9.22	13.28	16.59	19.18	21.33	23.11	24.55	25.86	26.99	28.01	28.86
All rated	1.61	3.21	4.66	5.90	6.92	7.80	8.52	9.14	9.70	10.22	10.69	11.08

Source: S&P's "Annual 2005 Global Corporate Default Study and Rating Transitions"

Table 9 One year transition matrix for a "AA" grade bond

	AAA	AA	A	BBB	BB	B	CCC	Default
AA	0.58	90.30	6.12	1.94	0.86	0.15	0.04	0.01

Time homogeneity

The assumption is that the transition is constant every year also means every year a bond in a credit grade has same transition path as the bond in the same grade in the previous year (see Table 9). All AA bonds have the same transition in the years $1, 2, \ldots, n$. Therefore, any bond that is rated as AA will have 0.58% probability of being upgraded to AAA. This will be true in the years 1998, 1999, 2000,…, etc.

Using the assumptions about the simple, time-homogeneous Markov model, stochastic processes can be specified completely in terms of transition probabilities. In practice, it is difficult to implement this time-homogenous assumption. However, we believe that some transitions can be usefully modelled as a Markovian chain for several years. For example, municipal bond ratings can be adequately described as being time-homogeneous Markovian on an annual scale over a period as long as five years. For transitions over a longer period, a different model should be used. As a practical matter, we conclude simply that the estimated TM should be updated regularly and then used to forecast for up to five years.

In practice, we use transition probabilities for two main reasons:

❏ in the trading book, to price credit sensitive instruments adjusting it through the risk-neutral transition probability;
❏ in the banking book, to measure the credit risk of portfolio losses of loans.

For the time non-homogenous state, the transition probability matrix remains Markovian because the basic requirement for the distribution across states in the next period depends only on the distribution in the current period. This means that transitions are "memoryless". History does not matter to the current state. In the credit rating context, this is equivalent to saying that, given a current rating for the credit, the likelihood that the credit will move to any other rating level, or that it will keep its current level, is independent of its past ratings history. This seems to go against the widely held view that future downgrade is perhaps more likely if the credit had experienced a downgrade in the previous period than if it had experienced an upgrade or no change in rating.

A Markovian chain is an approximation for a credit rating transition. The default probability is the key element that inhibits credit ratings from being Markovian. Default is an absorbing state. If this is also assumed to be one of the Markovian states then all assets would eventually default, which means credit is not exposed to the probability of loss but an actual loss.

Transition matrices across the risk segments
Rating transitions display significant differences across the risk segments (corporate, retail, structured finance, etc). The ratings in each market clearly have a different propensity to change. The logical implication of this is that identical ratings in different markets neither have the same meaning nor do they imply the same level of risk. These differences in ratings stability have important implications for investors.

The different level of rating transition is also a measure of diversification and credit rating volatility. Empirical studies have made the following findings.

❑ There is no time homogeneity – either across the risk segments or within a risk segment.
❑ In the short and medium term (1–3 years), rating appears to be more stable in structured finance than in corporate finance.
❑ In the long term (>3 years), structured finance has more volatility than the corporate bonds.
❑ Within the structured products, collateralised debt obligations and some Assets Based Securitisation (ABS) products exhibit much greater ratings mobility than other structured products.
❑ Structured finance has a strong path dependency pattern. A TM for structured finance tends to result in lower grades for conditional downgrading or in higher grades for conditional upgrading.
❑ Although the average pattern of ratings mobility in structured finance stays relatively constant over the time period and across the geographical area, such a pattern masks important heterogeneity across sectors.

Cycle-dependent TMs

We have seen that a time non-homogeneity TM helps the modelling of the credit rating process in a more accurate way. However, we know that modelling time non-homogeneity is very difficult for the following reasons.

❑ This requires the collection of data for a very long period, including both economic booms and recessions. Truncated business cycle coverage can bias the estimates of a TM.
❑ When estimating the risk factor based on the economic conditions and estimating its impact on rating transition, the questions that keep resurfacing are whether to use one or more of such variables, which variables to be used, whether the chosen variable can be forecasted and the how can we avoid the scarcity of data during an economic recession.

Trends in a transition matrix

The most important trends found in modelling a TM are as follows.

❑ *Applying a TM to retail assets*: Fair, Isaac & Co (FICO) has been propagating the use of risk management tools and techniques of corporate banking for retail banking. This includes the use of TMs and value-at-risk (VaR) modelling techniques.
❑ *Continuous-time transition matrix versus discrete-time transition matrix*: While discrete-time TMs assume a change in the credit rating of an obligor after a certain fixed period of time (generally one year), continuous-time TMs assume that the next change may happen any time (actually, this change occurs continuously). This modelling framework has attracted attention in the recent past. There are two advantages, namely that it is possible to estimate transition probabilities to a rare state and that it is also possible to estimate a TM for any period.
❑ *Correlation information*: TMs do not have correlation information. However, different TMs can be built for the entire economic cycle or for bad economic conditions. A TM under a particular economic variable indirectly captures the correlations or conditional probabilities. A number of studies have documented the fact that ratings TMs vary according to the stage of the business cycle, the industry of the obligor and the length of time that has elapsed since the issuance of the bond.

Table 10 One year transition matrix for a "C" grade bond

Start state	End state				Total PD
	A	B	C	D	
A	0.98	0.02	0.00	0.00	1
B	0.03	0.90	0.03	0.04	1
C	0.02	0.11	0.63	0.24	1
D	0.00	0.00	0.00	1.00	1

Table 11 Marginal and cumulative PDs

	Period 1	Period 2	Period 3
Marginal PD	0.24	0.1566	0.1028
Cumulative PD	0.24	0.3966	0.4994
Average PD	0.24	0.1983	0.1664
Survival rate	0.76	0.6034	0.5006

Converting TM into cumulative default rate

The TM gives the probability of moving from a given rating to a higher or lower rating, conditional on the rating at the beginning of the period. A general underlying assumption is that the transition follows a Markov process or those migrations across states are independent from one period to the next. This assumption has the following consequences:

❑ there is no carry-over effect;
❑ transition is a stochastic process;
❑ the conditional distribution given today's value is constant over time, ie, the PD distribution at the end of the period purely depends on the PD at the start of the period.

Cumulative default rate is the total default rate between the beginning of the time period and year T. Marginal default rate is the default rate during year T.

For example, let us find out the default probability for three periods (see Table 11) for an asset that has a PD of C in the beginning of the period (see Table 10), assuming time homogeneity. See Figure 11.

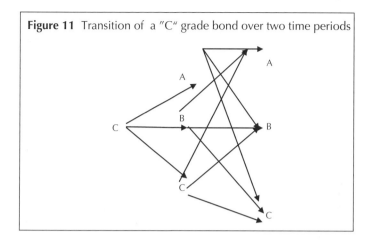

Figure 11 Transition of a "C" grade bond over two time periods

The PD for the first period = 0.24
The PD for the second period = $C \rightarrow A \rightarrow D = 0.02 \times 0 = 0$

$$C \rightarrow B \rightarrow D = 0.11 \times 0.04 = 0.0044$$
$$C \rightarrow C \rightarrow D = 0.63 \times 0.24 = 0.1512$$

The total default rate for the second period = 0.1556
The PD for the third period = $C \rightarrow A \rightarrow A \rightarrow D = 0$

$$C \rightarrow A \rightarrow B \rightarrow D = 0.02 \times 0.02 \times 0.04 = 0.000016$$
$$C \rightarrow A \rightarrow C \rightarrow D = 0$$
$$C \rightarrow B \rightarrow A \rightarrow D = 0$$
$$C \rightarrow B \rightarrow B \rightarrow D = 0.11 \times 0.90 \times 0.04 = 0.00396$$
$$C \rightarrow B \rightarrow C \rightarrow D = 0.11 \times 0.03 \times 0.24 = 0.000792$$
$$C \rightarrow C \rightarrow A \rightarrow D = 0$$
$$C \rightarrow C \rightarrow B \rightarrow D = 0.63 \times 0.11 \times 0.04 = 0.002772$$
$$C \rightarrow C \rightarrow C \rightarrow D = 0.63 \times 0.63 \times 0.24 = 0.095256$$

The total default rate for the third period = 0.10278.

Moody's and S&P's provide cumulative default rates. Figure 12 displays the cumulative default rates reported by S&P's: these rates are observable in practice. These rates are also used to build TMs. Credit migration or TMs characterise the past changes in credit quality of obligors (typically firms) and are a convenient summary of credit behaviour in a portfolio.

Assuming the TMs follow a Markovian chain process, a stochastic process can be modelled to build a one-year TM from a cumulative

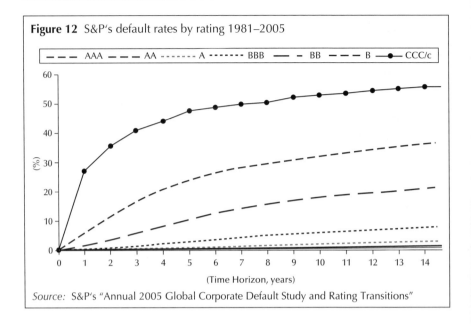

Figure 12 S&P's default rates by rating 1981–2005

Source: S&P's "Annual 2005 Global Corporate Default Study and Rating Transitions"

average default rate matrix. Analysis of the TM published by a rating agency helps in estimating the efficiency of the rating system. TM analysis requires high-frequency information for a long period prior to the defaults.

Criteria for analysing TMs

TMs are to be analysed on the following two criteria.

❑ Rating stability during the time horizon – few rating changes should occur during the time horizon:

  ❑ the frequency of rating changes;
  ❑ the frequency of large rating changes (ie, changes of >3 rating notches);
  ❑ the frequency of rating reversals (rating actions in the opposite direction to a previous rating action).

❑ Sensitivity to the changes in the environment. This is not in contradiction with the previous criteria; ratings should respond to any change in the company's situation. High- or good-quality ratings should be downgraded sufficiently early before any event of default. Therefore, an unstable TM should be observed for a higher than average default rate.

REFERENCES

Ederington, L., 1986, "Why Split Ratings Occur", *Financial Management*.

Beattie, V. and S. Searle, 1992, "Bond Ratings and Inter-Rater Agreement", *Journal of International Securities Markets*.

Cantor, E. and F. Packer, 1995, "Sovereign Credit Ratings", *Current Issues in Economics and Finance*, June.

Cantor, E. and F. Packer, 1996, "Multiple Ratings and Credit Standards, Differences of opinion in the credit ratings", *Reserve Bank of New York-Staff Papers*, April.

Morgan, O. P., 2002, "Rating Banks: Risk and Uncertainty in and Opaque Industry", *American Economic Review*.

Santos, J., 2003, "Why Firm Access to the Bond Market Differs Over the Business Cycle: A Theory and Some Evidence", *Federal Reserve Bank of New York*.

Credit Scoring – Statistical and Empirical Default Models

INTRODUCTION

Credit scoring is a method to assess credit risk. The meaning of "credit scoring" is to assign scores to the characteristics of debt and borrowers and historical default and other loss experienced as an indication of the risk level of the borrower.

Credit scoring is used on a stand-alone basis or as a part of a credit rating. When used on a stand-alone basis, credit scores are used to classify obligors/facilities into comparative grades or into accept/reject groups. If credit scores are used as a part of the credit rating, the credit rating is a measure of risk which may have an absolute default rate attached or may be a comparative grade.

In either case, credit scoring does not measure the quantum of credit risk/default rate. Traditionally, credit scores are not calibrated to the default rate or credit loss. Credit scoring is an intermediate step between measuring risk default rate and measuring risk factors. It is a summary indicator for credit factors.

History of credit scoring

Credit scoring as a method of credit evaluation has been used for more than 50 years. The first successful application of credit scoring was in the area of credit cards. The first retail credit scoring model for credit cards in the US was proposed in around 1941 based on the following parameters for scoring credit card applications:

❑ the applicant's job/position;

❑ the number of years spent in the current position;

❑ the number of years spent at the current address;

❑ bank accounts, life insurance policies;

❑ sex;

❑ the amount of the monthly instalment.

The increase in the US credit card business mandated a reduction in the decision time. In 1956, the Fair, Isaac & Co (FICO) was established to help consumer credit evaluation and in the 1960s computers were brought to process credit card applications. In 1963, Myers and Forgy proposed the application of a multivariate discrimination analysis for credit scoring. In 1975, with the passing of the US Equal Credit Opportunity Act I, credit scoring received complete acceptance. The Act outlawed the discrimination in granting credit by building credit scores based on race, language spoken, sex or age of the borrower.

Since the 1980s, credit scoring has been extended to personal loans. In the 1980s, linear programming and logistic regression were introduced into credit scoring.

Since the early 1990s, credit scoring has become the dominant method for assessing many kinds of consumer loan. Over the past four decades, automation has developed rapidly. Loan decisions are automated and made without the intervention or involvement of individual loan officers. At present, retail lending decisions are supported by a data warehouse. Meta data built as a part of the data warehouse helps in data mining for building, fine-tuning and validating parametric and non-parametric scoring models. Automation has also helped to reduce the time to market for building and updating credit score models.

Credit scoring models that were initially aimed at minimising credit losses by the ordinal ranking of obligors are now being developed to provide profit maximisation and cardinal measurement of credit risk. In the last decade, credit scoring has been extended to small business loans also from individual loans.

CREDIT SCORING MODELS

Credit scoring is a mechanism used to quantify the risk factors relevant for an obligor's ability and willingness to pay. The scores are

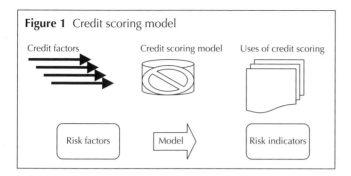

Figure 1 Credit scoring model

Credit factors   Credit scoring model   Uses of credit scoring

Risk factors   Model   Risk indicators

used either on a stand-alone basis or as a part of the credit rating process.

The aim of the credit score model is to build a single aggregate risk indicator for a set of risk factors (See Figure 1). The risk indicator indicates the ordinal or cardinal credit risk level of the obligor.

Models

Credit risk factors are assumed to be independent variables that best indicate the credit risk of the obligor, X_i, $i = 1, ..., n$. The aim is to assign a single aggregate risk indicator, also called the dependent variable Y, to the customer by measuring credit risk factors. Credit scoring is a process to discover or calibrate the credit risk measure Y from the values of credit risk factors.

There are two steps involved in credit scoring:

❏ model building;
❏ using model measuring scores for prediction.

To build a scoring model, past debtors are analysed to identify the credit factors or characteristics that relate to creditworthiness. The model is a tool to compare a borrower to past debtors with similar risk factors or characteristics whose credit performances are known, and to indicate by a single score number the future performances of current customers.

The input data is historical data on n credit customers with a known risk indicator. The input data is in the following form.

❏ Independent variables/risk factors: X_{ij}, $i = 1, ..., n$ customers, $j = 1, ..., r$ risk factors.
❏ Dependent variable/risk indicators: Y_i, $i = 1, ..., n$.

X_{ij} is the value of the jth predictor of the ith customer and Y_i is the known risk indicator of the ith customer. The risk values can either have two values (eg, default and non-default) or have multiple values (eg, different classes of risk levels). The scoring model is generated from the input data through various approaches and is applied to new customers to predict their unknown value of dependent variable Y. A credit decision is taken based on this prediction of Y. The prediction can either be the risk classes with two or multiple categorical values or a continuous score (eg, from 0 to 100%, which may be representing the default probabilities). If Y has continuous values, a credit decision is made by comparing the value of Y with a suitable threshold. For a given set of debtors we know the independent and dependent variable:

$X_{11}, X_{12}, X_{13}, ..., Y_1$
$X_{21}, X_{22}, X_{23}, ..., Y_2$
$...$

$...$

$X_{n1}, X_{n2}, X_{n3}, ..., Y_n$

Therefore, the credit scoring model aims at discovering the value of the unknown credit risk indicator given the values of the independent credit risk factors for a borrower:

$X_{u1}, X_{u2}, X_{u3}, ..., Y_u = ?$

Evaluating credit scoring: Credit scoring uses two data mining techniques: classification and clustering. Credit scoring, in common with other data mining techniques, is composed of three stages. From the data mining perspective the three stages are data preparation, data analysis and model evaluation (for a comparison of the data mining and credit scoring stages see Figure 2).

Model evaluation is one of the basic ingredients in credit scoring. Primarily there are two reasons for model evaluation:

❑ reconciling differences between model algorithms and different parameters for a particular algorithm;
❑ size of samples used to built the model (sample bias introduced) and pre-processing done on the inputs.

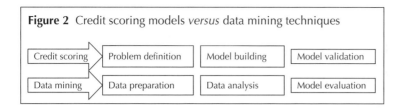

Figure 2 Credit scoring models *versus* data mining techniques

| Credit scoring | Problem definition | Model building | Model validation |
| Data mining | Data preparation | Data analysis | Model evaluation |

In this chapter we have discussed the challenges and issues relating to each stage – this includes input-related sub-processes, model algorithms and model-relevant sub-processes.

Objective of credit scoring

As with any other risk components, credit scores attempt to measure credit risk. However, owing to reasons of cost-effectiveness, the risk characteristics of retail portfolios are measured at portfolio level and not at the individual exposure level. However, the objective of credit score measurement cannot be different from risk measurement at the portfolio level. The credit scoring models should satisfy the following aims:

❑ the accuracy of estimates of credit risk and homogeneity of risk classes;
❑ the stability over time of risk classes and their properties;
❑ the dependence of the risk measurement on the business cycle;
❑ the stability of transition matrices; and
❑ the correlation of risks.

Issues that need to be addressed:

❑ type of variables and process of selecting variables;
❑ historical data;
❑ choice of statistical technique;
❑ forecasting horizon;
❑ stability of risk classes;
❑ frequency for tool revision; and
❑ the interaction between business cycles, forecasting and tool revision.

Most countries have one or more credit bureaux. Credit bureaux compile and disseminate reports on the creditworthiness of consumers and small businesses. Primarily, credit bureaux have two functions.

❑ Credit bureaux clean, process and make available the data relevant for assessing credit decisions.

❑ Credit bureaux collect the lending data from various types of lenders. They build credit scoring models to produce the relevant data and provide scoring models in addition to data.

Typically, a credit bureau collects and provides four types of data for consumer and small business lending.

❑ *Identification*: This includes a person's name (current and previous), addresses, social security number, date of birth and current and previous employers.

❑ *Existing credit*: Outstanding to banks, retailers and lenders (participating institutions). This includes account type, the date opened, the credit limit or loan amount, outstanding balances and the timeliness of payments on the account.

❑ *Public records*: Credit history from public records – bankruptcy filing, tax judgement, arrests and convictions.

❑ Number of enquiries with the credit bureaux for additional credit.

Various studies in the US have shown the benefit arising out of the services of credit bureaux in managing the credit loss. McCorkell (2002) has argued that using scorecards built with data supplied by credit bureaux results in 20–30% lower delinquency rates than lending decisions based solely on judgmental evaluation of applications for credit. In other words, for a given default rate, the use of scorecards yields an increase in the acceptance rate. Therefore, in the absence of credit bureau services for a given outstanding amount, the credit losses will increase. Alternatively, to maintain a given default rate, the outstanding amount has to be reduced drastically.

Scoring models – a classification process
The credit scoring is essentially a classification process. The process assigns obligors into one of the multiple categories or classes based on their credit risk factors. There are four types of classification processes.

❑ *Deductive*: Under this approach, the relevant attributes are given weights and aggregated. The choice, measurement and weights of the attributes are determined by experience. It is argued that this system is not fully objective and is based on past experience. In practice, deductive scores are losing their effectiveness and acceptance.

❑ *Empirical*: Here the selection of the relevant attributes and the calculation of the scores are based on the past credit data with the help of credit algorithms.

❑ *Inductive learning models*: The classification logic is extracted from discovering and analysing patterns found in prior solved cases.

❑ *Learning problem*: Sometimes, values of the dependent variable (credit risk) for the sample are not known. The models are then solved as a learning problem.

In making consumer credit decisions, the empirical scoring methods are not only a way of handling the large number of transactions, but it seems that they produce more accurate classifications than subjective judgmental assessments by human experts.

Classification versus clustering: Credit scoring primarily attempts to address the following two problems:

❑ definition of the classes to be predicted;
❑ construction of predictive function.

Classification aims at dividing the population into homogenous sub-groups with similar risk characterstics. Clustering is different from classification. Classification is driven by the externally defined classes (risk classes) while clustering segments the population on the basis of the data/attributes available within the data set.

If, for example, default as an attribute is available in the data set, then clustering can be used as a technique to build to clusters of "default/non-default".

If default is not one of the attributes available in the data set and we want to build a "default" class, it will be called classification.

Application versus behaviour scoring models
The most important questions that need to be answered during the lifespan of a credit are as follows:

❑ should the borrower be extended credit at all?
❑ what are the terms for the credit?

❑ is intervention required during the lifespan of the credit?

❑ what are the terms of the intervention?

Traditional credit scoring methods are being used to address the first two questions. Currently, new credit scoring models are also being established to address the last two questions. There are two kinds of credit evaluations during the credit process.

❑ *To make decisions on new credit applications*: This is to assess the level of the risk on the new application.

❑ *To supervise the existing credit*: This is to evaluate the level of the risk on the existing exposure to detect the default early.

For credit cards (unsecured revolving credit), in addition to application scoring, involving the identification of acceptable applications for extending credit, behaviour scoring, where borrowers likely to default or miss payments are identified, is also important. Some of the elements considered in credit card behaviour scoring are as follows:

❑ the average balance/total limit on the card;

❑ the average limit on the credit card – the total limit of the obligor for all the credit cards/number of credit cards the obligor owns;

❑ the average credit card transaction – the total value of the credit card transaction per obligor/total number of credit cards transactions in a month;

❑ the average card repayment – (net total income/12);

❑ gross balance carry over all the credit cards/net total monthly income;

❑ gross balance carry over all the credit cards/total credit line from all credit cards.

Application scoring: Application scoring models are the default or credit risk predictive models. Application models are designed to rank-order the risk of lenders' applicants by measuring the credit risk factors. In general, the application scoring models are automated decision models. Banks define a cut-off band or score under which the application is either assessed by an analyst or rejected.

Behaviour Scoring: Behaviour scoring models are used to help in the decision making process by forecasting future performance. The decisions supported are the limit, whether to market new products and how to manage the recovery. In addition to the application data, information on repayment is used for behaviour scoring. Two types of behaviour scoring models exist.

❑ Extension of the application models with information on the repayment behaviour.
❑ Probability models of customer behaviour using Markov chains. In this model, customers jump from one state to another. These models are divided into two types:

    ❑ Bayesian – customer own behaviour;
    ❑ Historical – sample of previous customers.

The payment behaviour is measured for the "performance period" under review. The performance period may vary from 1 to 5 years.

Risk factors

Risk factors generally either determine the ability and willingness to pay or measure the ability and willingness to pay.

As with any other credit decisions, the following "5C's" are the primary attributes of the obligor to be considered in credit scoring:

❑ character;
❑ capital or assets of the borrower;
❑ capacity to repay – ability to repay;
❑ collateral – willingness to repay; and
❑ conditions – economic conditions.

Data inputs for consumer lending

For different types of credits, different credit information is needed to assess their risks. Applications for lending and data from the credit bureau are the two primary sources of data inputs for consumer lending. For consumer credit and private customers, the data is relatively homogenous and easily obtained. Some examples of the predictors that are used usually are given in Table 1.

Credit bureaux work as an exchange for sharing credit risk factors. Each country has a different regulatory and legal regime. Sharing of credit variables is governed by privacy acts and fair lending acts. In many countries, only the default information is shared, this situation is referred to as a "negative only" model or regime. A sample of risk factors shared in a "negative only" regime is as follows.

Table 1 Some examples of commonly used predictors

Basic personal information	Age, sex, education
Family information	Marriage status, number of children
Residential information	Status, number of years at the present address
Employment status	Occupation, number of years in the current occupation
Financial status	Salary, expenses, other assets
Security information	Value and type of security
History of other borrowings	Other loans outstanding, repayment history, other applications for loans made

❑ Enquiries made:

    ❑ for the credit purposes;

    ❑ total enquiries;

    ❑ in the past 6 months;

    ❑ in the past 12 months;

    ❑ for the bank card purposes;

    ❑ total enquiries;

    ❑ in the past 6 months.

❑ The number of accounts outstanding for more than 60 days.

❑ The number of accounts charged off, collected, etc.

❑ The number of bankruptcies.

❑ The number of months since the most recent bankruptcies.

❑ Worst status on the bank card.

In the full model, information on account not defaulted or charged off is also shared. A sample of additional elements in a full model is as follows.

❑ Accounts opened, paid and closed:

    ❑ total;

    ❑ total in the past 6 months;

    ❑ total in the past 12 months.

❑ Average balances:

    ❑ all;

    ❑ revolving credit/bank card;

    ❑ finance instalment;

    ❑ for real estate.

Prohibited credit factors: Every country has some type of law to pro-hibit denial of credit on the basis of race, colour, religion, national origin, sex, marital status, age, etc. Therefore, credit scoring models cannot take one of these factors as risk for modelling.

In addition to the variables discussed so far, researchers have been working to further improve the performance of the applica-tion and behaviour scoring models.

Dunn (1991) found that the following three variables have a sig-nificantly positive effect on the default rates of the credit cards:

❑ the total minimum required payment to income ratio;
❑ the percentage of total credit line which the consumer has used;
❑ the number of credit cards on which the consumer has charged to the credit limit.

The minimum required payment to income ratio is more relevant for the consumer's ability to avoid default in the short term. The percentage of total credit line that has been used reflects a con-sumer's ability to avoid default by relying on additional credit to pay off old debt. The larger the number of credit cards the higher degree of default risk the obligor is exposed to.

The inputs identified in the preceding paragraphs are only illus-trative examples. The inputs (and the weights) used for one sample of customers may not be valid for another sample. The inputs differ according to the product, quantum and tenure of the loan and the population of the customer.

Data inputs for business lending
Variables should be able to distinguish between failing and non-failing obligors. There is still no general theory concerning primary variables that are relevant to distinguish between failing and non-failing firms. In general, cashflow-based analysis and ratios are considered the likely candidates for corporate credit scoring mod-els. However, some research has found that the financial ratios are not the relevant risk factors for small and medium-sized enter-prises (SMEs). The key sources of data inputs are summarised in Figure 3.

We have explained these factors in detail in the chapter "Credit Rating", although, although not all factors are considered relevant for SMEs.

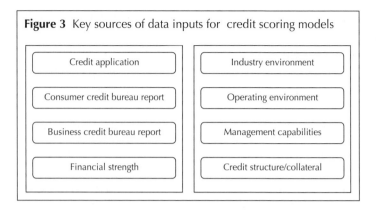

Figure 3 Key sources of data inputs for credit scoring models

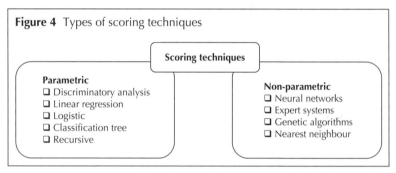

Figure 4 Types of scoring techniques

Types of models
Some of the ways to analyse the models are now discussed.

Scoring techniques

There are two approaches for studying scoring techniques (see Figure 4):

❑ parametric and non-parametric;
❑ classical statistical and inductive learning.

Parametric versus non-parametric

Parametric techniques: The parametric technique hypothesises a certain relationship between the dependent and independent variables. This relationship may be a statistical relationship or a theory-driven relationship. The parameters of the model are driven by the model algorithm. These techniques are referred to as model-driven approaches, parametric methods or theory-driven approaches.

Parametric techniques are difficult to implement for non-linear relationships between variables.

Linearity assumptions made by parametric models make it easier to build risk classes. We know that the relationship between a risk indicator and a risk factor is not linear. For non-linear models, parametric techniques split variables into classes/categories so that default risk is homogenous and a linearity assumption is valid.

Non-parametric techniques: This is a discovery-based approach in which a matching pattern or other algorithms are employed to determine the complex relationship in the data. Like the parametric approach this approach also assumes a certain underlying model, the model keeps changing with more data. These models utilise computational power to search and iterate until a good fit is achieved. Since they build a model automatically based on the patterns found in the data, they are also called non-parametric methods. This method seems to be more attractive if there is scant knowledge about the statistical properties of the data. Non-parametric approaches produce models that are large, idiosyncratic and difficult to interpret. There is always risk of over-fit and poor generalisation to new cases.

Classical statistical techniques and inductive learning

❑ *Classical statistical techniques* – regression techniques:

  ❑ discriminant analysis;
  ❑ logistic regression, also called conditional probability models.

❑ *Inductive learning*:

  ❑ recursive partitioning analysis;
  ❑ neural networks (NNs);
  ❑ expert systems;
  ❑ genetic algorithms;
  ❑ nearest neighbour.

Initially, the statistical methods were based on the discrimination analysis for general classification problems. Issues and concerns with this method led to a further refinement and the development of other regression methods. Classic statistical techniques and inductive learning techniques both have different assumptions concerning the relationships between the independent variables. Linear discriminant analysis is based on a linear combination of independent variables, logit analysis uses the logistic cumulative probability function and a genetic algorithm is a global procedure based on the mechanics of natural selection and natural genetics. The NN is a classification method.

Classification trees or recursive partitioning algorithms and expert systems classify the borrowers into groups, each group being homogenous in its default risk and as different from the default risk of other groups as possible. In classification trees, the borrower populations are split to ensure the difference in the average default risk between the two subsets is as large as possible.

Genetic algorithms are used to discover the "fittest" scorecards among populations of scorecards where fitness means correct classification. More recently, vector support machines have been used in classification problems. Superficially these machines resemble the linear programming approach.

Logistics are by far the most successful regression or classical statistical method. Discriminant credit scoring models have been used commercially for more than three decades and NN models and logit credit scoring models have been in use for more than 10 and 20 years, respectively.

The statistical methods provide statistical techniques to identify and validate the variables and models. However, both types of models have matured and are widely used. Each method uses a different variable, models different dimensions of the borrower and various experts have empirically found different levels of accuracy for each method (see Table 2).

Table 2 Statistical methods for credit scoring and their empirical level of accuracy

Author	Discriminant analysis	Logistic regression	NNs	Classification tree	Genetic algorithms
Srinivasan (1987)	87.5	89.3	–	–	93.2
Boyle (1992)	77.5	–	–	75	–
Henley (1995)	43.4	43.3	–	43.8	–
Desai (1997)	66.5	67.3	64	67.3	–
Yobas (1997)	68.4	–	62	62.3	64.5

Source: Adopted from "A Survey of the Issues in Consumer Credit Modelling Research" by L. C. Thomas (2005).

Accuracy is measured by comparing pairs of statistics such as acceptance rate, bad debt rate among the accepts, marginal bad debt rate, etc.

Challenges in classical techniques
These models broadly implement two types of approach to work with multiple variables:

❑ simultaneous procedure;
❑ stepwise procedure – choose the best variable.

The simultaneous procedure involves obtaining the model for the variables identified theoretically as economically important (eg, cashflow-based ratios). The stepwise procedure is to statistically fit the variables to produce a good discrimination model using and identifying the statistically important variables. The simultaneous model is relevant from an economic perspective and the stepwise model is relevant from a statistical fit perspective.

There are many challenges in the selection of independent variables. For the long-term validity of the model, variables should have more relevance from an economic perspective than from the statistical perspective. Variables empirically selected on the basis of statistical fit may lead to a sample-specific unstable model, which is over-fitted. Variables should be selected on the basis of the theoretical framework but no such theoretical framework exists. Empirical studies also throw up contradictory findings. For example, some studies have found the financial ratios relevant for small business and their models take financial ratios as variables, while other studies have found financial variables irrelevant and excluded them from their statistical models. There is also criticism against using financial ratios as accounting information as they are known for their deficiencies in accuracy, a lack of impartiality, errors and missing items. The sampling of firms is also an issue.

Which technique to use?
One of the important elements determining the choice of model is the forecasting horizon for firm failure. Based on the forecasting horizon, there are three types of firm failure:

❑ *sudden* – very short-term and sudden decline in performance;
❑ *acute* – rapid decline in the year before failure, with good performance before the rapid decline;

❑ *chronic failure* – poor performance for many years before failure.

For example, classic models also do not consider a time frame within which a firm is likely to fail. A logit model is fit for sudden failure and for acute failure only during the last year but is not fit for chronic failure.

The following factors determine the choice of the technique to be used for credit scoring:

❑ the types of variables (economic *versus* statistical variables);
❑ types of failures;
❑ the forecasting horizon;
❑ the sample size and historical data;
❑ the linearity or non-linearity of the variables;
❑ the sensitivity to extreme values.

Credit scoring models for risk segments
Different types of credit scoring models are used in various ways for each of the risk segments (see Figure 5).

Typically, for consumer and SME portfolios, the expected losses (ELs) tend to be higher (except for first mortgages) when compared to other exposures. However, correlations and therefore unexpected loss (UL) tends to be lower. This assumption has been the driving force for consumer lending models over the past four decades. The net impact of this assumption is that consumer and SME models do not consider correlation. ELs (the past lost experience) have been the driving force for model calibration and model building.

The factors to be considered, the relative weighting and even the statistical modelling used are different for consumer, SME and corporate segments. It should be noted that there are no correct models. The only differentiating factors are that a model "works" or "doesn't work".

Credit scores for consumer lending: Consumer lending means "homogeneous portfolios comprising of a large number of small, low-value loans to individuals and where the incremental risk of any single exposure is small". Lending includes credit cards, residential mortgages and home equity loans as well as other personal loans such as educational or auto loans.

Models have generally used empirical studies. They improve the prediction accuracy by appropriate variable selection. Authors agree

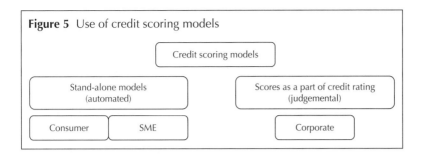

Figure 5 Use of credit scoring models

Credit scoring models

Stand-alone models (automated)

Scores as a part of credit rating (judgemental)

Consumer   SME

Corporate

for a need for a general theoretical framework for default but as yet there is still no general theory concerning primary variables that are relevant for distinguishing between defaulter and non-defaulter. These models are generally driven by the statistical fitness test. Even the variables are also identified from statistical significance.

Credit scoring for small businesses

The use of credit scoring for small businesses has changed the entire economics of small business lending. Banks can extend lending without face-to-face contact and based on third-party data. It has a major impact on the application processing costs, the collateral documents and relationship. The most important differences between small and large businesses are as follows.

❏ The cost for processing a lending is a good sum in terms of per dollar lent.
❏ Small business lending is driven by relationship.
❏ The quality of accounting statements and other disclosures in small businesses are not as good.
❏ The repayment experience is the most important input to the small business model. Surprisingly, financial statements and financial ratios are not particularly important inputs to the small business credit scoring model.

Although the models are similar to those of consumer lending, there are still a few notable differences.

❏ Small business is generally defined by the exposure (of say less than €1 million) and not by the legal structure of the business. There is a large variation in the legal structure of small businesses. The legal structure of an obligor has a major impact on the choice of credit risk factors and quality of financial statements.

❑ It is difficult to attain homogeneity with small businesses as the size of exposure, types of industry and the legal structure of obligors all vary substantially.
❑ Whether or not it is sufficient to only use the quantification method is still an unresolved question.

During the 1990s, credit scoring firms in the US developed products to evaluate the credit of small businesses, including trade credit. Over the past decade, small business loans have started to be evaluated using credit scoring.

For small businesses, credit scoring systems are statistical models that predict the likelihood that a firm will be seriously delinquent for 90 days or more in making loan repayment. The score does not predict a company's *ability* to pay, but rather its *willingness* to pay in a timely fashion. The quality of financial statements and legal structure makes it difficult to measure the ability to pay.

The willingness and ability of the business owner to repay personal borrowings could reasonably be assumed to correlate with the ability and willingness of the firm, controlled and managed by the owner, to repay its loans. Therefore, credit scoring models similar to the models developed for consumer lending can, in general, be applied to small businesses. FICO also supports this argument from the empirical analysis of a large database of small businesses loans.

On the basis of the use of scoring models, there are four types of credit scoring models for small businesses.

❑ *Specific model*: An application and behavioural model from a sample of businesses that most resemble the bank's actual behaviour.
❑ *Industry models*: These are application and behavioural models for a given industry.
❑ *Generic models*: These are application and behavioural models for predicting the likelihood of a company paying in a severely delinquent manner based on a sample of businesses from across all industry sectors.
❑ *Collection or payment models*: Payment performance is based on the owner's payment behaviour.

For consumer-related credit scoring, multivariate linear discriminant analysis has been the statistical method of preference for most analysts. This method is not found to be suitable for small businesses.

RMA (The Risk Management Association)–FICO (Fair, Isaac and Co.), in developing the Small Business Scoring System (SBSS), has chosen to use a regression model transformed to the logistic scale.

Credit scoring for corporate lending

Credit scoring is a part of the credit rating system in an internal rating system (see Table 3). However, the relative importance of credit scoring varies across the banks and across the portfolio. We have explained the rating process in the "Credit Rating" chapter. Some banks, especially smaller banks, use the ratings issued by external credit rating agencies. The rating agencies also use credit scoring as a part of the credit rating.

Treacy and Carey (1998) summarise their view on statistical models *versus* judgemental models as follows:

> *Many banks use statistical models as an element of the rating process, but banks generally believe that properly managed judgmental rating systems deliver more accurate estimates of risk than statistical models. For large exposures, the benefits of such human judgments may outweigh the higher costs of judgmental systems. In contrast, credit scores are often the primary basis for credit decisions for small lending exposures such as consumer credit.*

An example of a corporate rating model based on the empirical approach is Moody's Public Firm Risk model, which uses NNs as the main technology.

Table 3 Credit scoring *versus* credit rating

	Credit scoring	Internal credit rating	External credit rating
Model builder	Vendors or lenders	Lenders	External credit rating agencies
assets	Consumer and small businesses	Business loans for corporates	Large corporates, multi national companies and sovereign
Risk component	Class or scores for a particular risk segment	Number of grades specific to the institutions	Number of grades – common framework across the rating agencies
Users	Institution itself	Institution and supervisors	Investors, regulators
Method	Empirical	Empirical, expert judgment	Empirical, structural, expert judgment

Altman's Z-score model

Although experts have never used sound theoretical constructions to build the credit scoring models, they have generally concentrated on financial ratios. Experts have been working to discover the financial ratios that can predict corporate bankruptcy.

In 1966, William Beaver conducted a very comprehensive study using a variety of financial ratios and concluded that the cashflow-to-debt ratio was the single best predictor using univariate analysis. Edward Altman (1968) expanded Beaver's work to include multiple discriminant analysis (MDA).

Altman selected 33 publicly traded bankrupt manufacturing companies between 1946 and 1965 and matched them to 33 firms on a random basis for a stratified sample (assets and industry). The results of the MDA exercise yielded an equation that Altman called the Z-score, which correctly classified 94% of the bankrupt companies and 97% of the non-bankrupt companies one year prior to bankruptcy. These percentages dropped when trying to predict a bankruptcy two or more years ahead of its occurrence. The Z-score identifies the financial health of any business based on observable accounting and market ratios. The Altman model separates defaulting firms from non-defaulting firms based on the discriminatory power of a linear combination of financial ratios.

Altman had used the following five ratios in his Z-score:

❏ working capital over total assets;
❏ retained earning over total assets;
❏ earnings before interest and taxes over total assets;
❏ market value of equity over book value of total liabilities;
❏ sales over total assets.

Altman's Z-score is available in various forms. One of the ways to calculate the Z-scores is to use the following equation:

$$Z = -(X_1 + X_2 + X_3 + X_4 + X_5)$$

where, $X_1 = 1.2 \times$ (Current liabilities)/book assets value; $X_2 = 1.4 \times$ (Retained earnings)/book assets value; $X_3 = 3.3 \times$ (Operating income before depreciation)/book assets value; $X_4 = 0.6 \times$ (Market capitalisation)/book value of liabilities; $X_5 = $ Sales/book assets value.

Table 4 Z-score based cumulative default rates

Z-scores	1 year (%)	2 year (%)	3 year (%)	4 year (%)	6 year (%)	8 year (%)	10 year (%)
Z < −1.40	10.8	24.6	36.6	32.8	32.1	33.5	34.0
−0.84 < Z < 1.44	3.5	7.8	10.2	11.4	12.6	13.8	15.4
1.45 < Z < 2.28	0.7	1.8	2.9	4.2	6.3	7.6	7.7
2.29 < Z < 3.17	0.2	1.2	2.4	3.3	3.9	4.6	5.3
3.19 < Z < 4.47	0.2	0.4	0.6	0.9	0.9	1.4	1.9
4.50 < Z < 5.92	0.0	0.1	0.6	0.8	0.7	0.8	1.0
Z < 5.99	0.0	0.0	0.0	0.0	0.0	0.7	0.8

Source: "Benchmarking the rating agencies: In search of true credit rating transition matrix" by Edward Altman and Herbert Rijken (2002).

The calculation typically produces a Z-score of between −5 and 10, with a high Z-score implying a better credit quality and lower chance of bankruptcy. Z-scores are not directly interpreted as default probabilities and work as ordinal measures of financial health. Therefore, they cannot be directly used for valuation, quantitative risk assessment and capital allocation purposes.

However, different Z-scores mean a different credit quality and different default rates. This is supported by the studies conducted by Edward Altman and Herbert Rijken on the database of companies rated by Standard & Poor's for the period 1980–2002 (see Table 4). They computed cumulative default rates over a period of 10 years for different Z-scores.

Various versions of Z-scores have been published and used. Empirical studies in various countries by various experts have used different ratios to compute Z-scores and consequently some of the studies may also be conflicting.

Risk factors also vary based on the focus of the assessment. Factors for a short-term credit assessment focus are different from factors for medium-term and long-term focus. The liquidity ratio plays a major role for the short-term focus.

ISSUES IN CREDIT SCORING
Sampling bias
A scoring system is developed from a sample of obligors who have been granted credit. This sample is biased towards the applicants

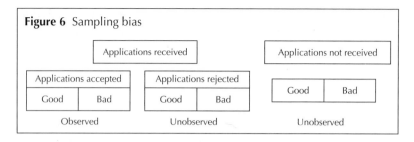

Figure 6 Sampling bias

who are eligible for credit; however, it is not unbiased when applied to the entire population of applicants seeking credit. The bias introduced in the scoring model is that the obligors who never applied for the loan and obligors who are rejected are not considered in developing systems to separate good classification from bad (see Figure 6).

Adjustment to the sampling population for the rejected applications is called reject inference.

Reject inference

There are three types of reject inference. One is to draw the sample from the representative of the whole population. The reject inference techniques are ideal but expensive to implement (the tracking of rejected applications is very expensive because data is simply not available).

The second is to draw the sample from the accepted population but extend the pattern of accepted application into the rejected application. Therefore, the information about the rejected applications can be built up and integrated by modelling some variable of the accepted application. However, various authors have argued that this method has unrealistic assumptions.

The last technique is to assume that the distribution of an accepted application is different from the rejected application. In such cases, inferences about rejected cases from an accepted

Table 5 Measuring application and behaviour performance

Months	1	2	3	4	5	6	7	8	9	10	11	12
Application scoring behaviour		Measure initial performance										

application will be incorrect. Various researchers have concluded that portfolio performance can be improved significantly by including the data on rejected applications.

For different scoring models, different techniques are used.

Time horizon

The time horizon is an important determinant in credit scoring. The time horizon has a different meaning in application and behaviour scoring. One of the important factors to be considered in the time horizon is that the time horizon must be consistent with the nature of data.

The time horizon is the time between the two snapshots in the scoring process (see Table 5). For application models this is the time period between the acceptance of the application and the measurement of the good/bad classification. For behaviour models the performance is measured during the two time horizon.

For the application scoring model: The time horizon represents the time between the application and the good/bad classification. In general, this period is taken to be 12–18 months. Any shorter time horizon will result in underestimation of the bad rates and a larger time horizon will introduce a population drift.

For the behavioural model: Changes in the performance are measured in the behaviour model. The performance variables are in addition to the application variables. The initial behaviour is measured during the "measure performance" period. Behaviour is again measured during "measure outcome" period. Changes in the variable during the two periods is called the behaviour model. For example, for a mortgage loan, pre-payment behaviour is measured for every sub-part portfolio every year. Sub-portfolios are built on the basis of the number of years completed. It is observed that pre-payments are higher for the sub-portfolio one, two and three years

13	14	15	16	17	18	19	20	21	22	23	24	..

Measure outcome between
12 and 18 months from lending
Measure outcome/performance

old bonds. After the first three years the pre-payment is tapered-off to say 1.5% pa.

Good *versus* bad

In most scoring model systems the performances of the risks are typically split into two categories: good and bad. The definition of "good" and "bad" risks varies and depends on the aim of the scoring and the risk segment. For business credit scoring models, the examples of "bad" definition may be bankruptcy filing by a company, bond default, bank loan default, etc. For consumer credit scoring models, bad risks may include the number of months of missed payment, amount over the overdraft limit, current-account turnover or functions of these and other variables.

Risk classes

Under the risk classes, the customers are identified into more than two classes based on the degree of "goodness". Bad is generally classified into one reject class.

The definition of risk class also varies in behaviour scoring and depends upon the purpose of the scoring. For behaviour scoring the classes are typically divided on the basis of the number of consecutive missed payments. Table 6 summarises a typical example.

Impact of default definition on predicting power of the existing credit scoring model

The default definition that is traditionally used by banks includes bankruptcy, restructuring and delay in payments, and this definition is not the same tight definition given by Basel. One of the important questions which needs to be answered after acceptance of the Basel II Accord is the impact of a tighter definition of default on the prediction power of the credit scoring model. Evelyn and Hayden (2003) argued that not much of the predictive power is lost if the bankruptcy model is used to predict the credit loss events of rescheduling and delayed payments instead of alternative models specifically derived for this definition. Therefore, the models being

Table 6 Risk classes

Consecutive missed payments	0	1	2	3
Class	R	R_1	R_2	R_3

used by the banks are not outdated for the tighter definition of default as provided by Basel II.

Incorporating economic cycles into the credit scoring models

Credit scoring models generally have several years of time lag between the transaction collected and used to build a credit scoring model. For example, a model used in 2006 was probably built in 2005 by using data collected in 2001–2003. Therefore, the model actually represents the macroeconomic conditions as they existed during that time. Thus, scorecards are to be constantly re-developed. There are two ways to address this problem. The first is to build scoring models suitable for various types of economic conditions. However, the data for different types of economic conditions may not be available or the models need to be built on the older data. The other approach is to build the scoring models using Bayesian learning networks incorporating Markov chains. In this method the impact of the economic variables is incorporated into the scoring models. However, these are not included in the scores themselves, but are built into the dynamics of the way the score changes. Incorporating economic cycles into the credit scoring method enables the building of robust credit scoring models.

Profit maximisation

One of the other challenges outside the scope of this book is to model profit maximisation into the credit scoring model. This is also called profit scoring. A borrower that is expected to generate profit (over a certain amount or by percentage) is good and the borrower that is going to lose money is called bad (or less than certain amount or percentage). Therefore, under profit maximisation, the definition of good and bad is changed and linked to profit threshold instead of default threshold.

In the last couple of years, pricing and profit maximisation models (Oliver 2001; Keeney and Oliver 2003) have started appearing. These models seek to identify price-setting policies as a function of default probability that maximise the overall profit.

Champion *versus* challenger

A challenger system run on the subset of the existing customers and their performance is measured with the existing "Champion" system. Many of the vendors of credit scoring software have an inbuilt facility for building challenger models. Challenger models enable the user to test out alternate policies and scoring models on the chosen samples. This facility allows minor adjustments to be made on the models continuously. Continuous minor adjustment also enables an elongation of time for complete rebuilding of scores.

Credit risk portfolio models

While credit scoring has been used for more than four decades, the current retail credit risk modelling at US banking institutions has been relatively slow in development and is characterised by a wide divergence in approaches.

There seems to be no incentives to develop retail credit risk portfolio models in the past. This is probably due to two reasons.

❑ The retail sector has never had to manage its risk in value.
❑ No mark-to-market is available for credit risk. The only channel available for mark to market is the securitisation framework. However, it is very difficult to extract any valuation or risk measures from the securitised framework. The reasons for such a minimal development of the retail portfolio model or adoption-only default risk models is the non-availability of the following information or data, which is useful for benchmarking, pricing, etc:

   ❑ agency credit rating and market data on pricing loans;
   ❑ option pricing models in retail credit;
   ❑ benchmark models to model "loss distribution". CreditRisk+ is the only model that can be used for retail credit.

The scoring models and the associated data provide an excellent raw material to develop and implement retail credit portfolio models. With a desire to implement the internal ratings based (IRB) approach, banks with large retail portfolios have started recognising the value of credit risk modelling for a wide variety of business purposes including allocation of economic capital. Given the advanced scoring models developed by the banks and historical data available with each bank and within the industry through credit bureaux, risks in the retail credit portfolio will be modelled earlier than in other portfolios.

The Basel II risk-bucketing approach requires PD and LGD to be estimated for homogenous sub-portfolios. Building a homogenous portfolio for corporates and SMEs is very difficult compared with retail portfolio using credit scoring. This bucketing exercise helps in reducing the importance of the idiosyncratic risks in the overall risk (as compared to systemic risk). However, this does not mean that systemic risk is also diversified away. Retail portfolio primarily includes low-value loans to individuals and small businesses managed on a pooled basis. Banks have developed application and behaviour scoring models. While application scoring models collect information during the loan application process, using behaviour models, information is refreshed periodically and this information will also help in collection and provisioning.

Most of the retail portfolio risk models being used are default risk models. The Basel Accord regulates risk measurement on the basis of risk segments. In terms of Basel II, credit scoring contributes to the internal assessment of credit risk or is a part of the internal rating system. So let us understand the impact of the Basel II Accord on the retail segment. According to the Accord, retail exposures can be divided into the following four types.

❑ *Unsecured revolving exposures*: An outstanding amount fluctuates within the limit. For an exposure to be classifed as *unsecured revolving exposure*, according to IRB requirements, the exposure must be unconditionally cancellable and should have low volatility of loss rates relative to its average level of loss rates. This includes credit cards and overdraft lines with cheques facility.
❑ *Residential mortgages*: This includes first and subsequent liens, term loans, lines of credit and legally binding commitments to lend.
❑ *Small business*: "Small business" applies to small loans of any kind, to individuals or companies for business purposes. For small business loans, the total exposure to a single borrower is limited to US$1 million, on a fully consolidated basis, although supervisors may allow amounts above the limit.
❑ Other not falling into the previous three categories.

The new Accord impacts the retail credit in the following three ways.

❑ Separate risk segments are to be built on the basis of risk characteristics. For each segment, the bank needs to consider historical data covering a minimum period of five years. The portfolio segmentation must involve product type and default character-istics and should segregate delinquent from non-delinquent exposure.

❑ Validation and verification of PD and LGD. The PD and LGD estimates should be valid for the next 12 months. This means scores should be adjusted for the economic cycles.

❑ Still no consumer portfolio models have been developed. Basel has developed the regulatory capital formulae using the one-factor Merton corporate model.

One of the most important criticisms of the regulatory capital computation formula proposed by Basel for the retail segment has been that the formula is not concave in some regions of PD. However, little of the work done in this field has covered retail credit risk portfolio models.

Retail products exhibit higher default frequencies, and higher ELs, than commercial products, and the economic capital to cover ELs is often lower for retail products. This result stems from empirical evidence suggesting that default and loss correlations are often lower for retail products than for commercial loan products.

Addressing the non-availability of historical credit risk, credit scores or dependent variable

In practice, it is a common problem that historical scoring data is not available. One of the solutions is to manually classify the historical borrower into a risk class without performing a complete scoring process (due to the non-availability of a significant amount of the data). This may be the best practical solution.

Advantages and disadvantages of credit scoring

Advantages of credit scoring

To improve the effectiveness of the credit decision process, credit scoring provides the following benefits:

❑ an objective decision criteria;
❑ a high degree of predictive accuracy;
❑ reduced time and cost through automation;
❑ the ability of models to handle a large number of cases efficiently and consistently.

Disadvantages of credit scoring

The following criticisms are often levelled at credit scoring models.

❑ Models are often non-transparent and are frequently not understood explicitly.
❑ The "reasonableness" of the scoring models is often suspicious.
❑ Credit scoring only considers quantitative inputs. As the exposure gets larger, the qualitative inputs also become an important driver for credit risk.

SCORING TECHNIQUES
Discriminant analysis

Credit decisioning is the decision of whether to give or refuse credit to the new application. Credit scoring helps in this decision and the aim of building credit scoring systems is to find out the best rule for decision making. Discriminant function analysis is a statistical technique used to determine which variables discriminate more effectively between two (or more) groups. Specifically, the question is whether or not the groups are significantly different from each other with respect to the mean value of a given variable(s). If the mean for a variable is significantly different to the value of the different groups, then this variable discriminates between the groups (see Figure 7).

Ultimately, the purpose is to build a classification system to classify applications into the groups. A scoring system does not individually identify a good performer from a bad performer; it classifies an applicant in a particular "good/bad odds" group.

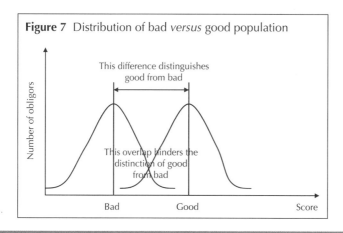

Figure 7 Distribution of bad *versus* good population

93

This classification does not mean that none of the "good" applicants will default or will turn into "bad" or *vice versa*. However, it does provide an acceptable bad/good ratio for the obligors classified as "good". (Actually, this is the default rate that the bank has to recover through pricing. The volatility in this ratio is to be provided through economic capital. Therefore, this ratio determines the pricing and required capital.)

A practice which has been common in the consumer credit industry is to divide the cases into three classes of risk (good, bad and indeterminate), but use only the "good" and "bad" classes to train the scoring models with the models then being used to classify new applicants as either good or bad risks. For example, "good" risks might be those borrowers who have never missed in payment, "bad" risks might be those who have missed for three or more consecutive repayments and "indeterminate" might be those who have missed either one or two consecutive repayments.

In general, there are three types of discriminant analysis.

❑ *Univariate normal case*: This considers only one variable as the determinant between good or bad. The distributions in both the "good" and "bad" classes are normally distributed.
❑ *Multivariate normal case*: Common covariance. More than one variable is used as the determinant of "good" and "bad". The distribution in both "good" and "bad" classes is normally distributed. The covariance of "good" and "bad" is the same.
❑ *Multivariate normal case*: Different covariance. More than one variable is used as the determinant of "good" and "bad". The distribution in both "good" and "bad" classes is normally distributed. The covariance of "good" and "bad" differs.

Some studies refer to MDA instead of multivariate. However, their analysis basically works in the same way. MDA, as the name implies, tries to discriminate between different groups and uses multiple variables. In this case it discriminates between the group of bankrupt businesses and the group of businesses. MDA is the most widely used statistical method, followed by logit models.

The most common application of discriminant analysis is to identify variables that discriminate between groups. This leads to the construction of a model of how to best predict which group a given variable belongs to. A model can be built step by step, where all variables are reviewed and evaluated at each step to determine which contributes the most to discriminating between groups. It can also follow a block-wise approach.

Most of the scoring models are based on the empirical approach. This aims to improve prediction accuracy by the appropriate selection of variables. There is little in the theoretical construction to support the choices in the variable selection; the variable selection is purely based on the statistical fit.

Discriminant analysis tries to derive the linear combination of two or more independent variables that will best discriminate between *a priori* defined groups, which in our case are failing and non-failing companies. This is achieved by the statistical decision rule of maximising the between-group variance relative to the within-group variance. This relationship is expressed as the ratio of between-group to within-group variance.

Discriminant analysis is very effective in ensuring that the variables in every group follow a multivariate normal distribution and the covariance matrices for every group are equal. However, empirical experiments have shown that failing firms, in particular, violate the normality condition. In addition, the equal group variances condition is also violated.

In order to do this, the MDA will take into account various samples from both groups. MDA tries to separate these two groups based on the financial ratios of each sample. Each ratio becomes a variable, X, and gets its own coefficient, V. This leads to the following formula, which calculates a sample's Z-score:

$$Z = \sum_{i=1}^{n} V_i X_i$$

where there are n different independent variables (risk factors such as financial ratios), of which X_i is an example, and V_i is its coefficient.

The average Z-score, $\overline{Z}$, is given by the following equation:

$$\overline{Z} = \frac{1}{n} \sum_{i=1}^{n} V_i X_i$$

MDA is used to divide the members of the sample into different groups (in the case of good/bad credit scoring the sample is to be

Table 7 Odds ratio and Logit

p	$p/(1 - p)$	$\ln(p/(1 - p))$
0.01	0.01	−4.60
0.05	0.05	−2.94
0.1	0.11	−2.20
0.2	0.25	−1.39
0.3	0.43	−0.85
0.4	0.67	−0.41
0.5	1.00	0.00
0.6	1.50	0.41
0.7	2.33	0.85
0.8	4.00	1.39
0.9	9.00	2.20
0.95	19.00	2.94
0.99	99.00	4.60

divided into two) and the average of the Z-scores is calculated for each of the group:

$$\overline{Z}_g = \frac{1}{N_g} \sum_{p=1}^{N_g} Z_{pg}$$

The aim of the MDA is to maximise the sum of the square among the groups and minimise the sum of the square across the group, which are given by the following two equations:

$$SS_{among} = \sum_{g=1}^{G} N_g \left[\overline{Z}_g - \overline{Z} \right]^2$$

$$SS_{across} = \sum_{g=1}^{G} \sum_{p=1}^{N_g} \left[Z_{pg} - \overline{Z}_g \right]^2$$

The aim of the MDA is to maximise the ratio of SS_{among} and SS_{across}.

Logistic regression

Logistic regression aims to assign the dependent variable into 0, 1. Therefore, the logistic method is not a regression method but a method for predicting probabilities. Logistical regression is used to predict binary (ie, yes or no; good or bad) outcomes from independent variables by using the following log equation:

$$\ln \left[\frac{p}{1-p} \right] = \alpha + \beta X + e$$

Table 8 Relative risk

$\dfrac{Y = 0 \text{ when } X = 1}{Y = 0 \text{ when } X = 0}$	$\dfrac{Y = 1 \text{ when } X = 1}{Y = 1 \text{ when } X = 0}$
$\dfrac{X = 0 \text{ when } Y = 1}{X = 0 \text{ when } Y = 0}$	$\dfrac{X = 1 \text{ when } Y = 1}{X = 0 \text{ when } Y = 0}$

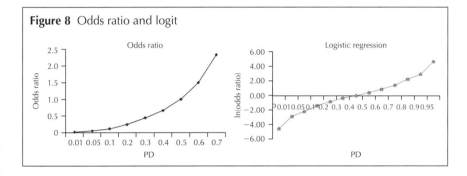

Figure 8 Odds ratio and logit

where p is the probability that the event, Y, occurs, $p = 1$, $p/(1 - p)$ is the "odds ratio" and $\ln [p/(1 - p)]$ is the log of odds ratio or "logit" (see Table 8 and Figure 8).

The logistic distribution constrains the estimated probabilities to lie between 0 and 1. The estimated probability is given by

$$\frac{p}{1-p} = \ell^{(\alpha + \beta X)}$$

Rearranging the above expression, the probability of event Y occurring is given by

$$p = \frac{1}{1 + \ell^{(-\alpha - \beta X)}} = \frac{1}{1 + \ell^{-(\alpha + \beta X)}}$$

where:

❑ for $\alpha + \beta X = 0$, then $p = 0.50$;
❑ as $\alpha + \beta X$ gets really big, p approaches 1;
❑ as $\alpha + \beta X$ gets really small, p approaches 0.

An interpretation of the logit coefficient, which is usually more intuitive is the "odds ratio", is as follows. Since

$$[p/(1 - p)] = \exp(\alpha + \beta X)$$

$\exp(\beta)$ is the effect of the independent variable on the "odds ratio".

Logit regression investigates the relationship between binary probabilities with explanatory (independent) variables. The advantage of the logit method is that it does not assume multivariate normality and equal covariance matrices as discriminant analysis does. Logit analysis incorporates non-linear effects, and uses the logistical cumulative function in predicting a bankruptcy.

The odds ratio is sometimes called the relative risk (Table 8). For a dichotomous (yes/no) variable, the odds ratio represents the relative risk that the probability that Y is 1 when X is 1 *relative* to the probability that Y is 1 when X is 0.

Logistic regression models are conditional probability models. These models allow estimation of the probability of company failure conditional on a range of firm characteristics. In logistic regression, the non-linear maximum likelihood (ML) estimation procedure is used to obtain the estimates of the parameters of the logit model.

Logistic regression is used to predict a dichotomous outcome, such as 0, 1 or good/bad. In the ordinary least square (OLS) analysis, the error variance (residuals) is assumed to be normally distributed. For dichotomous distribution, variance is assumed to be logistically distributed.

Instead of finding the best fitting line by minimising the squared residuals, as in the ordinary least square approach in the logistic method, the ML is estimated. ML is a way of finding the smallest possible deviation between the observed and predicted values (similar to finding the best fitting line) using calculus.

Probit regression assumes that the errors are distributed normally rather than logistically. With probit regression, one assumes that the dichotomous dependent variable actually has a continuous theoretical variable underlying it. The two approaches usually yield the same substantive results. Researchers tend to choose the approach with which they are most familiar, with some researchers preferring logistic to probit regression because the odds ratios can be computed. Probit finds the measure of the goodness of an applicant, but whether the applicant is actually good or bad depends on whether the score is greater or less than a cut-off level, C. This cut-off level is not assumed to be fixed but is a variable with standard normal distribution. Probit and logit models are also called binary choice models.

Such an objective gives rise to several questions about the properties of the score made available:

❑ the accuracy of estimates of probability and homogeneity of risk classes;
❑ the stability over time of risk classes and their properties;
❑ the dependence of risk measurement on the business cycle;
❑ the stability of transition matrices (TM);
❑ the correlation of risks.

With logistic regression there is no standardised solution and, to make things more complicated, the unstandardised solution does not have the same straightforward interpretation as it does with OLS regression. OLS uses R^2 to gauge the fit of the overall model while the χ-square test is used to indicate how well the logistic regression model fits the data.

Historically, a difficulty with logistic regression has been that one has to use ML estimate to weights. This requires non-linear optimising techniques using iterative procedures to solve and is computationally more intensive than the linear regression.

Neural networks
Unlike the classical techniques that can be implemented without software (although it may be cumbersome and time consuming), NNs are software applications. However, most of the solutions are independent of platform and software, which means they are designed to work well with all the major workflow and decision engine tools.

NNs emulate human reasoning, modelling non-linear solutions. In NN technology, one or more middle layers of variables or nodes are used in the regression. NNs are artificial intelligence algorithms that allow for some learning through experience to discern the relationship between borrower characteristics and the PD, and to determine which characteristics are most important in predicting default.

The advanced NNs involve statistical modelling techniques, fuzzy logic, Bayesian learning, genetic algorithms and natural language processing, etc.

Conceptually, NNs are an improvement on MDA. Although the technology used is more sophisticated than that required for conventional statistical methods of analysis, the results of the methods themselves are identical: a numerical score expressive of the outcome in question.

NN method is more flexible than the standard statistical techniques, since no assumption is made about the following factors:

❑ the relationship between risk factors (independent variables) and risk indicators (dependent variables);
❑ the distribution of independent or dependent variables;
❑ model errors;
❑ correlations among the variables.

NNs and classification trees are sometimes called expert systems as they are automated procedures with learning abilities. Expert systems also include systems where the learning of human experts has been incorporated into a set of rules, some of which have been developed using an inference engine from data presented to the system.

Banks have been using NNs to benchmark their internal ratings or scores to the agency ratings. For this task, the bank trains their NN model with financial, industry and management information (independent variable) and the rating grade (dependent variable) published by rating agencies. Once the model is trained, it can be used to predict the rating for the "out of the sample" obligor.

A NN consists of a large number of processing elements called neurons and the connections between them. Neurons are functions used to map independent variables onto dependent variables through a function $y = f(x)$. A NN finds the best possible approximate function for a non-linear relation. This approximation is coded in the neurons using the weights associated with each neuron. Weights get modified as the NN searches through the examples of correct input–output associations. With sufficient training, NNs are able to mimic the desired input–output model or relation. The ability to generalise from specific examples is the basic feature of the NN.

There is a node for each layer of input variables in a NN. The output has a single node for single-layer input and multiple nodes for multi-layer NN.

There are two types of NN: single-layer and multi-layer NNs. In single-layer networks the input variable is directly converted into the output (see Figure 9). In multi-layer networks there is

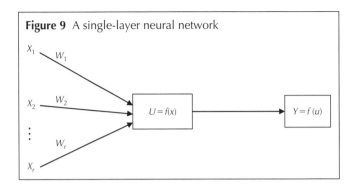

Figure 9 A single-layer neural network

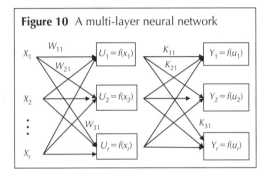

Figure 10 A multi-layer neural network

an intermediate hidden layer of nodes for each input node (see Figure 10).

Earlier in this chapter we have already shown that the input data is in the following form.

❏ Independent variables/risk factors: X_{ij}, $i = 1, \ldots, n$ for a customer, $j = 1, \ldots, r$ for a risk factor.
❏ Dependent variable/risk indicators: Y_i, $i = 1, \ldots, n$.

The input data layer is $X_1, X_2, \ldots, X_r$.

Single-layer neural networks
A single-layer neural network is generally represented by Figure 9. The risk factors relevant for the computation of risk are the inpus to a neural network based credit scoring. Examples of risk factors are given in the Table 10.

Table 9 Score card

Characteristics	Attributes	Points
Payment history	0–5 months	10
Number of months since most recent	6–11 months	15
derogatory public record	12–23 months	25
	24+ months	55
	No record	75
Outstanding debt	No facility	30
Average balances on revolving credit	US$1–99	65
	US$100–499	60
	US$500–999	35
	US$1000 or more	25
Credit history length with the bureau	<12 months	12
	12–23 months	35
	24–47 months	60
	>48 months	75
Pursuit of new credit	0	70
Number of enquiries in last six months	1	60
	2	45
	3	25
	4 or above	20
Existing bank card lines	0	15
	1	25
	2	50
	3	60
	4 or above	50

The factors are converted into a neuron using different weights for different factors. The weights are called synaptic weights.

Weights can be negative or positive and in this application they are known as synaptic weights. The positive weights are known as excitory and negative weights are known as inhibitory.

$$U = f(x)$$

$$U = \sum_{q=1}^{r} w_q X_q$$

The activation function, Y, is given by

$$Y = f(u)$$

There are various types of activation functions. Function can be a logistic function or just a threshold function. The type of activation

Table 10 Risk factors in credit scoring for individuals

Personal information	Professional educational	Financial capacity	Existing facilities	Credit bureaux information
Age	Education	Monthly net income	Existing credit limit or exposure with the bank	Credit bureaux score
Number of dependents	Activity, profession, industry	All existing Equated monthly instalments (EMI)	Secured/ unsecured	The number of enquiries received by the bureaux
In the negative list	Number of years in the employment with the same employer	Total assets of the applicant	Collateral offered-age	Payment history
The number of years obligor is banking with the bank	For small business – number of years in the business	Total liabilities of the applicant	Market value of the collateral	
Average monthly credit/debit balance		EMI on the proposed loan	Loan-to-value ratio	

function generally determines the problem and complexities of non-linearity which can be managed.

Multi-layer neural networks
Please see Figure 10 for a multi-layer neural network.

The multi-layer neural network has multiple hidden layers. Each hidden layer has multiple neurons. Each neuron has different weights. The output from one neuron becomes the input to other neuron and the chain continues.

$$U = f(x_q)$$

$$U_k = \sum_{q=1}^{r} w_{kq} X_q$$

$$Y_k = f_k(u)$$

The computation of weights (w and k) is known as training. There are many methods for training, and one widely used method is the back-propagation method, which uses training pairs that are repeatedly presented to the network to minimise the error function. A training pair consists of the value of each of the input (independent) variables and a dependent variable (value of Y in the equation).

Three of the most important aspects of NN architecture are the number of hidden layers, the number of neurons in the hidden layer and the error function.

Hidden layers are used to train the network for non-linearity. Non-linearity can be introduced in the hidden layer and activation functions. The number of hidden layers depends upon the complexities of non-linearity. The hidden layer models the complicated non-linear relationships between the variables and transforms the input variables. Experts have argued that two hidden layers are sufficient for modelling credit scores. The number of neurons for each hidden layer can be found only after analysing the results of the training.

The output of credit scoring is a classification of obligors into groups – generally two. Therefore, only two outputs are obtained from a NN. The error function is used to minimise the errors in classification.

Advantages of neural networks
The advantages associated with NNs are as follows.

❏ NNs are able to analyse complex patterns quickly and with a high accuracy level as they can learn from examples without any pre-programmed knowledge.
❏ There are no restrictive statistical assumptions as there are in MDA. There are no distributional assumptions on the input data and the data need not be linear. Non-linearity is an important advantage over MDA. Non-numeric data can easily be included in a NN because of the absence of the linearity constraint.
❏ NNs are perfectly suited to pattern recognition and classification in unstructured environments with "noisy data" ie, incomplete or inconsistent. NNs tolerate data errors and missing values by making use of the context and "filling in the gaps". Consequently, a NN is able to work with annual account data, which is often inconsistent and incomplete. In addition, a NN can overcome the problem of autocorrelation, which frequently arises in time series data.

❑ The NN technique can clearly bring out "failure/non-failure" output.

❑ In general, NNs seem to be more robust – especially when sample sizes are small – and more flexible than other methods.

Disadvantages of neural networks
The disadvantages of using NNs are as follows:

❑ the black box problem cannot be "applied";
❑ requires high-quality data;
❑ variables must be carefully selected *a priori*;
❑ there is a risk of over-fitting;
❑ NNs require definition of architecture;
❑ they have long processing times;
❑ there is a possibility of illogical network behaviour;
❑ a large training sample is required.

Expert system

An expert system is a software application representing expert skills to make an intelligent decision about a processing function. Expert systems have four key characteristics:

❑ the system is based on a knowledge base;
❑ the system has the tools to maintain and expand the knowledge base;
❑ the system can draw conclusions;
❑ the system can explain (justify) its decisions.

Expert systems are composed of the following three parts.

User interface: It must ensure easy and efficient use, error detection and data capture.

Knowledge base: The knowledge base covers the data, facts and rules to process the data and facts. Rules are expressed in terms of mathematical and logical equation. Logic and syntax may vary. The main challenge is that not all knowledge can be expressed in terms of formal rules. The knowledge base has building blocks of "if–then" rules. It also has a search heuristic. The expert base is trained on the number of examples of failed and non-failed firms and there may be thousands of rules. In an expert system, the rules need not be sequential and the matching of facts to rules is complex. There are algorithms for such matching.

Decision machine: This has access to rules and generates the appropriate conclusions. The decision machine has a series of facts that are matched with the rules to provide recommended actions.

The first step is to build the knowledge base. This is built by discussing the behaviour and decisions made with experts. The knowledge can also be extracted by using NNs. Researchers have built expert systems with 65% to 98% prediction accuracy (system output agreed with the expert system output).

Advantages of using an expert system
Expert systems have the following advantages.

❑ The expert system can take qualitative variables.
❑ The system does not assume any specific statistical distribution of the data.
❑ The expert system builds a number of if–then rules. These rules can easily be applied in practice to new cases for a classification decision. In this way, expert systems are regarded as user-friendly.

Disadvantages of using an expert system
The following drawbacks are associated with the use of expert systems.

❑ It is very difficult to program the intuition or "knowledge base" of the expert system and to determine which heuristics must be used.
❑ The process of extracting the knowledge into rules is very time consuming and expensive.
❑ Systems are not flexible and cannot learn. Expert systems are unable to use inductive learning to adapt the if–then rules to changing situations (ie, changes in the knowledge base).
❑ The data input quality has to be good. Expert systems are not capable of working with incomplete, noisy data or input information with errors.

Fuzzy logic
This is an extension of the expert method. This method starts from a number of if–then rules, which are based on pre-existing, qualitative

knowledge on corporate failure and are determined by the decision maker. These if–then rules make a link between a number of conditions concerning predefined variables and the failure status. Next, the relevance of each if–then rule is tested on an estimation data set. Each rule is attributed a rating index between zero and one, which denotes the probability of correctness of a rule. In this rating index, a higher rating corresponds to a better rule.

Fuzzy-rule-based classification models are built using a fuzzy rules set with firms classified as failing or non-failing. The fuzzy logic model is intuitive based. The most important negative feature of the fuzzy rule model is that it strongly depends on the arbitrarily determined if–then rules.

Search for the best method

As stated previously, there are only two types of models: models that "work" and those that "don't work". In a nutshell, there is no better method. The question of "which modelling method produces the best performing failure prediction model(s)?" cannot be answered conclusively. Unfortunately, no study has systematically compared the performance of all the different models: different factors or constraints determine different aspects of modelling and performance issues and each model addresses different issues. The following factors are the determinants of the predictive power of the model:

❑ definition of default/failure;
❑ time horizon for prediction;
❑ theoretical basis for choice of variable;
❑ number of independent variables;
❑ data and sample quality;
❑ sampling bias – assumptions about distribution of "good" and "bad";
❑ over-fitting and out of sample prediction.

Before implementing or choosing any model for implementation, it is recommended that the model is evaluated on the factors discussed in this section and the findings of empirical studies on the effectiveness of that model considered.

Some of the empirical findings useful for finding the better method are given below.

Businesses do not fail in a short period: There are criticisms of the use of dichotomy models. Dichotomy models arbitrarily divide firms into either failure or non-failure. Business failure is never a well-defined dichotomy. The definition of failure also has an impact on the dichotomy status.

Variables selection should have theoretical reasoning: This questions the way in which the variables included in the models are selected. Very often researchers start by forming a very wide range of possible variables and then reduce this range to a limited number of variables using statistical techniques. The variables found to fit through statistical techniques may not have any theoretical basis. This results in models with sample-specific variables that fit the data set ie, used, but that are not suitable for "out of the sample" data sets.

Number of variables: The model complexities reveal that the logit method performs better with a smaller number of variables (only one). NNs perform better if the number of independent variables is larger.

No empirical studies to support overall superiority of models: A study by Altman *et al* analysed over 1,000 healthy, vulnerable and unsound Italian industrial firms from 1982–1992 and found that performance models derived using NNs and those derived using the more standard statistical techniques yielded approximately the same degree of accuracy. They concluded that although NNs generally offered no advantage over the standard methods it would still be advantageous to apply both methods to certain applications, in particular to complex problems in which the flexibility of NNs would be particularly valuable.

Prediction time horizon: NNs are specifically more effective for "early detection – three years before failure" of financial distress and hence for minimising type I errors. For short-term failure prediction, the logistic regression model seems to offer better performance.

Relative performance:

❑ NNs are better than logit models in extracting information from attributes for forecasting bankruptcy;
❑ there is little evidence that the NN approach dominates the conventional multivariate models, particularly in the case of "out of the sample" prediction;

❏ the relative performance of NNs, compared with traditional statistical models, depends on the sample size used (NNs have increased prediction accuracy when small samples are used).

CASE STUDIES
Credit bureaux scoring

The goal of using a credit bureau risk scoring system is to achieve superior predictive power. An individual's application is sent to one of the bureaux for scoring based on the contents of the application and previous payment history in their credit bureau report. The systems at the bureau rank current elements of a credit report to predict the customer's future credit payment behaviour. Due to the depth of information available at a bureau, the credit scoring by a credit bureau tends to have a better success rank ordering risk or bankruptcy than other systems.

Banks purchase these scores for use in applicant screening, account acquisition and account management strategies.

Bureau scorecards are generally revalidated every 18–24 months. The bureaux use approximately 100 predictive variables in their model. Examples of some of the typical variables used with the relative weights are given in the Table 9. The variables are previous credit performance, the current level of indebtedness, the amount of time credit has been in use, pursuit of new credit and types of credit available.

Scorecard vendors have risk scorecards in place at the major credit bureaux.

A credit bureau typically also provides the following type of information which is relevant for the credit risk.

❏ *Credit bureau bankruptcy scores*: Scores predict the likelihood that a customer will declare bankruptcy or become a collection problem.
❏ *Recovery scoring after charge off*: These scores help in reducing and managing recovery scores.

Business requirements for the consumer credit scoring model

Typical components for consumer credit approval systems are given in the Figure 11.

The application should be able to interface with the leading credit bureaux in the country and should be able to extract the required data from the credit bureaux.

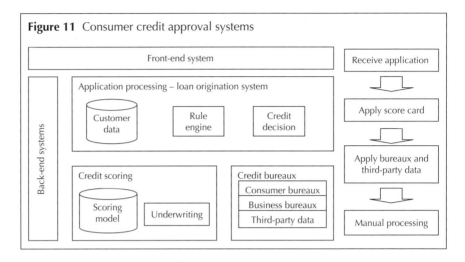

Figure 11 Consumer credit approval systems

Table 11 Some commercial bankruptcy and recovery credit scoring models and their scoring techniques.

Experian scorex	MDA
FICO	MDA, Logit
Equifax Predictive Science	MDA, NN
TransUnion	MDA, NN, Clustering

Credit scores should consider the following factors (Table 10).

❑ the ability to define behaviour scoring based on payments data;
❑ the ability to define collection rating;
❑ the definition of a challenger strategy for analysis and what-if analysis;
❑ the reporting of:

    ❑ cross analysis;
    ❑ score/rating migration;
    ❑ delinquency;

❑ the integration with application processing.

Table 11 gives some examples of the commercially available bankruptcy and recovery credit scoring models and lists their scoring techniques.

Applications available in the market vary dramatically in terms of their complexities and integration with other types of systems.

Different solutions are available for different types of solutions. Solutions also vary in their capacity to handle the number of applications.

An application is generally received through direct selling agents and submitted through various channels including the Web and agent branches. Applications are captured into loan origination systems through the Internet and other customer service channels or manually by an agent. Credit scoring is applied with automated acceptance or rejection. The second pass involves considering the credit bureau data and other data obtained from third parties. In the third pass, applications are processed manually.

The decision process

Decisioning is largely automated as the application passes through a series of models and decision trees. The models, decision trees and business logic are proprietary. The decision engine is a software application that allows analysts to programme business rules. A high percentage of applications are processed automatically.

Backend systems are the accounting and collection systems. These systems are required for recording the financial transactions.

REFERENCES

Altman, E., 1968, "Financial Ratios, Discriminant Analysis and the Prediction of Corporate Bankruptcy," *Journal of Finance,* **23,** pp 589–609.

Altman, E. and H. Rijken, 2002, "Benchmarking the Rating Agencies: In Search of the True Credit Rating Transition Matrix", Abstract published in FMA Annual Meeting in Denver.

Altman, E., G. Marco, and F. Varetto, 1994, *Corporate Distress Diagnosis: Comparisons Using Linear Discriminant Analysis and Neural Networks* (The Italian Experience).

Beaver, W., 1966, "Financial Ratios as Predictors of Failures," in Empirical Research in Accounting, selected studies, Working Paper.

Dunn, L. F. and T. U. Kim , 1999, "An Empirical Investigation of Credit Card Default".

Evelyn, H., 2003, "Are Credit Scoring Models Sensitive With Respect to Default Definitions? Evidence from the Austrian Market", University of Vienna.

Dunn, L. and T. Kim, 1999, "An Empirical Investigation of Credit Card Default", Working Paper, Ohio State University.

Henley, W. E., 1995, *Statistical Aspects of Credit Scoring*. Open University, Dissertation.

Keeney, R. L. and R. M. Oliver, 2003, "Improving Lender Offers Using Consumer Preferences", Proceedings of Credit Scoring and Credit Control VIII, Credit Research Centre, University of Edinburgh.

McCorkell, P. L. 2002, "The Impact of Credit Scoring and Automated Underwriting on Credit Availability," in Thomas A. Durkin and Michael E. Staten (eds): *The Impact of Public Policy on Consumer Credit* (Boston: Kluwer Academic Publishers).

Myers, J. H. and E. W. Forgy, 1963, "The development of numerical credit evaluation systems", *Journal of the American Statistical Association*, pp 799–806.

Oliver, R. M., 2001, "Risk Based Pricing – Some Early Results", Proceedings of Credit Scoring and Credit Control VII, Credit Research Centre, University of Edinburgh.

Srinivasan, V. and Y. H. Kim, 1987, "The Bierman-Hausman credit granting model: A note", *Management Science*, **33(10)**, pp 1361–2.

William F. T. and M. S. Carey, 1998, "Credit Risk Rating at Large U.S. Banks", Published in Federal Reserve Bulletin.

Yobas, M. B., J. N. Crook, and P. Ross, 2000, "Credit Scoring Using Neural and Evolutionary Techniques", *Journal of Management Mathematics*, **11(12)**, pp 111–25.

4

Market Information Based Models

INTRODUCTION

The measurement of credit risk is about modelling the borrower related information to establish a relationship between the credit risk measures and observable data. Borrower information is divided into three types.

❑ *Financial and other information provided by the borrower*: Information made available by the borrower under various legal and regulatory filings, and compliance requirements; financial statements including information voluntarily disclosed; or information submitted along with the credit application or contained in public documents such as prospects for issuance of bonds. This information is of two types: financial and borrower demographic. This information is used to build the credit scoring models.

❑ *Market data*: This second source includes the prices for bonds, equity and derivative instruments traded on the exchanges or dealt privately and rates published by the trade bodies and agencies.

❑ *Published credit rating*: The third source of information is the credit risk indicators or the credit ratings measured and published by the rating agencies.

However, none of the information available to the public is complete information.

Experts have made an attempt to build the models for defaultable securities using some of the borrower related information. Not all characteristics are used. In the market information based models use the market data like bonds, share price and credit derivative price (CD) are used. Such models can be broadly classified into the following two types.

❑ *Structural models*: These models utilise all the three sources of information. The term structural is derived from the financial structure of the firm, which means that information contained in the financial statements and market data is at the heart of these models. The Merton model is an example of a structural model.
❑ *Reduced form models*: These models use reduced or minimal information. The models use bonds, CD and equity prices and credit rating information and may or may not use financial and other information of the obligors. The Jarrow–Turnbull model is an example of a reduced form model.

Typically, each model has restrictive assumption and parameters. As the model becomes sophisticated, such assumptions and parameters are relaxed. Both types of models have evolved over last decade and are still evolving. Each of the models is in a dynamic state and with empirical findings, the restrictive assumptions and parameters are challenged on a continuous basis.

In this chapter, we have attempted to identify and understand the building blocks for each type of model and to identify the restriction imposed on the building blocks and how those restrictions are challenged and relaxed. This approach will help us to build a general framework for the market information based credit risk models and to understand and use the ever changing and evolving market information based models in the business environment.

BUILDING BLOCKS FOR MARKET INFORMATION BASED MODELS
Processes in the market information based models
Market information based models relate the market prices to the credit quality or changes in the market prices to the changes in the credit quality. The change in credit quality is represented through changes in the assets value, changes in credit rating or transition matrix and changes in the recovery. The forces for changing the credit quality are extracted out through a default process.

A firm's assets are the primary source of information to satisfy the lender's claims, the asset values of the firm determine the default time. Asset values of a firm keep changing owing to the business operation cycle and they also depend on the market expectation of the firm's ability to generate future cashflow. However, it is very difficult to measure and model asset values. Assuming the equity prices reflect the true value of the firm, equity prices and accounting information disclosed in the financial statements are a good source and proxy for asset values.

There are two approaches to the recovery rate process. Recovery rate as a function of asset value – default declared when the asset value falls below a certain level. In this case the expected recovery rate is assumed to be constant. The second approach is for lenders to allow a firm to continue as long as the expected recovery rate is likely to be higher than the targeted recovery rate. However, the expected recovery rates for different classes of lenders, in various bankruptcy regimes, in different economic cycles is not the same. If the recovery rate is assumed to be constant, then the default trigger is known.

The default process is driven by the default intensity.

Since this is a paradigm to value financial instruments, the risk-free interest rate curve plays an important role in valuation. A risk-free interest rate process can either be assumed to be a constant or a stochastic variable.

The following variables are used in market information based models:

❏ capital structure from financial statements;
❏ prices and changes in equity prices, bond prices and credit spreads;
❏ transition matrices (TMs);
❏ recovery rates;
❏ interest rates.

These variables are determined by the four processes shown in the Figure 1. All the four processes keep determining asset value and default threshold. Ideally, both the sides, ie, asset value and default threshold, should be modelled and equilibrium estimated. This is extremely complex. Existing models do not estimate both sides. Models assume that one of the variables (eg, the default threshold) is

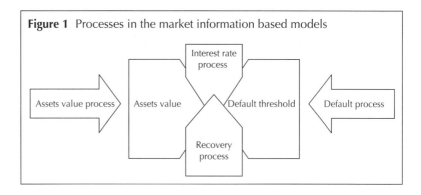

Figure 1 Processes in the market information based models

known and the other variable is unknown (asset value) or exogenous (not dependent upon the first variable).

For each of the known and unknown variables, various assumptions are made to make it measurable with the available data.

The four processes identified above continuously interact with each other under a given set of debt covenants (see Figure 2). These processes continuously change bond value, credit quality and push bonds to default. However, all the four processes are never modelled. In general, structural models focus on the asset value process and reduced form models focus on the default process. The recovery process and interest rate process are common to both types of models. However, underlying assumptions for modelling recovery and interest rates may be different in the two models.

Under each of the processes, various assumptions are made in the underlying process and debt covenants for analytical tractability, to build a closed form solution, to manage incomplete information available and to make solutions simpler.

Asset value process
The asset value of a firm keeps changing continuously owing to the operational cycle and changes in the environment.

❑ The passage of time changes the asset value at a given rate of value change, which includes profits generated, dividend and interest paid.
❑ Changes occur in the values of asset stock due to volatility in asset prices.

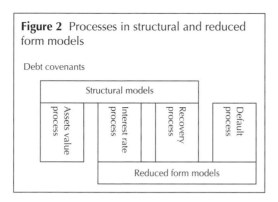

Figure 2 Processes in structural and reduced form models

Debt covenants

- Structural models
  - Assets value process
  - Interest rate process
  - Recovery process
  - Default process
- Reduced form models

❏ Sudden internal and external events generate changes in the assets value.

When assets value is modelled from the equity prices, the asset value process assumes that a firm's value follows a random process similar to that used to describe generic stocks in equity markets.

Asset values are modelled through asset drift due to the passage of time, asset volatility and asset value jumps. Asset values are measured indirectly from the firm value and contingent claims issued against the firm's assets. The firm's value is measured from equity prices.

Default is declared when the asset values touch the default threshold. Recovery rate, leverage, capital, dividend payout, default costs and other debt covenants impact asset values and the default threshold. There are two approaches to the default threshold under the structural models approach: either treat the default threshold as a function of the asset value or as an external variable that is independent of asset value.

❏ *Default threshold dependent on asset value*: Models based on an endogenous default threshold include Leland and Toft (1996) and Anderson *et al* (1996). The endogenous default models allow a borrower to decide when to default. This means default is declared if asset values touch a certain (low) level.

❏ *Default threshold as an external variable that is independent of the asset's value*: Models based on exogenous default threshold include Longstaff and Schwartz (1995), Collin–Dufresne and

Goldstein (2001) and Huang and Huang (2003). The exogenous default threshold is driven by the expected recovery rate.

Risk-free interest rate process

Interest rate modelling is an important input to all types of model. Interest rates are assumed to be constant in simpler models. This assumption is relaxed to model interest as a stochastic variable. An assumption about the constant risk-free interest rate process is to reduce the computation complexities of the stochastic variable. There are empirical studies against the use of the constant, as well as the stochastic, approach. Ogden (1987) found that the Merton model generally underestimates spreads and argues that the Merton models shortcomings are due to a lack of stochastic interest rate modelling. Lyden and Saraniti (2000) compared Merton Model and the Longstaff–Schwartz (1995) model and found that both models underestimate yield spreads, although they argue that allowing for stochastic interest rates has little impact on the under-estimation.

The Leland–Toft (1996) model does not allow stochastic interest rates, arguing that the stochastic modelling of interest rates is responsible for reducing the estimated credit spreads – one of the drawbacks of structural models.

Recovery process

The recovery process is a process to distribute the remaining assets after default to the bond holders. The recovery process is governed by the bankruptcy rules and seniority of the debt. Both of these aspects are very difficult to model. Therefore, most of the existing models price risky debt only prior to default. The recovery rate is either assumed to be a constant or a fixed proportion of the face value of the debt.

There are two approaches to measure recovery.

❑ Assets recovered from the defaulted firms.
❑ Some of the markets (such as the US) have bankrupt bond markets with sufficient liquidity for about a month after default. Prices of defaulted bonds represent the market expectation about recovery. In general, in the defaulted bonds market sellers are the banks and investment companies who were holding bonds as their portfolio and purchasers are the firms specialising in the recovery. Since

bonds are purchased by the firms specialising in the recovery, the selling price represents the best recovery possible.

With the development of a defaulted bonds market, the recovery process is modelled as a stochastic process. Recovery value is computed as a percentage of face value or market value or treasury value.

The economic cycle has a major impact on the prices for the defaulted assets and the defaulted bond prices. In the recent past, the recovery process has also been linked with the economic variables.

Default process

The default process is a process of changes in the credit quality. There are two approaches to address changes in the credit quality.

First, credit quality changes due to changes in the asset values of a firm. Therefore, changes in the firm's asset value reflect the changes in the credit quality. Asset values going below a certain threshold reflect the default rates. Such a threshold is generally assumed to be constant.

The other approach is to extract out the default rate or changes in the credit quality from the market prices of equity and bonds. The (expected) default rates and (expected) recovery rates drive the changes in the equity and bond prices. The default rate is assumed to be a process driven by intrinsic default rates (in the economy or industry) and is therefore either constant or driven by changes in the economic variables (similar to conditional defaults). Examples of economic variables include share price index, rating, economic state variable, etc.

Modelling with incomplete information

A firm's management always has better information than both the shareholders and the bond holders. The efficiency and depth of equity markets mean that equity prices typically represent better information than the bond price. Bond prices do not provide credit quality changes information completely. Empirical studies have proved that the credit spread not only represents credit quality but also the liquidity premium and tax impact. A proportion of influence of each of these onto the credit spreads keeps changing over time. Therefore, it is very difficult to segregate credit quality information, liquidity premium and tax impact in credit spread. The

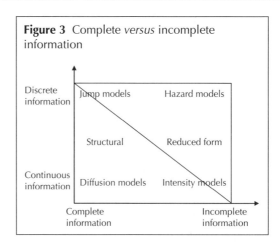

Figure 3 Complete *versus* incomplete information

financial and accounting information available at discrete points in time makes the financial statements less useful compared with the market information.

Therefore, in practice, complete information on the credit quality at an acceptable frequency is not available. All these issues are addressed by modelling the reduced or incomplete information through a reduced form modelling approach.

Information can be analysed on the two dimensions (see Figure 3) frequency and completeness. Though time is a continuous variable, information dissemination is always discrete. The frequency may vary from few minutes (equity price) to three months (quarterly results). The disseminated information is generally the least information since it is driven by the regulatory disclosures. The frequency is also a determinant of information completeness.

Complete and continuous information
Complete and continuous information is possible only under one of the following two assumptions.

❏ Only management of the firm can be assumed to have continuous and complete information.
❏ It is assumed that during the intervening period of an accounting disclosure there is no change in the leverage, debt position or asset value other than what has been disclosed or proposed as a change in the disclosure made. It also assumes that assets, debt and leverage quality and position are completely and correctly

disclosed. Frequently, none of these assumptions are valid. A bond investor in a publicly traded firm can better estimate the level of the firm's assets by observing prices of its equity and debt. On the other hand, the bond holders in a privately held firm have to rely on periodic accounting reports received at discrete dates.

Standard geometric Brownian models are based on the assumption of complete and continuous observation of the firm's asset value.

Complete and discrete information

Most of the structural models assume complete information. However, information is available at discrete time intervals. In practice, the complete information assumption in structural models is an approximation designed to facilitate a simpler way of capturing the various economic nuances of the firm's operations. A modeller can only have as much information as the market.

A criticism of the structural model has come from empirical studies showing that a structural model fails to correctly model short-duration (less than one year) bonds. This means that the existing bond valuation fails to consider information available at an interval of one year. This failure is due to the discrete intervals when the information is available.

It is often difficult to distinguish between the discrete information and incomplete information. Discrete information is incomplete until the arrival of the next discrete information. Immediately after the arrival of the information, discrete complete information becomes incomplete information.

Discrete or incomplete information introduces surprise into the bond and asset values. The surprise elements are modelled as jumps in the bond or asset values.

Structural models usually assume a known default threshold, and therefore a predictable arrival of default. This implies zero short-term credit spreads for the firm's debt, which is not consistent with the short-term spreads seen in practice. This shortcoming of structural models is corrected by introducing jumps into the asset value. If all claims against the firm's assets are publicly traded, then asset value is observed from equity prices. However, management does not disclose many claims that are not traded.

The main distinguishing characteristic of structural models and the reduced form models is the link structural models provide between the default rates and the firms' fundamental financial variables: assets and liabilities.

Compared with discrete and complete information, the reduced form approach assumes incomplete information. Assumption of incomplete information results in reduced form models with strong in-sample fitting properties but poor out-of-sample predictive ability.

Incomplete and continuous information

For a publicly traded firm, with good trading volumes through the equity values, the firm values are known continuously and, if bonds are also traded, the liability on traded claims is also known continuously to the market while the management of the firm has complete information on the asset value, claim's value and the firm's value. This is the best approximate example of incomplete continuous information.

Reduced form models use market prices of the firms' defaultable instruments (such as bonds or credit default swaps (CDSs)) and equity as the source of information to extract the arrival of default. Unfortunately, the credit spread information does not have credit quality as the only determinant. Therefore, the changes in bond prices do not represent only the credit quality changes.

Although easier to calibrate, reduced form models lack the link between credit risk and the information regarding the firms' financial situation or the asset value. Studies have proved that spreads do not represent the credit quality or credit quality changes fully. The information available with the bond holders or equity holders is not complete. Furthermore, all the debt is not traded in the market. Management keeps borrowing and repaying without the knowledge of the market. This results in incomplete information to the market and bond holders. Information may be continuous for the instruments traded in the market. However, a large portion of debt is never traded; in practice, the information is neither continuous nor complete.

The fundamental difference between the structural and the reduced form approach lies in how the models specify the timing of default. While structural models assume that default occurs when an exogenously modelled asset value hits some lower boundary, the reduced form models use an exogenous default intensity process to specify the default time.

Incomplete and discrete information

Structural models introduce jumps into the asset value to model discrete and surprise information. CreditMetrics models the incomplete information of default threshold instead of asset value. It assumes multiple default barriers for different rating classes.

Reduced form models specify an exogenous intensity of arrival of the default time which makes default an unpredictable event. To model discrete information, default intensity itself is not assumed to be constant but is linked with hazard rate in terms of changes in the economic variables or transition metrics, etc. The hazard rate itself keeps changing and varying.

Modelling for the incomplete information differs between the two types of the models even if both adopt the jump diffusion framework. Structural models converge to the reduced form model if they model incomplete asset information with a known or unknown default threshold. However, such models are not common.

Giesecke and Goldberg (2004) propose developing a structural model with incomplete information about the default threshold. Duffie and Lando (2001) assume accounting information to be noisy, thereby making information on the default threshold incomplete, and instead propose a "hybrid" model. Hybrid form models are a further refinement of structural models and rating-migration-based models are a further refinement of reduced form models.

Modelling the asset value process

The asset value of a firm keeps changing owing to the passage of time and volatility in the asset value observed in the market. Being a financial instrument, asset value is also impacted upon by the changes in the risk-free interest rates. In addition to the asset price volatility, asset values change owing to surprise events. Asset value modelling broadly has two building blocks – modelling diffusion processes, which include drift and volatility, and modelling surprise events through jumps.

Diffusion process

Diffusion models assume the asset values of a firm follow geometric Brownian motion. This model has two building blocks: the changes due to the passing of time, called drift, and asset value volatility. The model is represented as follows:

$$\Delta V = Value\ [Drift(time) + Volatility(Wiener\ motion)]$$
$$dV = \mu V dt + \sigma V dW$$

where W is a Wiener process and σ is the instantaneous asset volatility.

The drift primarily comes from interest accumulation (risk-free rates plus risk premium) and becomes reduced owing to payout such as dividend, coupon payments, etc.

Therefore,

$$dV = (r + \lambda - \delta)V dt + \sigma V dW$$

where r denotes the risk-free interest rate, λ denotes the asset risk premium and δ denotes the asset payout ratio (eg, reflecting the dividend and coupon payments).

Simple assumptions are simultaneously an advantage and a major weakness for the asset valuation model. Empirically, it is found that diffusion models systematically underestimate credit spreads. From a mathematical point of view, it is convenient to assume that a value process of the company is a continuous variable. In practice, we know that values are impacted upon by sudden events at discrete points in time.

Asset volatility

Asset volatility is related to, but different from, equity volatility. A firm's leverage has magnifying effects on its asset volatility. Since asset value volatility is generally not observed, structural models rely heavily on the observed equity return and equity volatility.

Under Merton's assumptions, the value of equity is a function of the value of the firm and time, so it follows directly from Itô's lemma that

$$\sigma E = \left(\frac{V}{E}\right)\frac{\partial E}{\partial V}\sigma V$$

The KMV is the name of the firm model utilises the relationship of equity value with asset (firm) value and equity volatility with asset volatility to measure implied default probability by measuring the distance to default. An asset's volatility directly determines its distance to default:

$$Distance\ to\ default = \frac{Market\ value\ of\ assets - Default\ threshold}{Market\ value\ of\ assets \times Asset\ volatility}$$

Jump-diffusion process

In pure diffusion models, the probability for a solvent company to default within a small interval of time is negligible. However, in reality a company may face sudden financial distress. Defaults may also be driven by shocks, which are not captured by the continuous paths of Brownian motion. Experience has shown that diffusion models fail to accurately predict the value of short-duration assets.

Inclusion of jumps into the diffusion process allows modelling of unpredicted extreme events, raising the probability for a solvent company to default within a small interval of time with some degree of accuracy. Considering that jumps on financial markets are usually rare but of large sizes, models typically assume the following characteristics:

❑ one jump per day, at the most;
❑ jump size dominates daily return when it occurs.

Further jumps are categorised into two types:

❑ jump due to the default;
❑ jump without the default trigger or a sudden change in the credit spread. The jump intensity is proportional to the jump in the credit quality. This can be upward or downward.

This helps to measure various jump parameters and their implications for the equity returns and credit spreads. Compared with the diffusion-only model, the jump model has four additional parameters:

❑ jump intensity;
❑ probability that the jump is upward;
❑ jump size parameters for upward or downward jump;
❑ jump risk premium.

Jump diffusion models were originally proposed by Merton (1976) to model stocks and options. In this type of model, the dynamics of the firm's value are driven by two random processes:

❏ a continuous diffusion component, as in Merton's original structural model;
❏ a discontinuous jump component modelled as a Poisson process.

Compared with a pure diffusion model, the inclusion of jump risk increases credit spreads, especially for shorter maturities. The resulting term structure of credit spreads is flexible and hence the model provides a powerful tool to fit a real spread curve.

The firm value of a company may be subject to sudden major changes, owing to external shocks or other unpredicted events. A pure diffusion model cannot capture such an incident, as all trajectories of a geometric Brownian motion are continuous. There are two ways to incorporate jumps:

❏ by superimposing a jump process (Poisson process) on the Brownian motion

Firm value = Diffusion modelling + Jump modelling

❏ by modelling the asset value as a discontinuous Levy process.

Merton (1976), Zhou (1997) and Kijima and Suzuki (2001) model jump risk through the addition of a Poisson process to a Brownian diffusion model (see Figure 4).

Levy process
The Levy process has been used in equity models and fixed-income models. The underlying distribution of the Levy process is flexible and can model the asymmetry and fat tail behaviour of a leptokurtic distribution. Since the Levy process itself models the jumps in the asset prices, there is no need to build jumps artificially into the asset value process. However, owing to complexities in modelling the Levy process, it is still not used extensively as a modelling tool but can be used to provide a qualitative insight into the behaviour of the asset pricing process. Hilberink and Rogers (2000) used the Levy process to address the problem of "zero short spreads" by extending the Leland–Toft (1996) model and allowing only the downward jumps in the asset pricing.

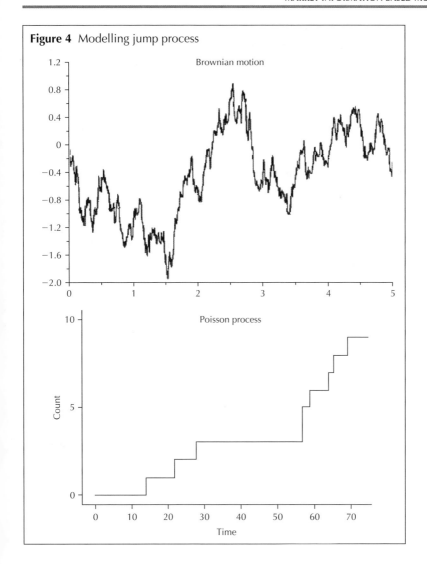

Figure 4 Modelling jump process

Risk-free interest rate process

The interest rate process is a stochastic process for the following features:

❏ discount bond prices;
❏ instantaneous forward rates;
❏ short rate.

Short rate is the internal rate of return of a discount bond with an infinitely short period of time to maturity.

Term structure is built as an output given some assumptions about how an overall economic equilibrium is achieved. Different assumptions about short rate lead to different interest rate models.

Vasicek assumes that the short rate follows a Markov process, that the bond price at maturity is determined by the short rate process and that the bond price is dependent only on one variable, called the short rate. It also assumes the short rate reverts to its long run mean rate. Cox, Ingersoll and Ross (CIR) model short-term volatility as being proportional to the square root of r to prevent it from becoming negative. Longstaff and Schwartz (1995) extend the CIR model by assuming the short rate to depend on the state variables.

These models are further changed to incorporate other time-varying parameters to the short rate process to provide enough degrees of freedom to also match the term structure of volatilities.

An overview of various interest rate models is given for the following reasons.

❏ Stochastic interest rates models are also extended to model credit spreads by incorporating the changes in credit quality and recovery rates into the interest rate models or the factors that lead to changes in the credit quality and recovery rates.
❏ Understanding interest rate models is an important ingredient towards understanding structural and reduced form models.

As with any other stochastic model, interest rates (r) are modelled as a function of drift and volatility,

$$dr = \mu r dt + \sigma r dW_1$$

where σ is the interest rate volatility, μ is the interest rate drift and dW_1 is the Brownian motion driving the interest rates. An overview of the interest rates models is given in Table 1.

The Leland–Toft (1996) model does not allow stochastic interest rates, arguing that previous studies have found that stochastic interest rates reduce the estimated credit spreads, and this problem of small credit spreads is precisely one of the drawbacks of structural models.

Correlation of interest rates with other risk processes
Correlation between the firm's value and the interest rate is modelled through the correlation coefficient ρ as

Table 1 Interest rate models

Interest rate process	Drift + volatility
Merton (1973)	$dr = bdt + \sigma dB$
Vasicek (1977)	$dr = \alpha(\gamma - r)dt + \sigma dB$
CIR (1985)	$dr = \alpha(\gamma - r)dt + \sigma\sqrt{r}dB$
Hull and White (1990) (extended by Vasicek)	$dr = (\theta(t) - \beta r)dt + \sigma dB$
Hull and White (1990) (extended by CIR)	$dr = (\theta(t) - \beta r)dt + \sigma\sqrt{r}dB$
Black et al (1990)	$d\log r = \theta(t)dt + \sigma dB$
Heath et al (1992) A model that applies forward rates to an existing term structure	$dF(t, T) = \mu_F^{\Pi}(t, T)dt + \sigma_F(t, T)dB^{\Pi}(t)$

Table 2 Correlation of interest rate process

Assets value	Default processes
Longstaff and Schwartz (1995) have modelled the correlation of the assets of the firms with the changes in interest rates. Their model explains why bonds with similar credit ratings but in different sectors or industries have widely differing credit spreads. Correlation between the assets' returns and interest rates is approximated by the correlation between the equity returns and the changes in the interest curves	In practice, the term structure and default process are interrelated. However, the two can also be assumed to be independent. This assumption is relaxed by some of the models

The Markov process for credit ratings is independent of the level of spot interest rates

For investment grade debt, this is a reasonable first approximation in the historical probabilities, although for speculative grade debt the accuracy of of the approximation deteriorates |

$$dW \times dW_1 = \rho dt$$

where dW is the Brownian motion driving the firm's value.

Correlation of interest rate with assets value and default process is explained in the Table 2.

Baselines for risk-free interest rates

There are two risk-free interest rate baselines available:

❑ treasury yields;
❑ interest rate swap yields.

In the case of treasury yield the counterparty is the government and for interest rate swaps the counterparty is a bank. Both have different credit risks. Traditionally, the treasury yield curve is used as a baseline for a risk-free interest rate term structure. However, this has changed over the past five years or so.

CDSs and interest rate swaps are instruments between the same players and therefore the swap yield curve is becoming a better proxy for valuation of CDSs.

This means the difference between both the yield curves is not constant. Experts attribute the difference between the two rates to the legal and taxation differences, demand and supply differences in each of the markets and unequal movements in underlying factors that affect each of the market segments.

Credit spreads

Credit spreads are the difference between yields on corporate debts that are subject to default risk and government bonds free of such risk.

There are two types of instruments available to measure the credit spread: bonds and CDSs (see Figure 5).

Reduced form models are a convenient framework to connect bond spreads with CDS premia. Using the risk-neutral default probability and no-arbitrage conditions, it is straightforward to establish the equivalence relationship between the two spreads.

If a firm's risk-neutral default intensity λ and risk-neutral expected fraction loss at default $(1 - R)$, where R is the recovery rate, are assumed to be relatively stable over time, the firm's CDS rate and its par-coupon credit spread would be approximately equal to the risk-neutral mean loss rate $(\lambda \times (1 - R))$ ignoring illiquidity effects. This is explained by Duffie and Singleton (1999).

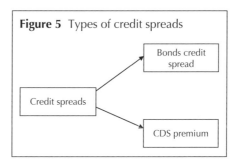

Figure 5 Types of credit spreads

Credit spreads

Bonds credit spread

CDS premium

However, they have shown diversion between spreads and risk-neutral mean loss rates. The illiquidity effects are pronounced if measured relative to treasury yield and as compared to interest rate swap yields.

Empirical studies have proved that the spreads for the same obligor between the two instruments are different. The reasons for such a difference are as follows.

❑ The two markets have a different definition of default.
❑ There is always a time lag between two markets to converge.
❑ Both the markets may not be pricing the credit equally accurately and efficiently.
❑ Price discrepancy is different for a different credit grade.
❑ The two markets may have different prediction power for the future credit events.
❑ In both the markets, liquidity premia are part of the credit spreads. Liquidity premia are different in the two markets.

Two instruments have different agreement terms and different liquidity in the market.

Amato and Remolona (2003) argue that spreads on corporate bonds tend to be many times wider than what would be implied by expected default losses alone. Similar views are also echoed by Huang and Huang (2003) who have shown that diffusion models typically explain only 20–30% of the observed credit spreads of firms.

During 1997–2003, the average spread on BBB-rated corporate bonds in the US with three to five years to maturity was about 170 basis points (bps) at annual rates. During the same period, the average yearly loss from default amounted to only 20 bps. The eight times wider spread on corporate bonds, therefore, cannot be explained from expected default losses alone.

Various propositions are put forward by experts to explain credit spread of more than eight times the expected loss (EL) from default. The role of taxes, liquidity premia, risk premia and an inability to diversify are some of the explanations given.

❑ In the US, corporate bonds are subjected to state income taxes while treasury bonds are not. The tax rates vary from 5–10%.

❑ Amato and Remolona (2003) argue that the difference between credit spread and default probability is due to the inability of investors to diversify the default risk in the corporate bond portfolio.

❑ The distribution of returns on corporate bonds is highly negatively skewed. Such skewness requires a large portfolio to achieve full diversification. A large portfolio of corporate bonds is difficult to build owing to smaller and shallow corporate bond markets. This leads to undiversified credit risk and wider spreads; as investors are not able to diversify their bond portfolio, they are asking for higher returns to compensate them for the higher/correlated risk they are taking.

Empirically it is established that taxes and liquidity conditions are the other major variables in addition to default expectation.

The importance of liquidity premia in the credit spreads can be established from the fact that observed credit spreads are higher for the highest quality corporate issue while it has a lower PD.

Default probability is a more important determinant of credit spreads for lower-rated investment-grade bonds. The credit quality explains around a third of variability in a pooled regression. When coefficients are allowed to vary at the level of the individual issue, explanatory power rises to 50% for a portfolio of lower rated investment grade bonds. Such a higher role of credit quality can be explained by idiosyncrasies playing a greater role in liquidity conditions and expected recovery rates for this type of portfolio.

Contrary to theory, recent empirical work suggests that changing default expectations can explain only a fraction of the variability in credit spreads. Empirically, the following features are established for credit spreads:

❑ negative correlation between credit spreads and the short-term default free interest rate;
❑ credit risk explains only a relatively small part of the credit spread;
❑ credit spreads are dependent on systematic factors;
❑ credit spreads are dependent on firm specific factors;
❑ credit spreads are dependent on liquidity and taxation.

There is considerable empirical evidence consistent with changes in credit spreads and changes in default-free interest rates being negatively correlated. Duffie (1998) fits a regression of the form

$$\Delta Spread = b_0 + b_1 \Delta Y + b_2 \Delta Term + e$$

where *Spread* is the time spread for bond with maturity T, ΔY is the change in the three-month treasury yield, *Term* denotes the difference between the 30-year constant treasury bond yield and the three-month treasury bill yield and $\Delta Term$ denotes the change in *Term* over the period.

Convergence of bond and CDS spreads: We know that a very wide definition of default exists in the bond markets. A very wide definition of default implies a higher risk-neutral default probability. The wider definition of default creates heterogeneity at default (as fractions of their respective principals) in the market values of the various debt instruments of the obligor. Duffie has shown that, by ignoring heterogeneity and assuming frictionless markets, CDS rates are very close to the par-coupon credit spread of the same maturity as the default swap.

Measuring CDS premium

The CDS rate is the annualised premium rate, as a fraction of notional. Using an actual 360-day count convention, the CDS rate is thus four times the quarterly premium. As there is no initial exchange of cashflows on a standard default swap, the at-market CDS rate is, in theory, a rate for a contract for which the net market value is zero.

Uses of credit spreads

The most important uses of credit spreads are risk-neutral probabilities of default. Reduced form models, represented by Jarrow and Turnbull (1995), Duffie and Singleton (1999), Madan and Unal (1998) and Hull and White (2000), typically treat default as a random stopping time with a stochastic arrival intensity. The credit spread is determined by risk-neutral valuation under the absence of arbitrage opportunities.

Risk-neutral probabilities of default are used for:

❏ building the credit term structure yield curves;
❏ bond rating changes;

❑ changes in the bond spreads;
❑ relationships between bond spreads and treasury yields.

Credit spread modelling under reduced form
Under structural models, the spread goes to zero with time to maturity going to zero. In the reduced form models, the default event is unpredictable – it comes without warning. There is always short-term uncertainty about the default event, for which investors demand a premium. This premium, expressed in terms of yield, is given by the intensity.

Calibrating reduced form models: Typically, all reduced form models compute the pricing. Therefore, reduced form models are directly calibrated to the market prices. There are three types of prices available in the market: bond spreads, CDS spreads and equity prices. Reduced form models are calibrated on to market prices.

Testing of structural and reduced form models: The credit spread in the bond markets and CDS market on average should be the same. Structural models are estimated with equity data and the reduced form model is estimated with bond data. CDS data for testing ensures a neutral ground on which the success of the different models can be evaluated.

Debt assumptions
Each model makes various assumptions about debt covenants to make the model simpler for mathematical modelling and such assumptions are then relaxed analytically. We will now discuss some of the assumptions that can be made.

Leverage and capital structure
Firms adjust their debt load to maintain a stable leverage ratio. In general, the leverage ratio is assumed to follow a mean-reverting process. The target leverage, in tune with intuition, depends on the current level of interest rates.

The KMV model places a greater weight on short-term obligations, probably for the purpose of better predicting the firm's default probability within one year. The logic of this approach is that debts due in the near term are more likely to cause a default. The KMV model measures the numerator of the leverage ratio as short-term debt plus half the value of long-term debt.

Time of default
Default can be expected to happen only on the following three dates:

❑ on the date of maturity;
❑ on the coupon payment date;
❑ any time if the asset value falls below the default threshold instead of occurring just at the maturity date of the debt or on the coupon date.

Default threshold
Black and Cox (1976) consider the default threshold as a safety covenant, which acts as a protection mechanism for the bond holders against an unsatisfactory performance. This will allow them to take control of the firm if the asset value reaches the default threshold.

Asset payout
Dividend and interest payments represent asset payout. An increase in the asset payout increases the default probability. The leverage ratio has an impact on the asset payout ratio. Various assumptions are made about asset payout. A zero-coupon bond is an assumption of no interest payment. No dividend payment is generally assumed.

Debt maturity
Debt maturity may not have a significant impact on the risk measurement except for the short duration where models undervalue the spreads.

❑ For infinite maturity or perpetual bonds, simpler models can be used. If a model assumes infinite maturity, it delivers analytic tractability but makes it impossible to capture the empirical regularity that borrowers are less likely to default over a given horizon if they are to repay the debt principal over a longer period of time.
❑ Short-duration assumptions are made to reduce agency costs and moral hazard.
❑ Finite maturity – for long duration – generates higher firm value.

Bankruptcy costs, agency costs and tax advantages
The simple assumption is to ignore such costs.

Interest payment on the debt

Coupon payments have an impact on the dynamics of bond pricing and firm valuation. On the coupon date, if coupon is paid, the firm's value decreases by the coupon paid. In general, there are two types of bond.

❑ *Zero-coupon bonds*: The holder of a zero-coupon bond receives the face value if the company survives until maturity, otherwise the investor receives the recovery fraction. Zero-coupon bonds are generally modelled on the basis of continuous compounding.

❑ *Coupon bonds*: Coupon bonds are reduced to several zero-coupon bonds of different maturity. Thus, even coupon bonds are treated as zero-coupon bonds.

Default triggers

Various types of covenants trigger default. We will now discuss some of the different default triggers.

❑ There are safety covenants in the issued debt that allow the creditors to close down and liquidate the firm if the value of its assets should fall below a certain level. Default occurs as soon as

$$Value\ of\ the\ firm \leq Default\ trigger$$

❑ A firm continues to operate until it pays back its debt. Here a default can only occur at the maturity of the outstanding debt or at coupon dates before that.

❑ The default trigger is not constant and is time dependent. For example, the covenant may prescribe that a default is triggered if the firm's value falls below the discounted value of the assets outstanding, ie, when the firm's value is worth less than a similar but default-free investment. Default occurs as soon as

$$Value\ of\ the\ firm \leq Value\ of\ the\ similar\ default\text{-}free\ bond$$

❑ Exactly modelling the covenants. However, generally this is not done as modelling covenants exactly requires assumptions that are very difficult to handle mathematically in other parts of the model.

Defaultable claim

A defaultable claim consists of the following features.

❑ A promised contingent claim X at time T in the event of no default.
❑ The cashflow stream received during the period prior to default.
❑ Two types of recovery process:

   ❑ recovery payoff at the time of default, if default occurs prior to or at time T;
   ❑ recovery payoff at time T if the default occurred after or at the time T;

Recovery payoff may be different for different seniority classes.

Information on the deteriorating repayment conditions
need to be disclosed

This information may be disclosed in private or in public. However, for the following reasons, this information is assumed to be incomplete whilst modelling:

❑ financial information is published after a time lag and only a limited set of information is available as public information;
❑ information from the management may be intentionally misleading.

Modelling default threshold

Black and Cox (1976) and Longstaff and Schwartz (1995) assume the default threshold to be a deterministic function of time. Others, including Nielsen and Santa-Clara, assume that the default barrier is a random barrier. Implicit in all these models is the assumption that the barrier, whether it is fixed or random, can be observed by investors. In each case, it follows that the distance to default (DD) is observable and default is predictable, while in practice the default barrier is random and non-observable.

A random and non-observable default threshold opposes the paradigm of modelling asset value through a diffusion process. In a diffusion process, the assumption is that the firms diffuse towards default. When modelling a random default threshold, the assumption is that default occurs only as a surprise. The PD is denoted and measured by two popular terms, "default intensity" and "hazard rate". As asset values keep changing, the default probability also keeps changing. The building blocks for modelling default thresholds include the following.

❑ The default threshold is modelled as an exogenous jump process. The default time is the first jump time of a Poisson process with deterministic or stochastic (Cox process) intensity.
❑ Since default requires a jump, the default-free market data does not have or provide default information. Market prices of defaultable instruments are the only source of information about the firms.
❑ These models take a purely probabilistic approach and do not attempt to explain default economically. The default intensity is typically modelled to depend on state variables such as interest rates and market indicators.

The model framework is very similar to that of interest rate models. The term reduced form model was coined by Duffie and Singleton (1999) for market information based models not considering the firm's value.

Modelling the recovery process

Bankruptcy results in the decline in the firm's value and expenses owing to the costs of the financial distress. Debt seniority is modelled through different recovery rates in the event of default.

The no-arbitrage condition requires a smooth switch between the pre-default and post-default value of debt (see Figure 6). Expected recovery rates drive this switch. If the bankruptcy cost is zero, to attain a given recovery rate borrowers will wait longer (smaller trigger) before starting the bankruptcy proceedings. If their expectations about the recovery rate go up, the bankruptcy process starts earlier and the pre-default value is higher.

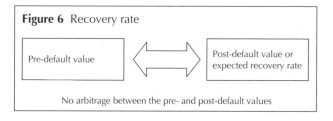

Figure 6 Recovery rate

Pre-default value ⟺ Post-default value or expected recovery rate

No arbitrage between the pre- and post-default values

In general, there are three approaches to model the recovery rates.

❑ *Recovery of face value* (RFV): The recovery rate is an exogenous fraction of the face value of the defaultable bond.
❑ *Recovery of treasury value* (RTV): Jarrow and Turnbull (1995) consider a recovery rate to be an exogenous fraction of the value of an equivalent default-free treasury bond.
❑ *Recoveries of market value* (RMV): Duffie and Singleton (1999) consider a recovery rate to be an exogenous fraction of the market value of the bond just before default. Of the three models above, the RMV specification has gained a great deal of attention in the markets where bankrupt bonds are traded.

Post-default valuation of firm

❑ If the value of defaulted firm is V and the cost of liquidation is K. The post-default value of the debt is given by

$$Post\text{-}default\ value\ of\ the\ debt = V - K$$

❑ Accept the terms of a new debt contract. If the value is assumed to be $V - K$, then creditors are indifferent towards the liquidation or reconstructing.

Modelling the default intensity process
Intensity models assume that a firm is exposed to the default intensity owing to external and internal factors. The intensity is measured or inferred from default experience. In contrast to the asset-value-based models, the time of default in intensity models is not determined via the value of the firm, but instead is taken to be the first jump of a point process (eg, a Poisson process). The governing default intensity parameters are measured (associated with the default probability measure, P) and inferred from market data. Default is modelled as the first arrival of a point process with the inferred risk-neutral hazard rate.

There are three approaches to understanding intensity-based models:

❑ types of intensity process – the basic type is the Poisson process;
❑ risk factors driving intensity;
❑ closed form equations.

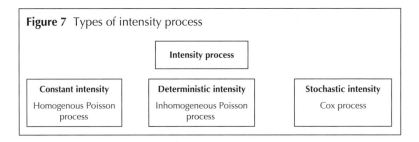

Figure 7 Types of intensity process

There are three types of intensity process (see Figure 7), dependent on whether the intensity is constant or variable. At a basic level, the Poisson process provides a convenient way of modelling default arrival risk in intensity-based default risk models. At the advanced level, default intensity is modelled as a stochastic Cox process.

Modelling the Poisson process
The Poisson process is used to model rare events or discrete countable events. The default rate is a discrete countable rare event. The time of default of a firm is modelled as the time of the first jump of a Poisson process. The intensity of the process is measured as λ. The following list gives some important properties of the Poisson process relevant for credit risk.

❏ The Poisson process has no memory. The probability of jumps in $(t, t + dt)$ is independent of the history. The next jump is not more likely if the last jump occurred a long time ago.
❏ The inter-arrival time is given by

$$Inter\text{-}arrival\ time = \lambda e^{-\lambda t}dt$$

❏ Two or more jumps at exactly the same time have zero probability.
❏ A Poisson process with intensity λ is a non-decreasing, integer-valued process with initial value. The primary differences between the Brownian process and Poisson process are as follows.

  ❏ Under the Brownian process: Jump size is decreased and probabilities are kept constant. A given probability is spread across the time intervals. This results in a diffused process.
  ❏ Under the Poisson process: The jump size is kept constant to a minimum of one default. This reduces the default rate for the time interval.

❑ The discrete Poisson process can be converted into a continuous rate of events by assuming a large portfolio of defaultable securities that are all driven by independent Poisson processes. Then the Poisson events occur almost continuously at a rate of λdt to the whole portfolio.

❑ The probability of at least one jump during a period (Δt) of $P_1 = 1 - e^{-\lambda \Delta t}$. For Monte Carlo simulation of the Poisson process, the random numbers are drawn from the interval $(0, 1)$ and it is checked if the number is lower than P_1. If it is, a jump is assumed to have occurred and a default count is incremented. Monte Carlo simulation may not give good results unless the simplest specification for a Poisson process is zero recovery and a default is triggered by the first jump of the Poisson process.

Intensity constant – homogenous Poisson process: For an intensity or hazard rate of λ, the PD is given by

$$P(T) = 1 - e^{-\lambda T}$$

Therefore,

$$\lambda = \frac{d(T)}{1 - P(T)}$$

where d is the density of default probability.

Deterministic intensity – inhomogeneous Poisson process: The PD is given by

$$P(T) = 1 - e^{-\int_0^T \lambda(t) dt}$$

where

$$\lambda(t) = h_i, \qquad t \in (T_{i-1}, T_i)$$

and h is constant and is calibrated from market data (ie, credit spread).

Stochastic intensity – Cox process: The construction of the Cox process intensity model is a two-step process. Variables or "risk factors", X, driving intensity are modelled in the first step. The second step is to model the risk factors into the intensity.

hazard rate or intensity $= f(\text{risk factors})$

and

$$\int_0^T f(X_s)ds \text{ is finite.}$$

Risk factors driving intensities or hazard rates
Existing models link default intensities to three types of risk factors: rating transition, stock prices and state variables.

Intensity based on state variables: Lando (1998) models default time as a Cox process. Lando assumes that the risk factor can be described by a Markov vector process and defines an intensity λ of the Cox process such that, conditional on the information generated by the state variable Y, an inhomogeneous Poisson process results with intensity λ $(t, Y(t))$.

Intensity based on rating migration: Credit ratings are the risk factors or indicators of the likelihood of default. They utilise historical transition probabilities for the various credit rating classes to determine the pseudo-probabilities (martingale, risk adjusted) used in valuation. There are two approaches to rating migration. Rating migration can either assumed to be constant or variable. Jarrow *et al* (1997) assume that the transition intensities are constant. Lando considers credit ratings as state variables and proposes the use of the Cox process framework to model default.

Including credit rating information in the bankruptcy process helps in modelling CDs in addition to the defaultable bonds.

Modelling rating transition: We know that

Price of defaultable zero-coupon bond = f(time, risk-free interest rate, issuer credit rating)

The following processes are examples of the approaches used for incorporating ratings into the default process.

Rating transition as a Markov process: It is assumed that the set of rating classes is $(1, ..., K)$ where the state K is assumed to correspond to the default event. It is assumed that the migration process follows a Markov chain, ie, the future evolution of ratings classes a particular bond not on the bond's history but only on its current rating, and is a stochastic process.

Given an initial rating of a defaultable bond, the future changes in its assessments by a rating agency are described by a stochastic process, referred to as the *migration process*. The Jarrow–Lando–Turnbull model incorporates the Markov chain dynamics of the ratings without allowing for stochastic spread dynamics.

Das and Tufano (1996) extend the Jarrow–Lando–Turnbull model to incorporate stochastic recovery rates and thus to have stochastic dynamics of the credit spreads within the individual classes. Their model is set up as a discrete-time approximation to a continuous-time model.

Lando (1998) modifies the Jarrow–Lando–Turnbull approach by introducing a conditional Markov migration process, which accounts for both the presence of different rating classes and for the postulated existence of the underlying state variables.

Some of the existing models assume the following simplifications.

❑ For a given credit class and maturity, the transition probability to default is assumed to be constant over time, and therefore the credit spreads also remain constant within every credit class.
❑ Two types of recovery rate assumptions – recovery rate is assumed to be zero or at default there is a cash payoff of recovery rate.
❑ A TM published by the rating agencies is used. The TMs published by the rating agencies do not represent the credit spread observed in the bond pricing. The credit spreads observed are higher. Therefore, the default probabilities are larger than the corresponding empirical probabilities.

Value of bond on maturity: Value is assumed to be either one or zero. The price of the defaultable bond is the risk-neutral expectation of its discounted payoff. The payoff in maturity is one for all rating classes, except in the last class (the default) where it is zero. Zero value assumes a zero recovery rate.

Transition time: There are two approaches with regards to the transition time.

❑ *Discrete time model*: The migration process and default time follow time-homogenous Markov chains.
❑ *Continuous time model*: The migration process follows the time-homogenous Markov chain. The migration process is driven by a time-dependent intensity matrix. At the default time T, the rating process reaches the absorbing state K.

Intensity based on stock prices: Default intensity depends on the stock price s. For example, the stock price is modelled as

$$ds(t) = \sigma s dz$$

where σ is the volatility and dz is the Brownian motion. The default intensity, λ, is given by

$$\lambda = \frac{a}{In(s/sl)^2}$$

where a and sl are constants.

Closed form equation: It is difficult to model default probabilities with stochastic intensity models. The affine intensity framework developed by Duffie and Kan (1996) provides a powerful class of intensity models that admit closed form solutions for default probabilities. The affine framework takes advantage of the Cox process.

STRUCTURAL MODELS

Structural models are based on the capital structure of a firm. Every firm has equity and debt in the capital structure with limited liability for equity holders, and they hold and churn assets continuously to service the capital by way of dividends, interest and repayment of debt. Default occurs when assets are unable to serve as debt ie, when assets are unable to generate sufficient funds for paying interest and repayment of debt.

There are various ways to implement the structural approach. However, all of these methods use arbitrage-free pricing methodologies.

The Merton model is based on the limited liability rule, which allows shareholders to default on their obligations while they surrender the firm's assets to the various stakeholders in accordance with the prescribed priority rules. The firm's liabilities are thus viewed as contingent claims issued against the firm's assets with the payoffs to the various debt holders. Default occurs at debt maturity whenever the firm's asset value falls short of the debt value. The loss rate is endogenously determined and depends upon the asset value, volatility and risk-free interest rate. Longstaff and Schwartz (1995) allow bankruptcy to occur at a random default time. Default is triggered the first time the value of the assets falls to the default trigger. They model interest rates as a stochastic process.

Empirical findings on structural models

Some of the empirical findings on structural models have included the following points.

❏ Estimation or calibration of credit spreads reveals that the predicted credit spread is far below observed credit spreads (Jones *et al* 1984). A similar conclusion is drawn by Huang and Huang (2003), who suggest that structural models tend to systematically underestimate the credit risk in the corporate bond market.
❏ Structural variables explain very little of the credit spread variation Huang and Huang (2003).
❏ Pricing error is very large for corporate bonds (Eom *et al* 2004).
❏ More flexible regression analysis suggests that the explaining power of default risk factors for credit spread is still very small (Collin-Dufresne 2001).
❏ Temporal changes of bond spread are not directly related to expected default loss (Elton *et al* 2001).
❏ The forecasting power of long-run volatility cannot be reconciled with the classical Merton model (Campbell and Taksler 2003).
❏ Asset volatility and leverage is an indicator of credit risk. Bonds with lower rating (speculative grades) have higher asset volatility and higher leverage.
❏ A major challenge for structural models is to raise the average predicted spread relative to the Merton model, without overstating the risks associated with volatility, leverage or coupon.

Landmarks in structural models

Some of the landmarks in modelling concepts for asset-value-based models include the following points.

❏ According to Black and Cox (1976), default can occur at any time until the maturity of the debt and the default threshold changes over time.
❏ Geske (1977) relaxed the condition of constant equity and debt to allow firms to issue equity and service debt.

❑ Longstaff and Schwartz (1995) developed a two-factor model to value risky debt, extending the one-factor model of Black and Cox (1976) in two ways:

❑ incorporating both default risk and interest rate risk,
❑ allowing for deviations from strict absolute priority.

An important feature of this model is that firms with similar default risk can have different credit spreads if their assets have different correlations with changes in interest rates. It also models stochastic risk-free interest rate processes.

❑ Leland and Toft (1996) argued that the debt value (and yield spread) is dependent on the company's capital structure. They proposed a process to optimise capital structure by modelling the impact of leverage on the debt value and considered the impact of bankruptcy costs and taxes on the structural model output. They assumed that the firm issues a constant amount of debt continuously with fixed maturity and continuous coupon payments.

❑ The Ericsson–Reneby model allowed valuation of finite maturity coupon debt with bankruptcy costs, corporate taxes and deviations from the absolute priority rule. It also modelled the re-organisation threshold.

❑ A mean-reverting leverage ratio process was used by Collin-Dufresne and Goldstein 2001).

Correlation in structural models

Cyclical impact on the PD is incorporated by modelling systematic risk factors into the stochastic asset diffusion process. In addition to the changes in the asset volatility, the cyclical factors also change the default point and liquidation costs. However, the impact on the asset volatility is more pronounced. In addition, the downward shift in asset values is greater for the high-volatility/low-credit quality firm, demonstrating that the pro-cyclical impact on the PD is stronger than for the low-volatility/high-credit quality firm. The KMV model – a type of structural model – has a three-level approach to the systemic factors:

❑ the first level incorporates a composite market risk factor;
❑ the second level includes an industry and country risk factor;
❑ the third level contains regional factors and sector indicators.

Recovery rates

Structural models have two different approaches to the recovery rates. Classic models like Merton assume recovery to be endogenous and models like CreditMetrics and Nielson et al (1993) recovery rate is exogenously determined. This assumption makes it possible to model default rate not only at the time of the maturity but any time before maturity. There are two approaches to recovery rate-constant and stochastic. It has been empirically established that LGD rates increase by 20–30% under distress conditions. Models with stochastic recovery rate can factor in the distress conditions in the recovery rate.

Shortcomings of structural form models

The main observed shortcomings of structural form models are as follows.

❑ All relevant credit risk elements, including default and recovery at default, are assumed to be a function of the structural characteristics of the firm: asset volatility (business risk) and leverage (financial risk).
❑ Structural form models require a firm's asset value as an input, which is a non-observable variable.
❑ These models do not consider credit rating changes that occur quite frequently for default-risky corporate debts.
❑ Structural form models assume that the value of the firm is continuous in time. Therefore, they assume default to be predictable and with no "sudden surprises".

Merton model

The Merton model is the first structural or asset valuation model. It is a structural model since it models the capital structure of a firm using assets values modelled from equity prices. Merton defines the default for a firm equivalent to an event that the total value of the firm's assets is less than its obligations at the time of debt maturity. In this case, the owners would hand the firm over to the creditors rather than paying back the debt. PD is measured as equal to the probability of observing this event. Therefore, under Merton model, risky debt of a firm is valued as an option on economic firm value and extends the Black–Scholes model to value risky debt.

Merton modelled a firm's asset value as a stochastic process, such as geometric Brownian motion, and assumed that the firm would default if the asset value, A, falls below a certain default boundary, X. The extent of randomness of this process characterises the riskiness of the firm. The value of the firm's assets is unobservable in practice. The default was allowed at only one point in time, the date of maturity. The model attempts to use the market observable equity value to determine both the economic firm value and debt value of a firm.

This model assumes various simplifying assumptions.

❑ *Only two types of claims – equity and debt*: Debt is zero-coupon bonds of the same seniority and maturity. This allows the balance sheet of the firm to be modelled precisely as required to fit into the Black–Scholes option pricing framework. On maturity date, debt holders get paid in full before equity holders receive anything.
❑ *No changes in bonds or equity until the maturity*: No new issues of bonds or equity, no repayment of bonds or equity, no cash dividends and zero coupon also means no interest payment.
❑ *Market is perfect*: There are no transaction costs and no taxes. All assets are infinitely divided, any investor believes that they can buy and sell an arbitrary amount of assets at the market price, borrowing and lending can be done at the same rate of interest and short sales of all assets are allowed.
❑ *Constant interest rates*: The term structure of interest rates is deterministic and flat.
❑ *Capital structure*: The Modigliani–Miller theorem applies for the value of the firm, ie, the value of the firm has an unchanging relationship to its capital structure.
❑ *No jumps*: The probability distribution of the firm's assets is described by a lognormal probability distribution. The value of the firm, Vt at a time t, follows an autoregressive process, ie, all information needed to predict the future dynamics of the firm value is contained in its past development. In particular, the value of the firm is not subject to any external shocks. Figure 8 shows the difference in credit spread between the Merton model and a Jump model.

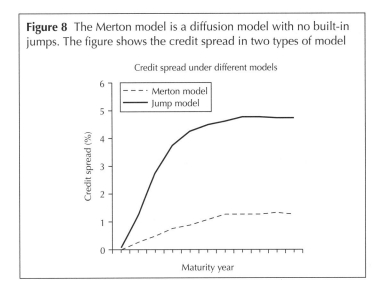

Figure 8 The Merton model is a diffusion model with no built-in jumps. The figure shows the credit spread in two types of model

Value of risky debt

At the time of debt maturity the value of the firm is V_T

❑ Debt holders get paid in full before equity holders receive anything.

❑ Debt holders receive the following:

   ❑ Face value F of the debt if $V_T \geq F$;

   ❑ V_T, otherwise.

This is equivalent to saying the value of the firm's debt is equivalent to the value of a zero coupon bond of the face value F and value of the European put option on the assets of the firm that matures at time T with strike price of F

Value of risky debt = Value of risk-free debt − Value of the Put option

We know that value of risk-free zero-coupon bond $= e^{-r(T-t)}F$ where r is the risk free interest rate

From the Black–Scholes formula

$$\text{Value of put option} = e^{-r(T-t)}F \cdot N(-d + \sigma\sqrt{T-t}) - VtN \cdot (-d)$$

where
$$d = \frac{1}{\sigma\sqrt{T-t}}\left[\ln\left(\frac{1}{L}\right) + \frac{1}{2}\sigma^2(T-t)\right]$$

and

$$L = \frac{e^{-r(T-t)}F}{Vt}$$

$$\text{Value of risky debt} = e^{-r(T-t)}F \cdot \left[N(d - \sigma\sqrt{T-t}) + \frac{1}{L} \cdot N(-d) \right]$$

Given value of L in the equation, clearly the value of risky debt declines when leverage L increases.

Let R denote the yield on the risky bond; therefore,

$$\text{Credit spreads} = R - r = -\frac{1}{(T-t)} \ln\left[N(d - \sigma\sqrt{T-t}) + \frac{1}{L} \cdot N(-d) \right]$$

This expression is used to derive the term structure of credit spreads. From this equation credit spreads are the function of

❑ Time to maturity $(T - t)$;
❑ Firm risks or the volatility of the firm value σ;
❑ Leverage Ratio L of the firm;
❑ Some of the implications from the equation mentioned in the previous paragraphs (see Table 3 also):

  ❑ When the put option is deep out of the money $Vt \gg F$, the PD is low and corporate debt trades as if it is default free.
  ❑ If the put option trades in the money, the volatility of the corporate debt is sensitive to the volatility of the underlying asset.
  ❑ If the default-free interest rate increases, the spread associated with corporate debt decreases. The magnitude of this change is larger for higher yield on the debt.
  ❑ As the maturity of the zero-coupon bond tends to zero, the credit spread will also tends to zero according to the model.

Advantages of the Merton approach
The Merton model addresses the deficiencies in the credit scoring or accounting information based credit analysis. It incorporates the market information or perception about the company into the credit analysis. The information contained in the market prices or equity prices is inherently future oriented and forward looking. Everyone is interested in the measurement of credit risk in the future.

Table 3 Valuation of defaultable securities

Equity returns	Higher returns indicate lower PD. Higher skewness also indicates more positive returns
Equity volatility	Higher equity volatility means higher assets volatility which in turn means assets value is more likely to hit the default threshold leading to a higher PD. Higher equity kurtosis means higher equity volatility
Expected recovery rates	Higher recovery rates reduces the default rates
Jump intensity	A jump has two components: mean and variance. Mean increases the assets value returns and variance increases the volatility. Net effect depends on the quantum of mean and variance
Leverage and default cost	Both increases the default threshold (assets size for default)
Dividend payout ratio	Higher rate reduces the assets value therefore increasing the credit risk

Compared with any other information source, the market information is the best information set of the future. Furthermore, with market information, default probabilities can be individually assessed on a day-to-day basis for each public company.

❏ The analysis can work as an early warning signal.
❏ It is possible to measure and compare risk on a cardinal scale in addition to an ordinal scale.

Problems when using the Merton approach

❏ A firm's value and volatility is not directly observable in the market. Only the equity market prices and equity volatility are observable.
❏ In particular, it is unrealistic to assume that volatility and riskless interest rates remain constant over time.
❏ The interest spreads (over a riskless interest rate) are assumed to be constant. If spreads followed the dynamics given by the model, they would tend to zero as the maturity of the bond approaches. In practice, experience shows that this is not the case. Spreads are usually bounded away from zero over all the time to the maturity of the bond. This observation has stimulated the development of many other default models for corporate bonds.
❏ Corporations have complex liability structures. In the Merton framework, it is necessary to simultaneously price all the

different types of liabilities senior to the corporate debt under consideration.

❑ Default can only occur at the time of a coupon and/or principal payment. However, in practice, payments to other liabilities other than those explicitly modelled may trigger default.

Hybrid model – the KMV model

KMV is the firm which started commercial implementation of the Merton model. KMV was acquired by Moodys in 2002. The KMV model is a valuable tool to make the Merton model operational and it has turned into a useful tool for practitioners. It is a tool used to bridge the gap between accounting information and market information. The most important contribution of the KMV model is to measure the PD at a daily frequency for public companies and the model allows the use of an estimate of the PD as an early warning signal. It enables complete automation and quantitative analysis of the PD.

KMV model process

❑ Measure the default point. The book value of the firm's total liabilities is generally taken to be the face value of the firm's debt.

❑ Using equity value and volatility, estimate firm's value V and firm volatility. Historical estimate of equity volatility is estimated from either historical stock returns data or from option implied volatility data. Volatility should be scaled to reflect the horizon of the forecast.

❑ Compute the distance to default. Distance to default is the number of standard deviations of the firm's value by which the firm's value is expected to exceed the default point F:

$$\delta = \frac{Vt - F}{\sigma V}$$

All the term have same meaning as given in the Merton model. Vt is the value of the firm, F is the face value of the debt and σ is the volatility in the firm value. Estimates of current assets and the current standard deviation of asset growth ("volatility") are calibrated from historical observations of the firm's equity-market capitalisation and of the liability measure. The default risk of a firm increases as the book value of its liabilities approaches the value of its assets, until such a time as the market value of the assets is no longer sufficient to repay its liabilities and the firm ultimately defaults on the payments of its obligations. The firm's obligations are therefore viewed as contingent claims (options) issued against the firm's assets, with payoffs to the various debt holders specified by the seniority of the claims and other covenants. The model basically uses these two non-linear equations, to translate the value and volatility of a firm's equity into an implied PD.

❑ Using the KMV database, identify the proportion of the firm with the same DD that actually defaulted within a year. This is the expected default frequency (EDF).

❑ This model accommodates five different types of liabilities, short-term liabilities, long-term liabilities, convertible debt, preferred equity and common equity, and combines market asset value, asset volatility and the default point term structure.

❑ It computes a DD term structure. This term structure is translated to a physical default probability, better known as an EDF credit measure, using an empirical mapping between DD and historical default data.

  ❑ It treats the firm as a perpetual entity that is continuously borrowing and retiring debt.

  ❑ By explicitly handling different classes of liabilities, it is able to capture richer nuances of the capital structure.

  ❑ It calculates its interim asset volatility by generating asset returns through a delevering of equity returns. This calculation is different from more common approaches that compute equity volatility and then delever it to compute asset volatility.

  ❑ KMV generates the final asset volatility by blending the interim empirical asset volatility as computed in the manner explained above together with a modelled volatility estimated from comparable firms. This step helps filter out noise generated in equity data series. The default probability generated by the KMV implementation of the VK model is called an EDF credit measure.

  ❑ A sector and seniority-wise constant LGD is assumed.

Merton versus *KMV*

While Merton assumes the firm's liabilities to consist of only two classes, a debt issue maturing on a specific date and equity, KMV allows liabilities to include current liabilities, short-term debt, long-term debt, convertible debt, preferred stock, convertible preferred stock and common equity.

Under the KMV model, dividend payments and cash payments of interest happen prior to the maturity of the debt unlike the Merton model.

In the Merton model, default was equivalent to the firm value being lower than the debt at the moment when the debt had to be repaid. KMV generalises the concept of default. Under the KMV model, default can happen even before the maturity of a particular debt issue. Equity, in this context, has no expiry date, but is modelled as a perpetual option.

In the Merton model, the firm value process is only modelled as a geometric Brownian motion for the purpose of the calculation of the unknown input variables.

KMV uses an empirical distribution to assign default probabilities to the stated DDs.

Weaknesses of the KMV model

Being an extension of the Merton model, the KMV model inherits all its severe structural problems.

❑ It requires some subjective estimation of the input parameters.
❑ It is difficult to construct theoretical EDFs without the assumption of normality of asset returns.
❑ Private firms' EDFs can be calculated only by using some comparability analysis based on accounting data.
❑ It does not distinguish among different types of long-term bonds according to their seniority, collateral, covenants or convertibility.
❑ KMV asserts to have done detailed research that has proved all results. However, it is proprietary in nature. The lack of a publicly available test is critical because the relationship between the DD and the estimated PD is so sensitive that small errors in the measurement of the DD or in the mapping between both quantities may lead to significant errors in the resulting default probability.

❑ The most critical inputs to the model are clearly the market value of equity, the face value of debt and the volatility of equity. As the market value of equity declines, the PD increases. This is both a strength and weakness of the model.

KMV model as reference for data
Moody's KMV provides, among other data, current firm-by-firm estimates of conditional PDs over time horizons including the benchmark horizons of one and five years. For a given firm and time horizon, the estimate of default probability is fitted non-parametrically from the historical default frequency of other firms that had the same estimated DD as the target firm. It is a readily available data set for essentially all public US companies, and for a large fraction of non-US public firms. Moody's KMV EDF measure is also extensively used in the financial services industry as a reference data covering over 26,000 publicly traded firms the world over. An alternative industry measure of default likelihood is the average historical default frequency of firms with the same credit rating as the target firm. However, as we have seen in the "Credit Rating" chapter, the rating does not exactly represent a measure of default probability.

REDUCED FORM MODELS
In contrast to structural models, the time of default is not determined via the firm's value or other such indices. The reduced form approach models PD and Recovery Rate (RR) independent of structural features of the firm, assets volatility and leverage. The reduced form approach views the credit event as a perfectly unpredictable event. The probability of the arrival of this event is in fact a hazard rate, also called the intensity rate since it represents the frequency of defaults that can occur in a given time interval. In general, reduced form models assume an exogenous RR that is independent from the PD. Both PD and RR are assumed to be stochastic.

The default event is modelled as a Poisson process, of which the default probability is governed by a stochastic hazard rate function. This hazard rate process is driven by an exogenously random variable. Default occurs when the random variable undergoes a

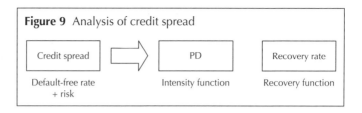

Figure 9 Analysis of credit spread

Credit spread		PD	Recovery rate
Default-free rate + risk		Intensity function	Recovery function

discrete shift in its level. These models treat defaults as unpredictable Poisson events. The time at which the discrete shift will occur cannot be foretold on the basis of available information. The random variable is generally inferred from market data, equity or credit spreads.

However, we know that credit spreads represent an incomplete information data set. Using the term structure of credit spreads for each credit class, the EL over the time $(t, t + dt)$ is estimated. EL means the product of the conditional PD and the recovery rate (see Figure 9) under the equivalent martingale (the risk-neutral) measure.

Default is modelled as a point process. Over the interval $(t, t + dt)$, the default probability conditional upon no default prior to time t is approximately $\lambda(t)dt$, where $\lambda(t)$ is the intensity (hazard) function. A stochastic change in credit spreads occurs only if default occurs.

The work of Jarrow and Turnbull (1995), Duffie and Singleton (1999), Hughston (1997) and Lando (1994, 1997) implies that, for many CDs, we need only to model the EL that is the product of the intensity function and the loss function. Hence, the reduced form models are extendable to produce a valuation of CDs.

Primarily there are three sources of data for reduced form models:

❏ equity prices;
❏ bond prices and credit spreads;
❏ CDS prices.

With these three sources of data available, the model should be built using one source of data and tested on the other sources. The power of the model is to discriminate default from non-default and valuation of the bond. Models should never be calibrated and tested on the same data used for testing.

Empirical findings on reduced form models

❑ Intensity models require the least amount of information. Credit risk is measured through risk-neutral default intensity. The value of the firm and the default-triggering barrier are not needed.

❑ Default is modelled as a surprise or jumps. The random time of default is unpredictable.

❑ Models use credit spreads. Credit spreads are more easy to handle than equity prices as an extension of the term structure models.

❑ The use of credit spread for valuation of defaultable claims is similar to the valuation of default-free contingent claims in term structure models.

❑ Current data regarding the level of the firm's assets and the firm's leverage or capital structure is (typically) not taken into account.

❑ Specific features related to safety covenants and the debt's seniority are not easy to handle in default intensity models.

Correlation in the reduced form models

The following list gives some of the approaches used to integrate the correlation into reduced form models.

❑ Duffie and Singleton (1999) model an intensity function with both idiosyncratic and systematic factors. The model can incorporate multiple systematic factors. The cyclical effect is observed in the correlated Poisson arrivals of randomly sized jumps in default intensities.

❑ Jarrow and Yu (2001) consider a doubly stochastic Poisson process. The default intensity depends on macroeconomic factors and an interdependence term linking firms across industries and sectors.

❑ Default contagions are modelled as an outcome of correlation. Default contagion means default intensities of related firms jump upwards.

Empirical studies have found that PD correlations fluctuate over time. Studies have found two contradictory findings – for some times the default correlations increase as credit quality improves, and for other times it is the other way round.

Reduced form models are subject to error because observed credit spreads incorporate noise and other factors, such as liquidity risk premiums, in addition to pure credit risk premiums.

Recovery rate in the reduced form models

Recovery rates are exogenously defined in the recovery models. Recovery rates are extremely volatile both across time and across industries. Reduced form models estimate the default intensity function using observed credit spreads on risky debt. The credit spread is defined as $PD \times LGD$. Thus, the reduced form models must find some way to disentangle the PD from the LGD in each observation of the credit spread. Some of the assumptions and empirical findings used under reduced form models include the following:

❑ LGD as a constant;
❑ LGD as a proportion to the bond value;
❑ independence of PD and LGD;
❑ LGD as bound rather than a single value.

Models like Chan-Lau has attempted to extract out recovery rate from the credit default swaps spreads as an upper and lower bounds.

Landmarks in reduced form models

Reduced form models require:

❑ a default intensity process;
❑ a recovery process;
❑ an interest rate process;
❑ correlation between these processes.

Table 4 shows the major landmarks in modelling reduced form models.

Advantages of reduced form models

Since reduced form models do not require parameter estimates related to the firm's unobservable asset value, these models are easier to implement. Furthermore, these models facilitate empirical estimation. Existing term structures of credit spreads can be fitted and a wide variety of dynamics for credit spreads can be accommodated.

Table 4 Reduced form models

Model	Default intensity (I) process	Recovery (RR) process	Interest rate (r) process	Correlation
Jarrow–Turnbull	Constant intensity	Recovery of treasury = constant fraction	Forward rate model	All three variables independent
Jarrow–Turnbull–Lando	Intensity depends upon rating – Cox process methodology	Recovery of treasury = constant	Forward rate	All three variables independent
Das–Tufano	Intensity depends upon rating	Stochastic RFV	Vasicek model	Correlation in recovery rate and interest rates
Madan–Unal	Intensity depends upon stock price	Stochastic RTV	Forward rate model	All three variables independent
Lando	Intensity depends upon state variable	RFV – depending upon state variable	Forward rate model – rate dependent of state variable	Correlation among all three variables
Duffie–Singleton	Hazard rate based on state variable	RMV depending upon state variable	Forward rate model – rate dependent of state variable	Correlation among all three variables

The default and probability migration models (eg, EDF in KMV and TM in CreditMetrics) measure credit risk over a one year risk horizon. Both are calibrated against historical annual default and credit migration rates. Both measures are insufficient to capture the short-term (eg, 10 day) credit risk characteristics.

Reduced form models are a more natural solution when we are interested in managing the short-term risk of a portfolio of tradable default-contingent claims. They allow direct calibration to the market prices of tradable assets for a risk horizon as short as the frequency at which market prices of tradable assets are observed and are fit for integrating market and credit risk.

Jarrow–Turnbull model

Robert Jarrow and Stuart Turnbull have developed a number of models under various assumptions and degrees of complexity to analyse the credit risks based on the credit spreads and other relevant market factors. Their focus is predominantly pricing of financial securities subject to credit risk. Their models do not focus on calculation of default probabilities.

Architecture of the Jarrow–Turnbull model

The Jarrow–Turnbull model has four building blocks.

❏ *Modelling default intensity*: Default intensity is modelled as a binomial process or an exponential process independent of previously defined market factors. The model assumes default to be an absorbing state.

❏ *Modelling credit spreads*: This is done through modelling market factors having an influence over the size of credit spreads. This is essential for modelling credit spread. Such factors can be the term structure of default risk free interest rates and an equity market index. The factors are generally modelled as correlated autoregressive processes such as the general Ito processes, geometric Brownian motion or the mean-reversion process. This model relates observed credit spreads to expected bond payoffs and the risk-free interest rate curve. Postulating the absence of arbitrage opportunities and a frictionless market, a risk-neutral world is assumed. This implies that expected returns and discount factors are equal for different investments and can directly be deduced from the default-free interest rate curve. Under these conditions, the models can be solved and risk-neutral default probabilities be derived.

❏ *Modelling recovery rate*: For the purpose of simplicity in the estimation, the recovery rate is assumed to be exogenously determined through equity or bond prices and a constant fraction of the bond's present value prior to the default. Credit spreads are assumed to be stochastic and independent of the recovery rate. Credit spreads are assumed to be dependent only on the stochastic structure of spot interest rates and the bankruptcy process.

❏ *Correlation*: Default rate, credit spread and recovery rate are assumed to be independent.

The Jarrow–Turnbull model presents convenient reduced form models of the valuation of contingent claims subject to default risk, focusing on applications to the term structure of interest rates for corporate or sovereign bonds.

This model works in a frictionless, arbitrage-free and complete market setting. Firms are allocated to credit risk classes. Default-free zero-coupon bonds and default-risky zero-coupon bonds in each of the risk class are traded, for all maturities and each with a face value of 1.

Default can occur at any time and comes as a surprise to all market participants. Default is modelled by a Poisson process, which stops at the first jump. Jarrow and Turnbull (1995) model default-free zero-coupon bond prices by specifying the default-free forward rate process as given by the Vasicek model. Jarrow and Turnbull (1995) assume that the Poisson default process under the equivalent martingale measure has intensity, λ_p. Default is assumed to be independent of default-free interest rates:

$$df_f(t, u) = \alpha_f(t, u)dt + \sigma(t, u)dW_1(t)$$

For default risky bonds, for $t \geq$ maturity,

$$df_r(t, u) = \alpha_r(t, u)dt + \sigma(t, u)dW_1(t)$$

and for $t <$ maturity,

$$df_r(t, u) = (\alpha_r(t, u) - \theta(t, u)\lambda_p)dt + \sigma(t, u)dW_1(t)$$

where $\theta(t, u)$ is the positive adjustment.

Jarrow and Turnbull (1995) derive the price of the default risky bond for $t <$ maturity as

$$v(t, u) = p(t, u)\{(\exp(-\lambda(u - t))) + RR(1 - \exp(-\lambda(u - t)))\}$$

where RR is recovery rate.

Jarrow–Lando–Turnbull model

This is a model for the term structure of the credit spreads. They model the bankruptcy or recovery process as a discrete state using a Markov chain in the credit rating. The model explicitly incorporates a firm's credit rating as an indicator of the likelihood of default. The model takes rating migration matrices as the input for determining the likelihood of default.

Interest rate modelling: This model can work on the top of any of the well-known term structure models.

Modelling transition probabilities: The model utilizes historical transition probabilities for the various credit rating classes to determine the pseudo-probabilities (martingale, risk adjusted) used in valuation.

Modelling recovery: Different seniority debt for a particular firm can be incorporated via different recovery rates in the event of default. Recovery rate is assumed to be constant. Therefore all changes in the credit spreads are completely dependent on the rating migration. It models the recovery of treasury. Recovery is obtained at the stated time though default might occur any time.

Correlation – between the interest rate process and default process: These two processes are assumed to be independent. This means the Markov process for the credit ratings is independent of the level of spot interest rates.

Default process: The bankruptcy process is modelled via a discrete state space, continuous time and time-homogeneous Markov chain in the credit ratings. The parameters of this bankruptcy process can be estimated from observable data.

Jarrow *et al* (1997) make the distinction between implicit and explicit (or historical) estimation of TMs, where implicit estimation refers to extracting transition and default information from market prices of risky zero-coupon bonds.

Duffie and Singleton model

The default-free term structure models are used by parameterising the model with default-adjusted interest rates instead of default-free interest rates.

In addition, the interest rates for defaultable bonds can also be modelled using the risk-free interest rates, hazard rate and recovery and credit loss. In this way, it is possible to model mean loss rates from historical information on defaultable bond yields. This is also called affine term structure modelling. The affine term structure model is applied for the following reasons:

- to show the flexibility of affine models in describing the basic features of yield and yield spreads on corporate bonds;
- to show that some of the models are incapable of capturing the negative correlation between the credit spreads;
- to develop a defaultable version of the Heath–Jarrow–Morton model;
- to price callable corporate bonds;
- to price CDs. To value a CD, a CD is assumed to be an option written on the counterparty. To price the put option, the correlation between the hazard rate (h) and the short rate is assumed and non-linear relationships are assumed between the option payoffs and the hazard rate and the default loss (L). This non-linear dependence allows the separate measurement of the hazard rate and fractional loss rate from the option price data.

The Duffie and Singleton model is a basic affine model. It attempts to answer the question "Is the price of a defaultable bond a good indicator of hazard rate and recovery rate?" There are two schools of thought. One assumes that the default hazard rate and the fractional recovery rate do not depend on the value of the defaultable bonds. The second assumes that the hazard rate depends on the value of the defaultable bond, especially in the case of swap contracts with asymmetric counterparty credit quality. The Duffie– Singleton model takes a middle path. It postulates a non-linear relationship.

The model treats default as an unpredictable event governed by a hazard rate process. Fraction default losses are measured in terms of loss in the market value that occurs at default.

It models a relationship between risk-neutral mean loss rate and par credit spreads. However, both rates diverge owing to illiquidity in the market. The model assumes an exogenous mean-loss rate process. For non-callable bonds, with hazard rate (h) and default loss (L), the mean loss rate is hL. A simplistic assumption is that the illiquidity of the security translates into a fractional cost rate of l, where l is a predictable process. Hence, the total mean loss rate of the security due to credit events and illiquidity is $hL + l$. This is equal to credit spread.

This model has also compared the loss of face value and loss market value assumption and concluded that both recovery assumptions generate the same similar par value spreads. However, loss of market assumption has more analytical tractability. It also suggests changes in the recovery parameters so as to compensate for the changes in the two methods.

REFERENCES

Amato, J. D. and E. M. Remolona, 2003, "The Credit Spread Puzzle", BIS Quarterly Review December 2003.

Anderson, R. W., S. Sundaresan, and P. Tychon, 1996, "Strategic Analysis of Contingent Claims", *European Economic Review.*

Black, F. and J. C. Cox, 1976, "Valuing Corporate Securities: Some Effects on Bond Indenture Provisions", *Journal of Finance.*

Black, F., E. Derman, and W. Toy, 1990, "A One-factor Model of Interest Rates and its Application to Treasury Bond Options", *Financial Analysts Journal.*

Campbell, J. Y. and G. B. Taksler, 2003, "Equity Volatility and Corporate Bond Yields", *Journal of Finance,* **58,** pp 2321.

Chan-Lau, J. A., 2003, "Anticipating Credit Events Using Credit Default Swaps, with an Application to Sovereign Debt Crisis", IMFWorking Paper.

Collin-Dufresne, P., 2001, "The Determinants of Credit Spread Changes", *Journal of Finance,* **56,** pp 21–77.

Collin–Dufresne, P. and R. S. Goldstein, 2001, "Do Credit Spreads Reflect Stationary Leverage Ratios", *Journal of Finance,* **56,** pp 1929.

Cox, J. C., J. E. Ingersoll, and S. A. Ross, 1985, "A Theory of the Term Structure of Interest Rates", *Econometrica,* **53,** pp 385–408.

Das, S. R. and P. Tufano, 1996, "Pricing Credit Sensitive Debt when Interest Rates, Credit Ratings and Credit Spreads are Stochast", *Journal of Financial Engineering,* **35,** pp 43–65.

Duffie, D., 1998, "Defaultable Term Structure Models with Fractional Recovery of Par", Graduate School of Business, Stanford University.

Duffie, D. and R. Kan, 1996, "A Yield Factor Model of Interest Rates", Mathematical Model.

Duffie, D. and D. Lando, 2001, "Term Structure of Credit Spreads with Incomplete Accounting Information", *Econometrica.*

Duffie, D. and K. J. Singleton, 1999, "Modeling Term Structures of Defaultable Bonds", *Review of Financial Studies,* **12,** pp 687.

Elton, E. J. *et al*, 2001, "Explaining the Rate Spread on Corporate Bonds", *Journal of Finance,* **56,** pp 247.

Eom, Y. U., J. Uelwege, and Y. Z. Yuang, 2004, "Structural Models of Corporate Bond Pricing: An Empirical Analysis", *Review of Financial Studies,* **17,** pp 499–544.

Geske, R., 1977, "The Valuation of Corporate Securities as Compound Options", *Journal of Financial and Quantitative Analysis*.

Giesecke, K. and M. A. Goldberg, 2004, "Sequential Default and Incomplete Information", *Journal of Risk*.

Heath, D., R. Jarrow, and A. Morton, 1988, "Bond Pricing and the Term Structure of Interest Rates: A New Methodology for Contingent Claim Valuation", *Econometrica*, **60**, pp 77–105.

Hilberink, B. and C. G. Rogers, 2000, "Optimal Capital Structure and Endogenous Default, Finance and Stochastics", *Finance and Stochastic*, **6**, pp 237–63.

Huang, J. and M. Huang, 2003, "How Much of the Corporate-Treasury Yield Spread is Due to Credit Risk?: A New Calibration Approach", Working Paper, Penn State University.

Hughston, L. P., 1997, "Pricing of Credit Derivatives", *IFR Financial Derivatives & Risk Management*.

Hull, J. and A. White, 2000, "Valuing Credit Default Swaps I: No Counterparty Default Risk", *Journal of Derivatives*.

Jarrow, R. and S. Turnbull, 1995, "Pricing Derivatives on Financial Securities Subject to Credit Risk", *Journal of Finance*, **50**, pp 53–85.

Jarrow, R. and F. Yu, 2001, "Counterparty Risk and the Pricing of Defaultable Securities", *Journal of Finance*.

Jarrow, R., D. Lando, and S. Turnbull, 1997, "A Markov Model for the Term Structure of Credit Risk Spreads", *Review of Financial Studies*, **10**, pp 481–523.

Jones, E. P. *et al*, 1984, "Contingent Claims Analysis of Corporate Capital Structures: An Empirical Investigation", *Journal of Finance*.

Nielson, L. T., and P. Santa-Clara, 1993, "Default Risk and Interest Rate Risk: The Term Structure of Default Spreads", Working Paper INSEAD.

Kijima, M., and E. Suzuki, 2001, "All Multivariate Markov Model for Simulating Correlated Defaults", Journal of Risk.

Lando, D., 1994, "Three Essays on Contingent Claims Pricing", PhD Dissertation, Cornell University.

Lando, D. 1998, "Cox Processes and Credit-Risky Securities", *Review of Derivatives Research*, **2,** pp 99–120.

Leland, U. E. and K. B. Toft, 1996, "Optimal Capital Structure, Endogenous Bankruptcy and the Term Structure of Credit Spreads", *Journal of Finance*.

Longstaff, F. A. and E. S. Schwartz, 1995, "A Simple Approach to Valuing Risky Fixed and Floating Rate Debt", *Journal of Finance*, **50**, pp 789–819.

Lyden, S. and D. Saraniti, 2000, "An Empirical Examination of the Classical Theory of Corporate Security Valuation", Barclays Global Investors.

Madan, D. and H. Unal, 1998, "Pricing the Risk of Default", *Review of Derivative Research*.

Merton, R., 1976, "Option Pricing when Underlying Returns are Discontinuous", *Journal of Financial Economics*.

Nielsen, L. T. and P. Santa-Clara, 1993, "Default Risk and Interest Rate Risk: The Term Structure of Default Spreads", Working Paper INSEAD.

Ogden, J. P., 1987, "Determinants of the Ratings and Yields on Corporate Bonds: Tests of Contingent Claims Model", *Journal of Financial Research*.

Vasicek, O. 1977, "An Equilibrium Characterization of the Term Structure", *Journal of Financial Economics*.

Zhou, S. 1997, "A Jump-Diffusion Approach to Modeling Credit Risk and Valuing Defaultable Securities", Working paper, Federal Reserve Board.

5

Credit Rating

INTRODUCTION

We should not forget that the basic economic function of (these) regulated entities (banks) is to take risk. If we minimise risk taking in order to reduce failure rates to zero, we will, by definition, have eliminated the purpose of the banking system.

Alan Greenspan, Ex-President FRB, 1996

In this chapter, we will attempt to identify the risk factors for the risk components we identified and measured in other chapters – default risk, recovery, EAD and maturity. In the "Probability of Default", "Market Information Based Models", and "Credit Scoring" chapters we have seen quantitative measurement of PD from idiosyncratic factors such as accounting ratios, or the market values of firm's securities such as bonds and equity; and macroeconomic factors such as petroleum price, interest rates, etc. As we know, the quantitative measurements are not sufficient to measure default risk. Quantitative factors need to be modelled, simulated, predicted, measured and interpreted under the light of qualitative and judgmental factors. *Measuring credit risk components or risk through a combination of quantitative, qualitative and judgmental measures is called credit rating.* Credit rating is the summary indicator for the risk of loss due to failure by a given borrower to pay as promised. Essentially, the experience, exposure and expertise of the rater play a major role in credit rating. Therefore, it is well accepted that rating as a measure of credit risk is not a precise measurement.

Credit rating is assigned either internally by the bank's staff using proprietary models and techniques or by the external rating agency using their proprietary methodology and models. The rating models and methodologies used by rating agencies are now partly disclosed in public. The Basel Accord endorses the rating measurements for risk and capital estimation and empowers the national supervisors to recognise the credit rating agencies locally.

Credit rating as a method to assess and measure credit risk is the traditional and most prevalent method of measuring the credit risk. Since 1909, external credit ratings have been provided by firms specialising in credit analysis. The first to offer ratings were Moody's in the US. The external ratings are the opinions on the credit quality of the issuer and issue. The credit ratings are the opinion statements about credit quality of the issue or the credit-worthiness of the issuer. External ratings offer bond investors access to low-cost information about the credit quality of the bond.

Companies with agency-rated debt tend to be large and publicly traded. The ratings are well accepted by the investment community, and extend not only to commercial firms but municipal, sovereign and other obligors. Worldwide, approximately 6,500 obligors have valid ratings out of which more than 3,000 are in the US. There were around 37 credit rating agencies outside the US in 2002. This number may have now doubled with the rating agency's role being endorsed by the Basel Accord.

Internal ratings refer to the ratings assigned by banks using a proprietary scale to their borrowers and exposures. Both internal and external ratings are moving from ordinal risk measures to absolute risk measures.

Convergence of internal and external rating methodology

The external rating methodologies are decades old. The rating agencies have gone through various economic cycles, are present in most countries and have rated debt across industries. All this has helped them refine their methodologies and transfer the learning across the economic cycles, countries and industries. With the endorsement from the Basel Accord, the agencies have further refined and extended their methodologies. For example, recovery ratings have been developed in the past three years to comply with the two-dimensional rating recommendations of the Basel Accord and ratings are provided with absolute loss and default rates.

The internal rating in some form has been in existence in the banking industry for a decade or so. However, owing to limited reach and experience, the internal rating methodology is not as developed as the external rating methodologies. The banks have been learning and adopting the methodologies from the rating agencies.

The internal and external ratings may differ with regards to the architecture, default definition, quantitative models and the measurement benchmarks used but broadly converge on the rating methodology. However, the methodologies continue to be different across the banks and rating agencies at a finer level for different industries and debt instruments, at the parameters and weights level.

Let us first study a generic rating methodology.

The analytical framework to measure credit risk

The analytical framework is divided into qualitative, quantitative, country and legal structures (see Figure 1). The rating process cannot be confined only to the various financial measures and quantitative frameworks. The qualitative framework generally includes business, industry, company environment analysis and their impact on the credit quality. Qualitative aspects are analysed through expert judgments. There are three approaches to the measurement of credit risk components:

❑ measure risk quantitatively;
❑ modify the quantitative risk measure with constrained expert judgment;
❑ modify the quantitative measure with expert judgment.

Nevertheless, the quantitative measurement of risk is common across the methods. The reliability or contribution of quantitative techniques in the risk measurement differs across the methodology. Furthermore, none of the methodology relies exclusively on the quantitative models. In the quantitative risk measurement approach, the bank may be relying 90% on the quantitative model. In constrained expert judgment, reliability on the quantitative model may be greater than 50% and in expert judgment it may be more than 30–40%.

Credit rating is an expert judgment method of measuring the risk components based on the quantitative models.

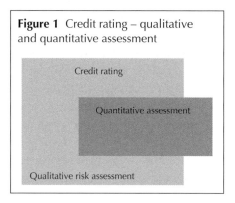

Figure 1 Credit rating – qualitative and quantitative assessment

Credit rating

Quantitative assessment

Qualitative risk assessment

We have discussed various types of quantitative models in Chapters 2, 3 and 4. Rating agencies have started selling quantitative model methodologies as software, ie, the private company models of Moody's. However, the actual models used by rating agencies are still a trade secret.

BUSINESS RISK ASSESSMENT METHODOLOGY

To build a generic risk methodology applicable for both the internal and external ratings, we suggest an assessment of seven factors:

❑ macroeconomic pressures;
❑ industry environment;
❑ operating environment;
❑ management capabilities;
❑ financial risk of the obligor;
❑ country environment;
❑ credit structure and willingness to pay.

Rating is an assessment of a firm's ability and willingness to pay in the future. Each of the factors recommended for assessment here are for the assessment of the impact on either ability or willingness to pay in the future. It is a road map to link historical experience to the future performance of the firm. However, the rating is not a forecast of future performance. Rather, rating aims at estimating variability of expected future performance based on a range of economic and competitive scenarios. Each of the risk factors are measured in terms of scores. However, there are no formulae for combining scores to measure the rating.

Some of the analysis explained here may not be in tune with accounting and other standards. For example, if a firm operates different businesses, the methodology believes that it is critical to analyse each type of business and asset class in its own right, while accounting standards such as the Financial Accounting Standards Board (FASB) and the International Accounting Standards (IAS) may require consolidation of business units. Analytical adjustments are made in computing financial ratios. For example, some of the companies borrow as well as hold investments and cash balances owing to tax arbitrage. In such cases, the debt ratios are computed on a net basis.

Macroeconomic environment

The impact of the factors, such as energy prices (especially the recent surge in the petroleum prices), basic raw material prices (mineral, ore, steel, cement, etc), economic growth, consumer confidence and demand, foreign exchange rates and volatility and fiscal balances or imbalances, cuts across the industries.

At any point in time the economic and business cycles impact each industry differently. It is a widely accepted fact that each industry has different default rates. The macroeconomic and business environments are analysed primarily for the following reasons.

❑ To understand, consider and measure the impact of the business and economic cycles on the industry in question. This enables identification of systemic factors and measurement of systemic risk.
❑ To understand and predict the government action, duties and taxes.
❑ To identify infrastructure bottlenecks – quality, pricing and supply of transport, electricity and water supply have an impact on the business.
❑ To identify the impact of the legal systems on creditors' rights, collateral and bankruptcy and property rights.
❑ To understand labour issues, such as labour regulations for hiring and firing, social security and welfare benefits.
❑ To understand industry structure or the operating environment for domestic, global and foreign firms and the protection provided by the Government.
❑ To understand price controls, especially the control of inflation and prices on mass consumption goods such as gasoline, electricity, food grains, etc.

We have discussed some of these issues in detail in "Sovereign Credit Rating" section in this chapter.

Industry environment

The industry environment has a major impact on the credit loss and credit risk. Credit loss and credit risk varies substantially across industries.

This fact has been recognised by the Basel Accord in its allocation of different risk weights to the exposures for different risk segments. (Please note that the risk segmentation according to Basel is not exactly based on industry.) Industry characteristics may or may not directly contribute to the rating. Best practice among rating agencies is for the industry characteristics to set the upper limit on the rating (regardless of financial characteristics).

Studying the industry environment helps in identifying the systemic and common factors to which firms are subjected. The primary aims of industry environment analysis are:

❑ to recognise the impact of systemic factors on the industry factors and the interaction of industry and macroeconomic factors;
❑ to measure and benchmark the firm against similar firms or the industry average;.
❑ to identify and benchmark the common idiosyncratic factors on the ability and willingness to pay.

Industry regulation

In each industry, government and regulators are empowered to intervene and manage the conflicting objectives, to protect public interest and to protect the long-term sustainability and competition in the industry. However, the quantum of such powers and the use of such powers differ. These intervening powers impact the measures of credit worthiness, default rates and risk premiums for each industry.

Each industry has different regulatory objectives. The banking industry is probably the most regulated industry, where the regulator aims at ensuring the safety of banking and the economic system. For telecom and utilities industries, price, supply, competition and quality are the regulatory objectives. For corporates in general, disclosures, transparency and accounting quality are the regulatory objectives.

The banking regulator is the most powerful institution with the necessary powers and resources to intervene in the working of institutions and financial markets to achieve the desired objective. The Long Term Capital Management (LTCM) hedge fund is famously known for the Federal Reserve (Fed) intervention and liquidity support in order to avoid "serious systemic instability". It is also said that "a bank fails when – and only when – its regulators decide that it should fail". The impact of the regulatory intervention on the credit risk measures is well recognised for the banking industry.

After banks, the municipals and utility industries are the sectors most influenced by regulation. The regulatory role varies substantially across countries. It can be in the form of direct support, sheltering and guaranteed returns or insulation from competition, facilitation to external capital, etc. Direct government support plays an important role.

Outside the financial intermediary industry, the regulators generally do not focus on the credit quality of the regulated companies. The regulation impacts indirectly through demand, margins and growth. With the fiscal deficit going up in all countries, the government intervention for nationalisation, etc, has also gone down the world over in the past decade or so. Currently, governments intervene indirectly through policies on privatisation and encouraging competition and foreign direct investment. For credit ratings it is important that the regulatory and governmental roles are consistent and transparent.

Industry strength/weakness
This emanates from exposure to technological changes, lead time, quantum of investment required and labour conditions in an industry. This analysis helps in confirming the future growth and investment required. Industry prospects such as growth, stability or decline determine the business prospects. The entry barriers, such as higher capital investment, long lead time and specialised equipments, determine the level of competition.

Demand for products and services
The three cycles experienced by an industry are growth, stability and decline. The very term "cycle" seem to imply regularity. In actuality, this is seldom the case. All cycles vary considerably in their duration, magnitude and dynamics.

Cycles are broadly divided into four types:

❑ business or economic cycles;
❑ demand-driven industry cycles;
❑ capacity- or supply-driven industry cycles;
❑ supply–demand cycles due to natural phenomena.

At any given point, it is difficult to know the severity and the stage in a cycle. This information may only be known with a lag, which may vary from a month to a year.

Additional complexities are introduced from a confluence of different cycles. Cycles are never simple and vary in their dynamics. They are also difficult to differentiate from secular changes in industry fundamentals. Sometimes secular changes are taken as cyclic and *vice versa*. There are two schools of thought on assessing the impact of industry and economic cycles: cyclic-independent assessment and cyclic-dependent assessment. (None of the approaches are acceptable. In a separate section we have made an attempt to explain how to achieve a balance between the two.) To provide the cyclic-independent rating the business strategy and financial policy should also be independent of the cycle. Furthermore, the public confidence level should also be constant. However, this is seldom the case.

Cyclic-dependent assessment, according to the critics, is also not acceptable. One reason is that rating through-the-cycle (TTC) requires an ability to predict the cyclical pattern – and this is usually difficult to do. In practice, there is no such thing as a "normal" cycle. Even when compared with historical cycles, the latest cycle will be either longer or shorter or steeper or less severe than repetitions of earlier cycles. Sometimes the cycles are so severe that they permanently alter the competitive position. The demand recession for fast-moving consumer goods in India in 2003–2005 is an example. The market share of the local branch of Uni Lever *versus* ITC company has changed during the period. ITC Company has emerged as a market leader in many product categories.

Industry structure and competition
The number of players in the industry, barriers of entry into industry (in the form of larger investment, marketing channels, etc),

commoditisation of products/services and the capacity being built by the players by way of fresh investment by the existing players are the key determinants of the industry structure. Firms build a competitive position from:

❑ marketing – image, product differentiation, service, other factors;
❑ technology – quality;
❑ efficiency – price and distribution capabilities;
❑ regulation – tax advantage, regulatory protection.

There are three levels of competition:

❑ regional;
❑ national;
❑ global.

The basis for competition determines the factors to be analysed for a given company.

Firm size: The size of a firm usually provides a measure of diversification and often affects competitive issues. Size is a major factor in marketing and sales cost sustainability and therefore the market size.

Market share analysis often provides important insights. However, large shares are not always synonymous with competitive advantage or industry dominance. The market share needs to be analysed *vis à vis* the industry structure and comparable participants, and in local, regional, national and international markets and product concentration perspectives.

Diversification factors: If a company works in more than one industry, each industry needs to be separately analysed and added according to their relative importance. The potential benefits of diversification, which may not be apparent from the additive approach, are then considered. Companies may be operating in different industries, but may not experience diversification benefits as they are similarly exposed to the economic cycles.

Relative industry risk
Despite all the disclosures and financial statements, the market demands a different or larger premium on the firms in certain industries. Furthermore, this premium may not be the same across different rating grades or groups of risk grades.

Industry traits have a major impact on the future cashflow generation and therefore on the financial strengths of a firm.

The relative performance of an industry in terms of credit risk also needs to be measured and this plays a significant role in the credit rating of a firm. The three most common indicators used to measure the relative industry default risk are:

❑ financial ratios – industry average, median leverage or coverage ratio;
❑ industry average market implied ratings;
❑ industry average ratings.

Empirically it is found that relative industry default rates fluctuate substantially over time; therefore, good industry risk indicators should also be dynamic. None of the factors identified above provide the entire measurement of industry rating *per se*. By combining these three factors, rating agencies have established a composite index to measure the relative credit risk performance.

Credit issues are weighed in conjunction with the risk characteristics of the industry to arrive at an accurate evaluation of credit quality.

Relative industry risks help portfolio managers to decide the industry allocations. Ideally the different relative industry risk should alter the risk at the firm level within the same grade. While the rating distribution for an industry will be different, in a relatively risky industry more firms will be in the lower grades.

The historical experience of rating agencies has shown a drift of industries on the relative risk. For example, the US telecommunications industry drifted from low risk in the 1980s to high risk in 1990.

The difference in default rates across the risk segments can be recognised through two ways – distribution of firms in various rating grades or absolute measure of the default rate of each rating grade.

It is well known that the average default rate is not the same across the industry. Figure 2 shows the average default rates in various industries in the US. Figure 3 shows the rating distribution of financial institutions (FIs) and corporates in 2005.

While the observed default rates of FIs are in tune with the rating grade, the market demands more premiums for a credit default swap (CDS) spread for FIs against corporates. Figure 4 shows the ratio of FI CDS spreads to similarly rated corporate CDS spreads.

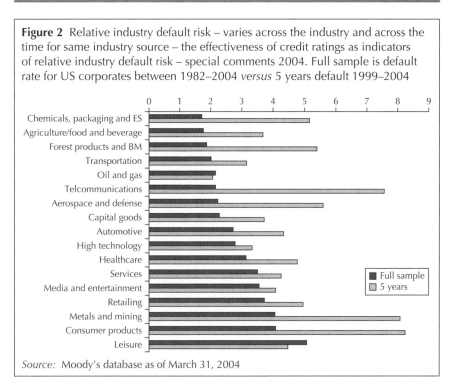

Figure 2 Relative industry default risk – varies across the industry and across the time for same industry source – the effectiveness of credit ratings as indicators of relative industry default risk – special comments 2004. Full sample is default rate for US corporates between 1982–2004 *versus* 5 years default 1999–2004

Source: Moody's database as of March 31, 2004

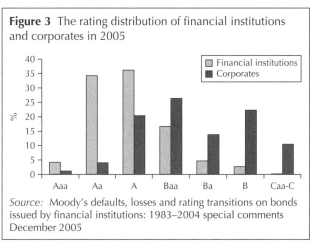

Figure 3 The rating distribution of financial institutions and corporates in 2005

Source: Moody's defaults, losses and rating transitions on bonds issued by financial institutions: 1983–2004 special comments December 2005

This may be due to the following reasons:

❑ irrespective of the historical default and loss experience, the market is expecting a higher credit loss among FIs;

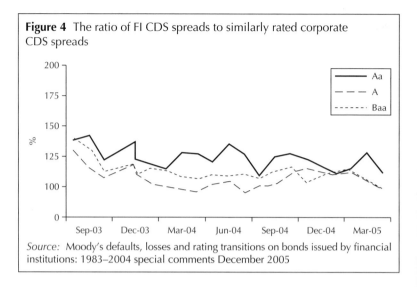

Figure 4 The ratio of FI CDS spreads to similarly rated corporate CDS spreads

Source: Moody's defaults, losses and rating transitions on bonds issued by financial institutions: 1983–2004 special comments December 2005

❏ the credit risk of a FI is more strongly correlated with market risk, and therefore requires a higher premium;

❏ the market is demanding additional diversification premiums since the market has a large exposure against FIs.

Building the impact of cyclicity into the ratings

The objective of the rating agencies is to provide an accurate relative (ie, ordinal) ranking of credit risk at each point in time and to achieve an optimal balance between rating timeliness and rating stability.

Impact of Economic Cycle on the Default Rate Estimation: the business cycle impacts the PD (see Figure 5). Economic upturn reduces the PD and economic down turn increases the PD.

Each rating grade has a number of exposures and associated default rate. The economic cycle actually changes the default rates. Rating is a process of classifying exposures into different grades.

The changing economic cycles can be factored into the rating grades by two ways: either change the associated default rates for each grade or allocate the exposures to different grades. We know that the default rates associated with a credit rating are generally fixed. Therefore, economic cycles change the number of exposures in a rating grade by pushing a large number of exposures upward (favourable cycle) or downward (adverse cycle); see Figure 5.

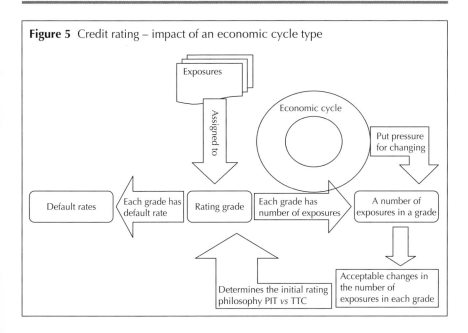

Figure 5 Credit rating – impact of an economic cycle type

Factoring impact of Economic Cycle into the Rating: There are two options. First, consider the forthcoming downturn in the economic cycle and assign the lower rating initially itself, so that for a larger economic variation no change is required in the ratings of the exposures (the number of exposures remain largely constant). This method is called the through-the-cycle (TTC) method. Second, consider the rating at a given point-in-time (PIT) and adjust the rating as and when the economic conditions change. This is referred to as the PIT rating method. TTC is equivalent to assigning the higher default rate *ab initio* factoring in the economic downturn and maintaining the transition to the minimum.

Analysing PIT and TTC Methods (Table 2): The rating objectives determine the rating method. Rating objectives are determined by time horizon.

❏ Short Term – current conditions (PIT rating assessment) ;
❏ Long Term – expected creditworthiness over the life of the loan or the entire credit cycle (TTC rating assessment) – long horizon.

Rating agencies usually assign grades from a long-term perspective and consequently their choices are considered to be approximately

Table 1 Rating approach under different economic conditions

	PIT will appear as	TTC will appear as
Under stressed conditions – bad economic and business cycle	Pro-cyclic	Fixed – stable – or uncorrelated
Under average economic and business cycle	Fixed – stable	Counter cyclical

those based on the TTC rating, although their short-term ratings may be equivalent to the PIT rating (see Table 1).

The difference between PIT and TTC ratings is whether changes in the economic environment surrounding borrowers are absorbed by rating migration as in PIT, or are incorporated into changes in the actual default rate of each grade as in TTC.

The choice of between PIT and TTC ratings, or a mixture of the two, depends on the length of time the FIs are exposed to credit risk. If the majority of a bank's loans have a long time before maturity, it is desirable to assign a grade considering creditworthiness over the whole period. However, it is difficult in practice to assess the future change in a borrower's conditions over the long term including the business cycle as in the case of the TTC rating. One alternative might be to adopt an approach in which ratings are based on recent conditions and to assess the degree of credit risk under the economic downturn by use of a stress test. This approach is based on PIT rating but takes account of TTC components in application.

FIs evaluate the creditworthiness of borrowers over three to five years. This indicates that their choice is somewhere between the above two types of ratings.

The TTC philosophy probably accounts for the considerable emphasis on a borrower's industry and its position within the industry. Therefore, whether or not the industry supply and demand conditions are more important than the overall business conditions or the cashflow is also a factor for consideration the selection of the approach.

Although we cannot judge *a priori* which rating method is better than the other, it is still very important for FIs to understand whether their own internal rating systems are more PIT-oriented, TTC-oriented or follow a mixed approach. In other words, they need to know how their systems are affected by the business cycle. This is because the assessment of an internal rating system using actual default rates per grade depends on whether PIT or TTC rating is used. With TTC rating, default rates per grade are expected to be stable and hence their stability is confirmed, whereas with PIT rating they will fluctuate over the business cycle.

Carey and Hrycay (2001) describe the TTC methodology as a rating assignment based on a stress scenario. The TTC rating is affected significantly by the economic and industry cycles. This makes agency ratings insensitive to the cycle. According to Cantor and Mann (2003), this is equivalent to extracting the permanent component from changes in the observed credit quality, on the basis of a forecasting analysis: "Even though an issuer might experience a change in its financial performance as a result of an adjustment in the macroeconomic environment, its rating may nonetheless remain unchanged if it is likely that its previous financial condition will be restored during the next phase of the cycle".

The middle of the road approach is best suited for the internal rating systems. It should be noted that TTC is very difficult to implement in practice, since it is difficult to agree and simulate on the worst conditions. Therefore, a consistency in approach as required by Basel may not be achieved. Furthermore, it is expensive to simulate and agree on the worst conditions. Therefore, this method may not be considered by banks for the retail or middle market segment but is considered only for large exposure. However, with the availability of automating tools and external processed data, the internal rating approach may adopt an approach with the TTC influence as large as possible.

Therefore, it is possible to analyse characteristics of ratings afterwards by decomposing total qualitative changes in loan portfolios over the business cycle into two parts: the PIT part, which can be explained by rating migration, and the TTC part, which can be explained by changes in the actual default rate.

The relative size of the PIT part or TTC part based on the above definition varies depending on each FI. In addition, some institutions have different sensitivities to the business cycle depending on the ratings. For example, the actual default rate has a high sensitivity to the business cycle in higher ratings (closer to TTC), while it has a relatively low sensitivity in lower ratings (closer to PIT).

Table 2 Rating approach PIT *versus* TTC

PIT	TTC
Current conditions	Expected conditions
Short term	Long term
❑ This approach is used for allocating economic capital, monitor loans and establish loan loss reserves	❑ Credit assessment for large lending
	❑ Potential stress conditions are also considered through covenants, loan amount, term, collateral and guarantee
❑ Rating is based on the current and most likely future outlook over a one-year horizon	❑ Generally used by rating agencies
❑ PIT ratings are required for measuring rating migration and inputs to the portfolio models	❑ Determine the rating at the worst point in the credit cycle. Therefore, rating stable over the credit cycle

Therefore, it is advisable that institutions consider these factors when verifying the validity of PD estimates.

Operating environment

One of the important factors to be considered along with the firm is to analyse an industry's life cycle and the growth or maturation along with similar measurements at the company level. The primary challenge is to identify the difference between the cycle and secular changes. The company environment or the idiosyncratic risks should be seen along with the industry risk. The competitive advantages of the firm differentiate it from other firms, and provide longer life and capabilities to sail through the business and industry cycles.

The relative position of the firm in the industry is measured in terms of the competitive advantage or position. The competitive position comes from:

❑ market position;
❑ business operation efficiencies;
❑ regulation – tax advantage, regulatory protection;
❑ management capabilities; and
❑ financial strengths.

Market position

Market share analysis often provides important insights. However, large shares are not always synonymous with competitive advantage or industry dominance. As mentioned before, the market share needs to be analysed *vis à vis* the industry structure and comparable partici-pants, and in a local, regional, national and international market and product concentration perspective. For any particular company, one or more factors can hold special significance, even if that factor is not common to the industry. The factors included are dependent on one product, one production location, one customer, etc.

❑ *Product portfolio*: Particularly the margins, value addition and the product diversification. If a company works in more than one industry, each industry needs to be separately analysed and added according to their relative importance. The potential benefits of diversification, which may not be apparent from the additive approach, are then considered. Companies may be operating in different industries, but may not be experiencing diversification benefits because they are similarly exposed to the economic cycles.
❑ *Brand equity*: The main question to be assessed is whether the brand equity is being maintained or eroded.
❑ *Supply–demand position*: Is the volume growing, stagnant or declining.
❑ *Distribution*: The relationship with the other participants of the supply chain.

Business operation efficiencies
Every firm develops business operation efficiencies in terms of costs for marketing, logistics, raw materials, machinery, location, volume, etc.

Management capabilities
Management plays a substantial role in the operational success of the firm, managing the business and industry environment and competition, and risk tolerance and risk management. In addition to their operational role, they have a role to play in assessment and management of business and industry risk, corporate governance, financial risk and policy. Some managers are so powerful that they change the macroeconomic environment such as the taxation regime. In every country, plenty of cases have existed, especially during the 1970s and 1980s, for influencing tax regimes for the benefit of the firm or for the disadvantage of the competitor. This continues to happen even now, but in a more subtle way.

In general, management is assessed on the following criteria:

❑ corporate governance;
❑ operational successes;
❑ risk tolerance and risk management;
❑ management consistency and credibility;
❑ corporate culture;
❑ organisation.

Corporate governance: Corporate governance works at three levels:

❑ disclosure, accounting and financial statement quality and adherence to the standards;
❑ alignment of interests of various stakeholders;
❑ importance of the borrower company within the group and the legal structures built.

Operational successes: Management is evaluated on the basis of the ability to create a healthy business mix, maintain operating efficiency and strengthen market position, delivering on past projections or maintaining previously articulated strategies when evaluating future growth plans and related financial projections.

Risk tolerance and risk management: These are leverages built in the past to implement aggressive financial plans.

Corporate culture: Corporate culture encompasses the following factors:

❑ the business model – organic *versus* inorganic growth;
❑ consistency in the business strategies;
❑ history of restructuring;
❑ legal, regulatory and tax disputes as compared to peers;
❑ constraints on the management enforced by the shareholders.

Organisational considerations: These include effects such as:

❑ succession planning and reliance on an individual or a few individuals;
❑ turnover at the senior level;
❑ professionalisation of the management;
❑ importance of the finance function in the organisation;
❑ quality of risk and performance management;
❑ shareholder-imposed constraints on management prerogatives.

A judgment of the quality of management is subjective. Financial performance over time provides a more objective measure. The credibility of the company and differences in the stated polices from followed policies are very difficult to measure.

The most difficult task in management evaluation is to distinguish "because of management" from "despite management" effects on the company track history.

ASSESSING FINANCIAL STRENGTH
Financial strength is assessed from financial statement quality, the interplay of business and financial risk and financial flexibility.

Accounting and financial statement quality
The rating process does not entail auditing a company's financial records. Therefore, the quality of the financial statements should be reliable. Analysis of the audited financial records begins with a review of accounting quality. Maintenance of the accounting quality and adherence to the accounting policies and standards is part of the financial policies.

The purpose is to determine whether ratios and statistics derived from financial statements accurately represent a measure of the company's performance and position relative to both its peer group and the company universe as a whole. Therefore, accounting policies, not only within the company but also across its peer group, have to be compared. In particular, the rating process is very much one of comparisons, so it is important to have a common frame of reference. Auditing provides a basic common frame of reference. However, as companies have different accounting policies and adopt different "generally accepted accounting principles", some of the issues remain. It should be noted that it might not be possible to completely recast the financial statements. An example of the items generally analysed for differences across the firms are as follows:

❑ income recognition;
❑ depreciation methods;
❑ inventory valuation;
❑ accounting treatment of the purchased goodwill;
❑ employee benefits – pension, medical, superannuation, leave, etc;
❑ off-balance-sheet items – derivative instruments, leases and project finance and receivable sales;
❑ consolidation basis.

Interplay of business and financial risk

A business risk profile determines the appropriate level of financial risk (see Table 3). The aim of a financial policy is to manage the interplay between business and financial risks. Financial risk management should flow from the business risk management rather than the other way round. The problem with financial policies and financial prudence is that it is not always adhered to, and neither is the pursuit of the highest rating in the best interest of the company.

Financial risk is measured largely through quantitative means, particularly by using financial ratios. However, financial ratios *per se* do not convey much meaning unless they are compared against the industry or benchmarks. Therefore, financial data quality and a common frame of reference are needed. Benchmarks vary greatly by industry, and several analytical adjustments typically are required to calculate ratios for an individual company. Cross-border comparisons require additional care, given the differences in accounting conventions and local financial systems.

Firms have started recognising the importance of financial risk and the interplay of business and financial risk. A firm with no financial risk is not optimal. An under-leveraged company deprives its owners of potentially greater value by the use of leverage.

Financial flexibility

Financial flexibility mainly emanates from the capital structure, good quality of assets, better profitability and cashflow. Business and environment play a significant role in financial flexibility. In addition, the following factors influence financial flexibility:

- absence of litigations;
- appropriate insurance coverage;
- covenants and restrictions flowing from the existing loan agreements;
- continuity and contingency plans;
- access to capital markets;
- ability to sell assets;
- over-reliance on bank borrowings or commercial paper;
- adverse maturity schedule for long-term debt;
- maturing preferred stocks;
- inability to access the debt market.

Table 3 Business and financial risk

	Capital structure	Profitability and asset value	Cashflow
Economic environment	The economic cycle and regulations has impact on the quantum and structure of the capital that can be raised	The level of competition therefore profitability is driven by the working of public sector, regulatory regime, duties and taxes	The general business environment determines the demand and supply level therefore the cashflow
	Country risk is driven by the economic and regulatory environment	Government intervenes to protect the public interest	Internal sources for investment are driven by the business environment
Industry environment	Some of the industry have adverse capital structure – telecom, Oil exploration etc	Growth, stability or decline and the pattern of the business cycle has impact on the assets quality and profitability	Competitive environment has impact on the future sales and cashflow projections
	In general, the rating is driven by the industry level rating ceiling	Firm size, maturity of the industry, marketing expenses, technology changes determine the profitability	Future growth prospects determine the level of investment and financing needed
Company environment	Management ability and vision drives the capital structures and financial risks	Large firms have substantial staying power. Real estate land and assets that have more value than their book value	Capital-intensive firms may have adverse cashflow
	Small firms have low equity, adverse capital structure and high business risks	Growing firms need a substantial investment in marketing setup therefore has adverse impact on the profitability	Competitive advantage has a positive impact on the cashflow projections
Country environment	Determines the quantum and mix of capital structure	Interest rate and foreign exchange volatility adversely impact the profitability	Economic volatility, growth prospects and banking systems depth impact on the cashflow projections

Assets quality and value

The correct valuation of assets is an important factor in the analysis of financial flexibility, leverage, profitability and cashflows. This is an important factor in merger analysis. Goodwill and intangible assets may require analytical adjustments in asset valuation. Knowing the true value of assets is a key to the analysis. Leverage as reported in the financial statements is meaningless if assets are materially under-valued. Market values of a company's assets or independent asset appraisals can offer additional insights. However, there are shortcomings in these methods of valuation.

Capital structure and leverage

Existing leverage works as a constraint on the future borrowings and the ability and willingness to repay. Furthermore, the traditional debt/equity ratio as a measure for leverage may not be sufficient for the following two reasons.

❑ Structural models consider the book value of debt and the market value of equity, so the debt/equity ratio is more accurately measured by the ratio

$$Debt/equity\ ratio = \frac{Total\ long\text{-}term\ debt}{Total\ long\text{-}term\ debt + Market\ value\ of\ equity}$$

❑ Short-term borrowing, which does not appear in the previous equation, is currently more common.

There is no correct measure of appropriate leverage. The asset mix is a critical determinant of the appropriate level of leverage. Assets with stable cashflow or market values justify greater use of debt financing than those with clouded marketability.

Adjustments are required for considering off-balance-sheet financing. Some of the examples of off balance sheet liabilities include operating leases, guarantees, factored, transferred and securitised receivables, contingent liabilities, debt of joint ventures and unconsolidated subsidiaries.

There is always a debate on whether preferred stock is debt or equity. In general, it is divided into equity and debt. According to accepted practice, the maturity date determines the portion to be considered equity and debt. A lower percentage is allocated as equity when it approaches maturity.

Profitability and fixed-charge coverage

Profitability and fixed-charge coverage are the critical determinants of credit protection. Earnings and profits should be viewed in relation to a company's burden of fixed charges.

The primary fixed-charge coverage ratios are:

❑ earnings before interest and taxes (EBIT) coverage of interest;
❑ earnings before interest, taxes, depreciation and amortisation (EBIDA) coverage of interest;
❑ earnings before interest, taxes and rent (EBITR) coverage of interest plus total rents.

The significant measures of profitability are:

❑ pre-tax, pre-interest return on capital;
❑ operating income as a percentage of sales;
❑ earnings on business segment assets.

While the absolute levels of ratios are important, it is equally important to focus on trends and to compare these ratios with those of competitors.

Various industries follow different cycles and have different earning characteristics. Therefore, what may be considered favourable for one business may be relatively poor for another. For example, the drug industry usually generates high operating margins and high returns on capital. Defence contractors generate low operating margins, but high returns on capital. The pipeline industry has high operating margins and low returns on capital.

Therefore, profitability measures across the industries may not provide a meaningful comparison. Comparisons with a company's peers and industry average may give an idea about the competitive strengths and pricing flexibility.

For correct measurement of profits, earnings and profits from non-core activities need to be segregated. Profits from non-core activities may not be sustainable. Adjustments may also be required for various other items, provisions, depreciation and other line items in profit and loss accounts.

Cashflow analysis

Principal and interest requires cashflow for servicing (see Table 4). The cashflow is the single most critical aspect in determining the credit rating. There is not always a strong relationship between profitability and cashflow. Cashflow analysis helps in understanding the level of debt servicing capabilities. The significance and importance of cashflow in determining the credit rating is higher for weaker credit rating grades.

Table 4 Cash flow analysis

From sales	From revenue
Cash from sales	Revenue
Less cash production costs	Less operating expense, depreciation, amortisation
Gross cash profit	Operating EBIT
Less cash operating expenses	
CFO (EBDIT)	Less net interest paid
Less cash paid for interests	Less tax paid
Plus miscellaneous cash income	
Less miscellaneous cash expenses	
Less cash tax paid	
FFO	FFO
Less current maturing long-term debt	Less/plus working capital
Cash after debt amortisation	CFO
	Less/plus non-operational cashflow
	Less capital expenditure
	Less dividend paid
Changes in the investment and fixed assets	Free cashflow
	Receipt from assets and investment sale
	Business investment
	Exceptional items
Cash from investment	Net cash inflow/outflow
Change in short-term and long-term debt	Change in short-term and long-term debt
Change in equity	Change in equity
Cash from financing	Cash from financing
Changes in cash	Changes in cash

Projected financial statement analysis

Financial statement analysis is central to appraising the capital structure, profitability, assets value and adequacy of cashflow. Financial statement analysis is performed for historical financial statements and the projected financial statements. The analysis yields an assessment of the difference between current or projected performance and liquidity, on the one hand, and projected debt service obligations on the other. In general, the larger the cushion, the more favourable the rating. A primary difference between internal ratings and external ratings is whether the financial analysis is keyed to a "base" (or "most likely") case or to a downside (or "stress") scenario. Banks assign ratings on the basis of the borrower's current condition and most likely outlook, whereas the rating agencies assign grades on the basis of a downside scenario.

As a part of the financial analysis the borrower's financial ratios are compared to prevailing industry norms or averages. Firms in declining industries (TV manufacturers) and highly competitive (mobile/PC manufacturers) are considered more risky, whereas firms with diversified lines of business are viewed as less risky. A related factor, the borrower's position in its industry, is also an important factor in determining ratings. Those borrowers with substantial market power or those that are perceived to be "market leaders" in other respects are considered less risky because they are thought to be less vulnerable to competitive pressure.

Projected cashflow

This focuses on the borrower's prospects of generating sufficient cash in future periods to pay interest on the loan being considered as well as to pay back that loan principal under a mutually agreed repayment schedule. Cashflow analysis is critical in reviewing whether a borrower has the ability to repay individual debt. A review of the borrower's cashflow statements can offer information about other sources of repayment as well as the borrower's overall financial condition and future prospects. Evaluating cashflow is the single most important element in determining whether a business has the ability to repay debt. The principal methods of computing cashflow available to service debt are as follows:

❏ the accrual conversion method;
❏ supplemental or traditional cashflow analysis;
❏ primary sources of cashflow:

   ❏ cash from operations (CFO);

❏ investment and financing as a source of cashflow:

   ❏ cash from additional equity contributed to the business;
   ❏ cash from the sale of operating assets;
   ❏ cash from additional borrowing;
   ❏ cash from liquidation of the business.

Cashflow from operations

Enterprises generate cashflow (see Table 4) from three sources: operations, financing and investment activity. Enterprises need cash for three purposes: to conduct their operations, pay their obligations and provide returns to their investors. The primary interest of the lenders is to assess the sources of cashflow available for the repayment. To substantiate this, the other two types of analysis are also performed, ie, the cashflow sources and cashflow priority.

Funds from operations are the most important barometer of the business. Interest and debt coverage is measured with respect to the funds from operations and free cashflow. These ratios are the key differentiator between the investment grade and speculative grades. However, better cash ratios are not always good. Sometimes the better ratios indicate a declining market.

Some of the ratios computed are funds from operations (FFO) or CFO *versus*:

❑ interest cover;
❑ fixed-charge cover;
❑ debt;
❑ debt service coverage;
❑ net interest cover;
❑ capital expenditure.

The methods to measure cashflow may vary from bank to bank and may also vary with the industry of the borrower. Depending upon the method to compute cashflow, the method to compute some of the ratios will also vary. However, such variations may not be large and they may only be in the details.

Cash from investment
Companies need investment for fixed and working capital and financial assets. The investment analysis is critical for fixed capital-intensive firms and growth companies. Rapidly growing companies need to finance the growing inventories and receivables. With the increased turnover, companies generally improve their working-capital management techniques. Therefore, it may be difficult extrapolating recent trends to compute the future cashflow.

Flexibility in the investment timing is an important dimension of cashflow adequacy. Investment may be discretionary or committed for ongoing large projects. Cashflow is also required for acquisition. Capital investment plans and acquisition plans give an idea about the future acquisition.

Cash from financing

A company's size and its financing needs play a major role in whether it can raise funds in the public debt markets. Similarly, a firm's role in the national economy – and this is particularly true outside the US – can enhance its access to bank and public funds.

Access to the common stock market may be primarily a question of a management's willingness to accept dilution of equity, rather than a question of whether funds are available.

Cashflow requirements for investment need to be assessed for cashflow generated from investment, operations and financing.

Case study of the financial strengths of a bank

To measure financial strength (Table 5) of a bank the following factors are analysed.

Profitability

❑ *Net interest margins*: This is the profit margin for interest earnings over their funding costs. The global trend of financial disintermediation primarily hurts the Net Interest Margin (NIM). The impact of every action should be measured in terms of their impact on NIM.
❑ *Fees and commission*: This is non-interest earnings. The importance and earnings vary in the line of business, banks and geography.
❑ *Trading income*: Trading represents probably the most volatile large category of revenues. Trading income volatility can, in particular, hurt the earning power of those institutions that do not have an offsetting cushion of stable earnings.

Table 5 Financial ratios – for period 2001–2003

Key financial ratios for various long-term debt ratings							
Three-year (2001–2003) medians	AAA	AA	A	BBB	BB	B	CCC
EBIT interest coverage (x)	23.8	13.6	6.9	4.2	2.3	0.9	0.4
EBITDA interest coverage (x)	25.3	17.1	9.4	5.9	3.1	1.6	0.9
FFO/total debt (%)	167.8	77.5	43.2	34.6	20.0	10.1	2.9
Free operating cashflow/total debt (%)	104.1	41.1	25.4	16.9	7.9	2.6	(0.9)
Total debt/EBITDA (x)	0.2	1.1	1.7	2.4	3.8	5.6	7.4
Return on capital (%)	35.1	26.9	16.8	13.4	10.3	6.7	2.3
Total debt/capital (x)	6.2	34.8	39.8	45.6	57.2	74.2	101.2

Source: S&P

❑ *Expenses and efficiency*: Banking is a people- and technology-driven industry and the cost of both these items is a major component in the total expenses. The efforts and success of cost containment and cost against the industry average is the guiding factor. However, cutting cost beyond a point impacts the competitive advantage. Cost/income and fixed cost *versus* variable cost are the better indicators to measure.

❑ *Risk profiles*: The banking industry now measures three types of risk – credit, market and operational – and provides for the expected losses (ELs) and unexpected losses (ULs). ELs are provided through the expenses, pricing and profitability. ULs are provided through capital.

Capital structure

❑ Bank capital should not be measured just in absolute terms. The bank capital is not to be seen in terms of usual leverage ratios but in terms of risk profiles. Therefore, capital is measured in terms of economic capital and not against the regulatory capital as a measure of protection and safety. In particular, this is required to be measured for securitised assets. The bank's capacity to generate the capital from internal sources is a major financial strength of the bank.

❑ Cashflow provides an ability to grow and diversify into new businesses so as to grow and protect the cashflow.

❑ Liquidity measurements show a bank's reliance on so-called confidence-sensitive funding markets. Core deposits and purchased funds ratios are the appropriate measures of the liquidity.

❑ For asset quality, the critical factor is the credit concentration and loan loss reserves. Provisions and non-performing assets and various related ratios are measured.

SOVEREIGN CREDIT RATING

The sovereign credit rating reflects the opinion on the future ability and willingness of sovereign governments to service their commercial financial obligations in full and on time. A sovereign rating is a forward-looking estimate of default probability.

Sovereign risk is important because the unique and wide-ranging powers and resources of national government impact the financial and operating environment of companies. Internationally, past experience has shown time and again that defaults by otherwise creditworthy borrowers can stem directly from a sovereign default.

Sovereign risk arises from two of the sovereign powers: one is to control and regulate the use of the available foreign exchange and the other is the power to tax and issue the local currency in potentially unlimited amounts (thereby building inflation in the country and eroding the purchase power and therefore value of the debt).

Foreign currency denominated debt is impacted upon by the power to regulate foreign exchange, while both types of debt, ie, domestic currency and foreign currency, are impacted upon by the power to tax.

According to Standard & Poor's (S&P's), there were 69 cases of sovereign default during the 20-year period from 1975 to 1995.

For most international debt issuers, the sovereign risk factor remains an extremely important consideration in the assignment of overall creditworthiness. During 2002, the Argentine government rationed the availability of foreign exchange to private-sector entities to the point that some of these entities defaulted on foreign currency debt obligations, despite many of these same firms having sufficient funds to meet these obligations in a timely manner if access to foreign exchange had been possible.

Sovereign risk is generally managed by providing external guarantees and structures and asset-based repayments.

Sovereign ceiling principle

Financially distressed governments are likely to impose exchange controls or otherwise interfere with the ability of domestic firms to service their external debt. Therefore, firms cannot get ratings better than the sovereign ratings on their foreign currency debt. This is called the sovereign ceiling principle. All of the major rating agencies apply the sovereign ceiling principle, with varying strictness. In most cases, the sovereign ceiling rating coincides with the foreign currency bond rating assigned to the corresponding national government. However, this ceiling is broken if the corporate is able to structure an offshore collateral arrangement under which funds would never enter the domicile country.

Sovereign ratings address the credit risk of national governments, but not the specific default risk of the issuers. A rating assigned to a non-sovereign entity generally has a ceiling of the sovereign rating of its main country of domicile, but may some times be higher. Foreign currency ratings may be higher when the non-sovereign entity has stronger credit characteristics than the sovereign and when the risk of the sovereign limiting access to foreign exchange needed for debt service is less than the risk of sovereign default. This can include a highly creditworthy private sector issuer located in a sovereign ie, a member of a monetary union with a higher-rated central bank, an issuer with a significant percentage of assets and business offshore and an issuer with a very supportive offshore parent.

Sovereign default and transition studies indicate that, compared with corporate ratings, sovereign ratings show more stability at most rating levels. In many instances, the sovereign default record is lower than the corporate default record. However, such comparisons are affected by the small sample size of rated sovereign defaults.

Understanding sovereign ratings

As with other credit rating processes, the sovereign rating process analyses both quantitative and qualitative factors (see Table 7).

The quantitative aspect of the analysis incorporates a number of measures of economic and financial performance and contingent liabilities of the government. The political and policy framework is factored in through qualitative factors.

The following factors are generally considered for local and foreign currency ratings.

❑ *Political risk*:

  ❑ stability and legitimacy of political institutions.

❑ *Economic structures and flexibility*:

  ❑ income disparities in the country;
  ❑ income level shares of various sectors in the national income;
  ❑ labour flexibility;
  ❑ asset size of the financial sector;
  ❑ gross domestic product (GDP) and economic growth.

❑ *Fiscal flexibility*:

  ❑ expenditure, revenue and deficit;
  ❑ debt and debt-servicing burden;

- ❑ salary, pension and other expenditure on government employees.

❑ *Monetary flexibility and the efficiency of the financial sector*:

- ❑ price behaviour and economic cycle stage;
- ❑ monetary policy instruments and their effectiveness;
- ❑ central bank independence;
- ❑ exchange rate stability.

For foreign currency rating the following additional factors are considered.

❑ *External liquidity*:

- ❑ size of FX-foreign exchange reserves;
- ❑ balance of payment;
- ❑ current and capital account convertibility;
- ❑ public and private sector external debt.

Credit rating of the sovereign

The sovereign rating is divided into local currency and foreign currency. The difference is due to differences in the capacity to generate resources.

The political, social and economic factors impact a government's ability and willingness to honour local and foreign currency debt in varying degrees. A sovereign government's ability and willingness to service local currency debt are supported by its taxation and fiscal powers to raise resources and its ability to control the domestic monetary and financial systems, which gives it potentially unlimited access to local currency resources.

However, to service foreign currency debt, the sovereign generates foreign exchange through exports and capital flow. This constrains the repayment capacity.

The rating process is to assess the government's economic strategy, particularly its fiscal and monetary policies, as well as its plans for privatisation, other microeconomic reform and additional factors likely to support or erode incentives for timely debt service.

The aim of foreign currency rating is to assess the external liquidity, the external debt burden and the other factors that can change these two factors so as to achieve the desired level of comfort for repayment of foreign currency debt.

Table 6 Sovereign credit rating in local and foreign currency for selected countries

Country	Local Cy	Foreign Cy	Country	Local Cy	Foreign Cy
Australia	AAA	AAA	Japan	AA−	AA−
Belgium	AA+	AA+	Korea	A+	A−
Brazil	BB	BB−	Mexico	A	BBB
Canada	AAA	AAA	Netherlands	AAA	AAA
China	BBB+	BBB+	Russia	BBB	BBB−
France	AAA	AAA	South Africa	A	BBB
Germany	AAA	AAA	Switzerland	AAA	AAA
Hong Kong	AA−	A+	Taiwan	AA−	AA−
India	BB+	BB+	UK	AAA	AAA
Italy	AA−	AA−	USA	AAA	AAA

Sovereign credit characteristics

A sovereign rating reflects the opinion on a central government's willingness and ability to service commercial financial obligations on a timely basis. Table 6 indicates the sovereign credit rating for selected of countries.

CREDIT STRUCTURES

The willingness to pay and the priority are driven by the credit structures or interests in the collateral and assets of the borrowers, support or guarantee or credit enhancement provided by third parties and covenants imposed (Figure 6). Based on the credit structures, debt securities are divided into the following categories.

❑ *Secured*:

  ❑ asset based bonds and loans;
  ❑ secured bank loans;
  ❑ secured bonds.

❑ *Unsecured*:

  ❑ unsecured debt (bank loans, bonds and other identified general claims);
  ❑ subordinated debt loans and bonds;
  ❑ preferred stock.

The risk mitigation by credit structure is driven by the following two factors.

Table 7 Sovereign rating

AAA and AA	A	BBB	BB	B	CCC
Stable and accountable political institutions	Some geopolitical risks	Possible social stress	Possible social stress	Changes in the government lead to economic disarray. Social stress.	Weak political institutions
Flexibility to respond to changing circumstances	Less diversified policies	Record of satisfactory economic performance in the past	Structural impediments to growth and high economic disparities	Variable economic performance, vulnerable to political and external influence	A clear danger of default
Efficient public sector with flexibility to implement counter cyclic fiscal policies	Less developed local debt market and therefore requires offshore borrowings	High tax and fees regime, spending pressure, need to borrow externally	Non-official Debt linked to or denominated in foreign currency	Untested macroeconomic stabilisation efforts. Short-term debt linked to foreign currency	May be on default on bilateral trade
Sustainable monetary and exchange rate policies	Less flexibility and moderate inflation	Less developed financial markets, limited monetary tools	Direct interventions by the central bank and variable inflation	Structural imbalances, shallow debt markets and variable and high inflation	Sharp currency depreciation, high inflation
Strong and diversified financial sector	Developing capital markets, some ongoing challenges	Evolving financial sector and contingent liabilities for government	Financial sector under stress during recession. Significant contingent liabilities for government	Underdeveloped financial markets with significant contingent liability for the government	Weak financial sector. Acute shortage of credit
Very high external liquidity	Moderate external debt	Moderate to high external debt	Moderate to high external debt	Moderate to high external debt	High fiscal and external debt

Source: Based on S&P methodology.

❑ How easy is it for the creditors to establish an interest in the credit structure or security provided?

❑ How easy is it to realise the security interest? In general, there are four options to realise the interest: receivership, liquidation, schemes of arrangement and supervision under the auspices of court.

As expected, secured creditors have historically achieved the highest recovery ratings across the portfolio.

Collateral valuation

Collateral value is the primary driving factor in the credit structure. There are two options in collateral valuation: going concern or distress sale. Under reorganisation and receivership, the going concern values can be applied. Under liquidation the distress-sale values can be applied. For valuation, not all collaterals are equal.

Collateral valuation has a major impact on the facility and recovery rating and, therefore, the overall rating of the issue.

Valuation as a going concern

The analysis method for the valuation of collateral or recovery as a going concern varies widely under different legal regimes across geographies and also under different financing preferences. Here, an attempt is made to discuss a general method applicable to North America. However, each bank has to establish its own methodology, including banks in North America. The method also varies for loans, bonds and preferred stocks. If a defaulted company's intrinsic or economic value is greater than its liquidation (claim on the company) value, then the company is recommended to attempt to reorganise and continue to operate. Recovery ratings will also differ with the going concern assumption or liquidation assumptions.

Bankruptcy is first reflected in the cashflow. A firm is technically bankrupt when it is unable to meet its operational and financial commitments. Covenants further trigger or aggravate the situation. The firms near the cashflow triggers are more prone to the risks of bankruptcy. To estimate future cashflows, the existing future cashflow should be discounted and this discount varies from industry to industry.

Valuation under distressed recovery or sale
This is a structural analysis for the securities currently defaulted or distressed. Distressed recovery is estimated under a default scenario. The following factors determine the distressed recovery.

❑ *Time horizon*: The recovery varies under different assumptions about the time horizon. The time horizon can be as large as the maturity period of the security or some period starting from the distress time. The time period varies between the two.
❑ *Assumed stress*: For defaults including default correlation and interest rates.
❑ *Total cashflow analysis* versus *separate interest and principal repayment*.
❑ *Discount factors*.

Credit support
There are two widely used credit supports:

(1) sovereign support provided by government and regulators;
(2) credit enhancement, typically provided by the parent to a subsidiary or by an originator in a structure finance transaction.

Sovereign support
Sovereign support is provided to banks and utility organisations under the premise of "too big to fail" or since failure of the bank and utility may have a profound impact on day-to-day life and national and regional economies. This support needs to be factored into the ratings. Public rating agencies have published the methodologies to consider sovereigns and other institutions providing support to ratings. Two factors need to be analysed:

(1) the financial ability of the potential supporter;
(2) the willingness, or propensity, to support the troubled institution.

The following mechanisms are to be factored into the "support" provided to banks:

❑ the rating floor, below which long-term debt rating for a bank cannot fall so long as it retains the support; therefore, the rating may exhibit less volatility;
❑ the support may not be the same for local currency and foreign currency; usually, the foreign currency support will be weaker than that of the local currency;

❏ the willingness to support the troubled institutions is based *inter alia* on the systemic importance of the bank, both domestically and internationally, and any special relationship with the sovereign (support provider);

❏ a majority shareholding by government does not automatically provide support.

Case study: Sovereign credit support for banks – Moody's bank rating methodology. Moody's divides the bank rating methodology into two stages. Using the usual corporate rating methodology based on the financial and structural strengths, a baseline rating is estimated. The external government and other institutions' support are estimated separately and a final rating is estimated. Since the support is generally available only for the domestic currency, two ratings are estimated separately for the domestic and foreign currency.

The support is never absolute but limited. This is owing to:

❏ the degree to which the authority is able to support an important bank may be limited due to a monetary regime that does not permit the creation of unlimited quantities of local currency;

❏ risk of a local currency deposit freeze.

Therefore, there is a ceiling on the highest rating that can be assigned to the local currency deposits of a bank domiciled within the rated jurisdiction.

Sovereign support is so important for the bank rating that the Basel Accord provides the following information in paragraphs 60 and 61:

> *There are two options for claims on banks. Under the first option, all banks incorporated in a given country will be assigned a risk weight one category less favourable than that assigned to claims on the sovereign of that country. However, for claims on banks in countries with sovereigns rated BB+ to B− and on banks in un-rated countries, the risk weight will be capped at 100%.*

According to paragraph 57 of the Accord, the claims on non-government public sector entities (PSEs) will receive similar treatment as claims on banks, sovereigns or corporates.

The treatment of PSEs, such as local authorities, administrative bodies and commercial undertakings, is summarised in Table 8.

Table 8 The treatment of PSEs

Treatment as sovereigns	Treatment as banks	Treatment as corporates
If the PSE has specific revenue raising power though tax, duties and fees	Strict lending rules but bankruptcy not possible due to their public status	Entity function as a commercial enterprise in the competitive environment
Specific arrangements with the central government to reduce the default risk	No revenue raising powers	Even though the government and local authorities may be the major shareholder or owners

Support from parents and subsidiaries

Credit quality is a zero sum game. If some credit worthiness is passed on to the subsidiary, the parent's credit worthiness is notched down. In general, a parent and subsidiary are viewed as a single unit. If the parent provides a material advantage to the subsidiary, it creates disadvantages to its own creditors. This will lead to passing some credit to the subsidiary and parent level issues are notched down. During liquidation, creditors of the parent company are entitled only to the residual net worth of the subsidiary.

Adjustment may also be necessary for recognising the impact of the group. This depends upon whether the company under review is one of the core companies in the group or not. In general, the credit rating of a core company is independently assessed and the rating of a group of companies is adjusted to be lower than the core company's ranking. If the ranking of the non-core company under review is very low and the ranking of the core company is quite good and there is a real control by the core company, the ranking of company under review may be improved (with the condition that its rating will be lower by at least one notch) by adding points in the credit ranking. The additional points will depend upon the type of control. In general, the following control is taken as a real control:

❏ controlling more than half of the directors;
❏ agreement to control financial, business and enterprise policies;
❏ financing half or more of the liabilities;
❏ controlling the decision-making body.

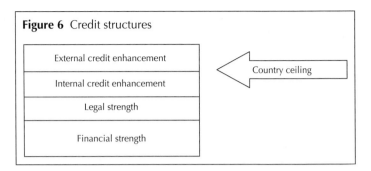

Figure 6 Credit structures

A group is generally defined in the companies act of the country. That definition may be taken as a rule of thumb. However, the litmus test is the real control and not the legal control.

Credit enhancement

According to paragraph 546 of the Accord, "A credit enhancement is a contractual arrangement in which the bank retains or assumes a securitisation exposure and, in substance, provides some degree of added protection to other parties to the transaction" and these can take various forms.

Case study: Structured finance transaction

Securitisation deals and structures closely depend upon collaterals values and the bankruptcy and legal structures. The structured financed rating is typically driven by the following factors (Figure 6).

❑ *Financial strength.*

  ❑ Credit quality of the underlying assets: the analysis focuses on the PDs amongst the assets, the chance of recovery in the event of defaults (including the length of any delay in recoveries) and the uncertainty associated with these default probabilities and recoveries.
  ❑ Debt service coverage ratio.

Legal strength or willingness to pay

This determines the following factors.

❑ Rights on the receivable: can true sale to Special Purpose Vehicle (SPV) be challenged by the creditors of the originators in the event of the originator's insolvency? Is there any ceiling on the interest rate to be charged on the consumer loans?

❑ Potential interference or delays in receipt of receivable, in particular during the insolvency of the originator.
❑ The legal structures implemented.

  ❑ Jurisdiction:

    ❑ the originator's jurisdiction;
    ❑ the offshore location in which a special-purpose entity is located;
    ❑ the jurisdiction in which the trust is located.

  ❑ Triggers:

    ❑ specified event triggers that either accelerate amortisation of the transaction or obligate the originator to repurchase receivables;
    ❑ that provide negotiating powers to creditors;
    ❑ that work as early warning signals;
    ❑ that trap all the cash on certain specified events:

      ❑ notice and acknowledgements – this enhances the right to receive; a notice and acknowledgement from the central government greatly reduces the sovereign risk of redirection of the payment to the central bank.

Collection procedure

❑ Collection procedures ensure that the receivables are received by the trustee to fund periodic debt service.
❑ If specification to retain all or a majority of flows for accelerated amortisation under certain trigger events.
❑ Provision of a reserve account.

Representations and warranty

About the receivable – all hindrance for the transfer of receivable to the trust are cleared and do not exist.

Credit enhancement

The structure of the transaction has a significant impact on the risk profile of the issued notes. A waterfall method is the recommended method to analyse the cashflow and the prioritisation. Credit worthiness of the parties providing external and internal credit

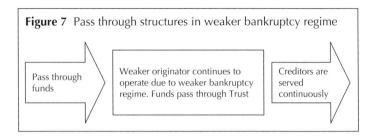

Figure 7 Pass through structures in weaker bankruptcy regime

Pass through funds

Weaker originator continues to operate due to weaker bankruptcy regime. Funds pass through Trust

Creditors are served continuously

enhancement is considered to evaluate their capabilities of adequately performing their role.

Some of the internal credit enhancement mechanisms are:

❑ reserve account;
❑ capture of excess spread;
❑ over-collateralisation;
❑ inclusion of subordinated notes.

Some of the external credit enhancement facilities are:

❑ hedging for interest rate;
❑ hedging currency.

Weak bankruptcy regimes

Weaker bankruptcy regimes are characterised by weaker creditor rights (see Figure 7). Weaker creditors rights mean the obligor will continue with the normal operations of their business during the bankruptcy proceedings and control the restructuring proceedings to a great extent. Ironically, some experts feel that a weaker regime is helpful for securitised transactions.

Covenants and credit quality

Covenants provide a framework regarding how the borrower will conduct their business and financial affairs. The stronger covenant package provides a greater degree of control to the lender. Borrowers typically seek the least restrictive covenant package. Imposing covenants on public bonds is not a common trend, but imposing covenants is an important tool for bank loan or private placement to provide the leeway to control the deterioration in the quality. For a public bond, a covenant does more harm than good for the credit quality. A tight covenant may accelerate the debt causing a default that might otherwise be avoidable.

Covenants generally do not play much of a role in the credit quality or credit ratings and they should not be analysed to a great extent. A covenant can, at best, be an expression of the management's intent, at least for the public companies. Neither can covenants insulate a subsidiary from the failure of the parent company. Some of the flaws in protecting the credit quality through the covenants are as follows.

❏ Covenants cannot alter the business adversity. Covenants do not address fundamental credit strength or augment cashflow.
❏ Enforcement is dubious. Courts do not enforce covenants, they award damages.
❏ In practice, lenders waive covenants for a variety of reasons. Lenders do not precipitate the default by adhering to the covenant. Instead they choose to change the interest rate or improve the security position.

Typical covenants
The basic factors present in all covenant documents include:

❏ financial and other information and frequency of its submission;
❏ default definition and cross-default and cross-acceleration clauses;
❏ modifications of loan conditions and covenants.

Transaction-specific covenants are governed by:

❏ level of credit quality – covenants increase in number and grow more stringent as the quality of the credit declines;
❏ private *versus* public debt – covenants are more stringent and finely negotiated for private debt;
❏ tenor – covenants appropriate to the tenor are imposed.

The following examples illustrate features of transaction-specific covenants.

❏ Preservation of the repayment capacity of the borrower. This is achieved through restricting new borrowings, prohibiting diversion of cash generated from operations and asset sales, restricting payments to shareholders, restricting asset sales and investment decisions.
❏ Preventing occurrence of credit damaging events without first repaying the debt.

❏ Negative-pledge clauses, cross-acceleration (or cross-default) provisions and limits on obligations are imposed to protect assets or the value of the assets for the creditors or to safeguard the priority position of the credit in the event of bankruptcy or default. These covenants preserve the value of assets for all creditors and – what is particularly important – safeguard the priority positions of particular lenders.

❏ Providing a steady flow of information for early warning signals and triggering the corrective actions.

❏ Bank loans also include covenants to pay instalments on time.

❏ In many cases, covenants can serve more than one function. Often, what is not included has more impact than what is included.

Structure for bank loan *versus* other debt

The main pertinent features of the bank loans can be summarised as follows.

❏ *Security*: Strong collateral is a secondary source of repayment. It differentiates between the secured and unsecured debt. The strength of the collateral is derived from the nature and quality of the collateral and the degree priority of secured lenders over the unsecured claimants. The strength varies with tangible and intangible assets, capital stock, common stock, accounts receivable, inventory, etc.

❏ *Bank loans are generally focused on the non-investment grade*: These loans are typically secured while investment grade bonds are usually unsecured. A lien on assets and/or common stock encourages lenders to invest in riskier credit profiles by providing an alternative source of repayment other than potentially volatile cashflows.

❏ *Priority in bankruptcy and valuation of collateral*: How collateral is valued, how it is likely to be treated in the bankruptcy and the difference between the two is very important. While valuing collaterals, values should not be treated as a going concern while the appropriate method according to the structure, financial strength and type of the collateral is to value it as a distress sale.

❏ *Effective covenant structure*: The terms are generally more stringent under bank loan agreements as compared with bond indentures. Financial covenants under bank loans provide early warning signals. Negative covenants typically prevent additional indebtedness.

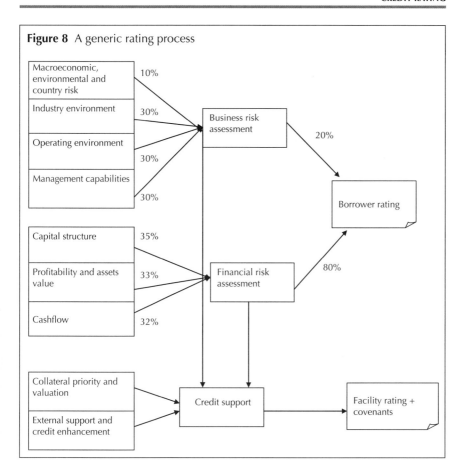

Figure 8 A generic rating process

We have discussed in detail the workings of credit risk mitigation as a credit support structure in the separate chapter, "Credit Risk Mitigation".

A GENERIC RATING PROCESS

This is applicable to both internal and external rating (see Figure 8). The rating systems are based both on the qualitative and quantitative assessments. There are various types of quantitative models. Financial strength is a major area of assessment through these models. In the final assessment of the borrower rating, there is no formal model and the weights are not fixed. In essence, systems are based on the general consideration and modelling, and vary for

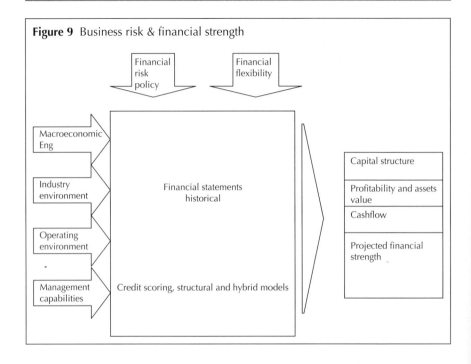

Figure 9 Business risk & financial strength

industry and macroeconomic conditions and for country. The weights indicated here are only samples to give an idea of the importance.

As a part of the internal rating process, banks do give weight to the opinion of the external rating agencies, if available.

Importance of business risk analysis

In most of the internal or external frameworks, the weight given to the business risk is generally one-third or one-quarter of the financial risk. However, analysis of the business risk (see Figure 9) is used in forecasting the future financial statements and analysis of the financial risk. Furthermore, the benchmarks and weights used for each industry, country and macroeconomic environment also vary. In a nutshell, the model used to measure the financial strength is different for different business risks.

Forecasted financial strength is also impacted upon by the financial risk policies and the quality of financial statements and accounts.

EXTERNAL RATINGS

Ratings agencies are firms such as S&P's and Moody's. They provide ratings, ie, opinions regarding the creditworthiness of issuers or capital markets obligations.

An external credit rating is the rating agency's opinion on the general creditworthiness of an obligor, or the creditworthiness of an obligor with respect to a particular debt security or other financial obligation. Ratings generally indicate an issuer's ability to meet financial obligations denominated in the issuer's domestic currency. Ratings in the domestic and foreign currency may be different.

Defining external rating

In the words of S&P's, "A credit rating is the opinion of the general creditworthiness of an obligor, or the creditworthiness of an obligor with respect to a particular debt security or other financial obligation, based on relevant risk factors".

In Moody's words, a rating is ". . . an opinion on the future ability and legal obligation of an issuer to make timely payments of principal and interest on a specific fixed income security". Moody's states that "ratings are intended to serve as an indicator or forecast of the potential credit loss because of failure to pay, a delay in payment, or partial payment".

It may be noted that a rating does not constitute a recommendation to purchase, sell or hold a particular security. A rating also does not mean a comment on the suitability of an investment for a particular investor. Rating is also not equivalent to an audit or audit-like process. No audit is conducted by the rating agency.

Since the issuance of the market risk amendment by the Basel Committee in 1996, banking regulators have been making increasing use of the external credit ratings. This has been further strengthened by recognising and incorporating external credit ratings in the standardised method for credit risk management in the Basel II Accord.

Rating agencies use quantitative and qualitative attributes of the obligor and analyse the credit structures. The proprietary rating models, the expertise of the rating in general, rating obligors of the same industry, assessment of similar credit structures, experience and benchmarks, judgment and the private information about the obligor play a major role in the determination of the credit rating.

Due to various pressures, criticism and expectation credit agencies have started disclosing their rating processes, rating change processes and rating trigger processes. Wider and sudden failures and criticism due to credit losses have also compelled issuers to make larger disclosures on the liquidity positions and rating triggers.

External ratings were originally ordinal measures of credit risk (they help rank firms by their quality of credit) but nowadays agencies ratings also represent cardinal default/loss numbers.

Basel has prescribed the following criteria to recognise a credit rating agency.

> *Rating Agency should have a rating methodology for each of the market segments. The rating methodology must be rigorous, systematic, and subject to validation based on historical experience and back testing. Ratings must be reviewed on the ongoing basis and ratings must be responsive to changes in financial condition.*
>
> *Rating agency must disclose assessment methodologies, including the definition of default, the time horizon, and the meaning of each rating; the actual default rates experienced in each assessment category; and the transitions of the assessments or transition matrix.*
>
> *Rating agency should be independent, should have skilled resources, international access and internal processes.*

Ratings from recognised agencies are regarded as unbiased evaluations and their ratings are widely accepted by market participants and regulatory agencies. FIs, when required to hold investment-grade bonds by their regulators, use the ratings provided by the recognised agency to determine which bonds are of investment grade. For issuer rating, the rating is an opinion on the obligor's overall capacity to meet its financial obligations. The opinion is not specific to any particular liability of the company and it does not consider the merits of having guarantors for some of the obligations.

External credit ratings represent credit quality and represent a summary indicator of the borrower characteristics and risk characteristics.

❑ *Borrower characteristics*:

   ❑ financial strength.

❑ *Risk characteristics*:

   ❑ default probability;

❑ loss given default (LGD);
❑ transition risk.

The overall credit quality within the same rating is comparable but may differ with respect to specific credit quality characteristics. This differentiation is more pronounced for bonds in different industry segments. Within a specific segment, the loss severity, financial strength and transition risk is generally uniform across the rating grades and the grades differ on relative default probability. However, this may also vary. Therefore, it is necessary for the user of the rating to understand the constituents of the credit quality and its variation across the industry and grades. This may also differ between the investment grade and speculative grade. For example, Moody's investment grade for corporates is driven by PDs and a speculative grade is driven by LGD. For structured finance, Moody's places an equal emphasis on LGD and PD. The rating can be analysed in two dimensions:

❑ obligor or issue under review;
❑ time horizon for validity of rating.

Types of external rating

Rating is a process to classify the issuers and issue into different categories or classes representing the likelihood of the company failing to pay its obligations. There are three main types of rating: issuer credit ratings, issue-specific credit ratings and specialised ratings.

Issuer credit ratings

This is the opinion on the obligor's capacity to meet its financial obligations. The opinion is not specific to any particular issue of the company, and it does not consider the availability of any guarantors or seniority. There are three types of issuer ratings: counterparty ratings, corporate credit ratings and sovereign credit ratings.

Both S&P's and Moody's publish issuer ratings that assess the creditworthiness of a firm, even if the company has no outstanding public debt. These issuer ratings reflect their opinions on an entity's ability to meet its senior (unsecured) financial obligations.

An issuer credit rating is issued for measuring the company's repayment ability beyond debt markets and under a variety of financial contracts, including swaps, forwards, options and letters of credit, for extension of credit lines, the provision of information to potential suppliers or customers and various other counterparty transactions.

Issue-specific credit ratings: An issue credit rating is a current opinion of the creditworthiness of an obligor with respect to a specific financial obligation or a specific financial programme. It takes into consideration the terms and conditions of the obligation as well as the creditworthiness of guarantors, insurers and other forms of credit enhancement. External ratings are those of debt issues, not of issuers. The ratings assigned to senior unsecured debt may be said to be closer to the issuer ratings since the debt defaults only if the issuer does. The country-specific factors such as political, economic and monetary risk factors are also analysed to consider country risk. The country risk is measured in terms of currency and not in terms of the place of issuance. There are six types of issues:

❑ equipment trust certificates;
❑ secured;
❑ senior unsecured;
❑ subordinated;
❑ junior subordinated;
❑ preferred stock and deferrable payment debt.

Specialised ratings: Specialised ratings incorporate an evaluation of covenants and collateral packages designed to mitigate the risk of loss, even if a default occurs. Specialised ratings are used for loans, private placement and other instruments. Loan ratings serve the syndicated loan and project finance markets and assess the lender's prospects of recovery after default by examining the value of any collateral or of other protective features commonly provided to lenders. Loans, private placements and other instruments such as secured bonds, if well secured and offering good ultimate recovery prospects, may have a higher rating than the issuer rating. Conversely, instruments that are subordinated to the senior debt of an issuer will normally carry a lower rating than the issuer rating. Bond and money fund managers use fund ratings to differentiate their bond and money funds from those of their competitors. The ratings provide investors with information on the credit quality and volatility of a fund.

Table 9 Recovery rating

1+	Highest expectation full recovery	100% recovery	Issue rating = Issuer rating + 3 notches
1	Full recovery expected	100%	Issue rating = Issuer rating + 2 notches
2	Substantial recovery	80–100%	Issue rating = Issuer rating
3	Meaningful recovery	50–80%	Issue rating = Issuer rating
4	Marginal recovery	25–50%	Issue rating = Issuer rating
5	Negligible recovery	0–25%	Issue rating = Issuer rating

Recovery ratings: An issue credit rating is an opinion of the credit-worthiness of an obligor with regard to a specific debt issue. Issue ratings take into account the ranking, payment terms and recovery prospects of the issue, and may be rated the same as the corporate credit ratings – lower for junior debt or higher for well-secured debt. However, in all cases, issue ratings remain anchored to the corporate ratings.

The recovery rating scale estimates the likely recovery of principal in the event of default and is de-linked from the corporate credit rating (see Table 9).

The meaning of external rating

Long-term ratings: Every borrower is responsible for meeting its financial commitments on a timely basis. Every borrower is exposed to the external cycles and risks. Ratings are graded on the basis of the likely impact the external factors are going to have on the financial strengths. Obviously, the lower grade borrowers have a smaller time horizon over which the current rating will be valid.

External credit rating ranks the credit standing of debt using coded letters (see Table 10). These ranks are "ordinal numbers" and not absolute values of the level of the risk. By contrast PD represents the absolute values of the likelihood of default over a given time horizon.

Short-term rating: Assessment of the likelihood of the timely repayment of obligations is considered in the short-term in relevant markets. Due to the short-term horizon, the external factors and cycle have almost no impact on the rating.

The short-term credit ratings by S&P's are as follows.

Table 10 One dimensional external rating

S&P	Moody's	Credit quality	Grade
AAA	Aaa	Very high quality	Investment grade
AA+	Aa1	High quality	
AA	Aa2		
AA	Aa3		
A+	A1	Good repayment ability	
A	A2	Susceptible to external	
A	A3	economic adversities	
BBB+	Baa1	Adequate repayment ability	
BBB	Baa2	More susceptible to external	
BBB	Baa3	economic adversities	
BB+	Ba1	Uncertain repayment ability	Speculative grade
BB	Ba2		
BB	Ba3		
B+	B1	High-risk investing	
B	B2		
B	B3		
CCC+	Caa1	Vulnerability to default	
CCC	Caa2		
CCC	Caa3		
CC	Ca-C	Bankruptcy likelihood	
C			
D			Default

❑ A-1: highest category. The obligor's capacity to meet its financial commitment on the obligation is strong.
❑ A-2: The obligor's capacity to meet its financial commitment on the obligation is satisfactory.
❑ A-3: Weakened capacity of the obligor to meet its financial commitment on the obligation.
❑ B: The obligation has significant speculative characteristics. The ongoing uncertainties could impact the obligor's capacities to meet its financial commitment on the obligation.
❑ C: Vulnerable to non-payment. Favourable business, financial and economic conditions are required to meet the financial obligation.
❑ D: Default.

The credit rating changes are implemented through the following three mechanisms.

❑ *Outlooks*: An outlook notation indicates the possible direction in which a rating may move over the next two to three years. Rating outlooks convey the analyst's opinion of the likely direction of the rating over the next 12 to 18 months:

❑ "positive" – may be raised;
❑ "negative" – may be lowered;
❑ "stable" – unlikely to change;
❑ "developing" – may be raised or lowered.

❑ Rating reviews are a formal rating action indicating that there is stronger chance that the rating will change within 90 days of going on review. Rating reviews are premised on certain well-defined triggers (eg, conclusion of negotiations with creditors, closure of certain transactions, additional financing, etc), which will determine the rating outcome on conclusion of the review. As a result, rating reviews should be viewed by investors as a signal that a potential rating change is imminent.

❑ *Credit watch*: A credit watch listing highlights the potential for near-term change in a credit rating. It signals to investors that further analysis is being performed.

Mapping risk categories

Since there are a limited number of rating categories, multiple firms fall in the same category and it is difficult to distinguish among the credit qualities using ratings alone. The objectives of valuation, risk assessment and capital allocation can be achieved only by mapping these measures to historical default probabilities. On the positive side, these ratings cover a large range of the major corporate market and cover a long history for each borrower. These ratings have internationally established credibility because of the long history of rating agencies and the extensive testing of their relative performance.

On balance, a close reading of Moody's and S&P's detailed descriptions of rating criteria and procedures suggests that the two agencies' ratings incorporate elements of PD and LGD but are still not precise EL measures. Risk tends to increase non-linearly on both bank and agency scales. Therefore, the risk from one rating grade to another does not change in the same proportion but exponentially. For example, on the agency scales, default rates are low for the least risky grades but rise rapidly as the grade worsens (see Table 11).

Table 11 Default rates for the various risk grades

S&P long-term rating	PD/Loss %	Moody's long-term rating	PD/Loss %
AAA	0.00	Aaa	0.00
AA+, AA, AA−	0.00	Aa, Aa1, Aa2, Aa3	0.03
A+, A, A−	0.07	A, A1, A2, A3	0.10
BBB+, BBB, BBB−	0.25	Baa, Baa1, Baa2, Baa3	0.13
BB+, BB, BB−	1.17	Ba, Ba1, Ba2, Ba3	1.42
B+, B, B−	5.39	B, B1, B2, B3	7.62
CCC, CC, C	19.96	Caa, Ca, C	NA

According to the Basel Accord, the supervisors are responsible for mapping the credit ratings to the credit risk or risk weights. The Accord recognises that the credit risk will vary according to the size and scope of the pool of issuers that each rating agency covers, the range and meaning of the assessments that it assigns and the definition of default used by the rating agency.

External rating process

Rating agencies do not have any government mandate and operate independently of any investment banking firm, bank or similar organisation. The recognition as a rating agency ultimately depends on the willingness of investors to accept its credit opinion. Rating agencies generally work on the principles of:

❑ independence;
❑ objectivity;
❑ analytic integrity;
❑ disclosure.

Rating agencies perform adequate qualitative, quantitative and structural analysis. The process may vary for a risk segment. However, in general, the process is as explained in this chapter.

The analyst team meets the issuer's management to review in detail the key qualitative, quantitative and structural factors. Documents are also reviewed for analysing structure and quantitative factors. Analysts also visit various offices and business places of the issuer to get an insight into the working and strengths of the issuer. In general, the following issues are discussed:

❑ the impact of the business environment;

❑ the industry environment and prospects and impact on various business of the issuer;
❑ comparison of the business segments with the competitors and industry norms;
❑ financial policies and financial goals with projected cashflow, balance sheet and income statements and capital spending plans, financing alternatives and contingency plans;
❑ accounting practices.

The analyst team recommends the initial rating and the rating committee at the rating agency discusses the rating with pertinent facts and reasoning. In general, the committee works on the basis of voting. The issuer can appeal against the decision before publication of the rating. Once the rating is assigned, it is disseminated to the public. The rationale behind the rating is also published.

The publication of the rating to the public at large is not done in all markets. In markets such as the US, the rating is published openly. In other markets, the rating is published to the public if it is requested by the issuer or it is disclosed to the parties indicated by the issuer.

After the assignment of rating, an ongoing review is maintained on the material factors regarding the issuer and issuer industry. The rating is reviewed formally at least once a year with the issuer management. The issuer is also expected to notify promptly the material, financial and operational changes that could affect the rating. If the change in circumstances warrants changes in the rating, a preliminary review is undertaken and the issuer may be listed for credit watch and the entire process of rating, as described in this section, is followed.

When a rating change appears necessary, the agency undertakes a preliminary review that may lead to a credit watch listing. The next step is a comprehensive analysis, including, if needed, a meeting with management and a presentation to the rating committee. The rating committee considers the circumstances, comes to a decision and notifies the issuer, subject to the appeal process noted above.

Public credit agencies assign ratings based on expert judgment as well as credit risk models. In fact, many credit agencies provide credit risk models as products to banks and FIs. Various risk management models play a more important role than before. The process of rating by credit agencies depends significantly on the analyst's background and experience. While certain tools can help, human efforts represent the largest part of the job.

External rating differences across agencies

Rating is an art as much as a science. Therefore, it is critical to ensure that rating is not confined only to quantitative measures. Each rating is subjective and subjectivity is at the heart of rating.

With rating being an art as well as a science, and there being no clear-cut formulae for rating, it is likely that the ratings will differ across the agencies. The question is of degree. As regulatory capital may rely significantly on the external ratings, significant differences in rating across the agencies may not be acceptable. Furthermore, this will encourage borrowers to shop for the best ratings.

In various studies, it is found that the difference between the ratings of Moody's and S&P's is minimal and very small. On the other hand, differences between Moody's or S&P's to other rating agencies in the US are comparatively larger. The difference is still large when compared with the Japanese credit rating agencies.

Problems with external ratings

Why are the default or transition probabilities derived from historical rating agencies and by the KMV model different?

Empirical studies have found external ratings to be slow in their response to the environment changes and changes in the ratings of issuers and issues. Therefore, the historical frequency of staying in a class is overstated from the true probabilities of maintaining the same credit quality and this results in changes to the median and mean default rates as measured from the rating agency's rate and true rate.

INTERNAL RATINGS
Introduction

Internal ratings are the summary indicators of the risk measures, derived internally by the bank using bank-specific methodology, models and data. Internal rating represents a measure of one or more of the following risk components:

❑ PD – default rate;
❑ LGD – loss given default;
❑ EAD – exposure at default;
❑ M – maturity.

Each of the components is measured separately. No systems have yet been developed to measure all components combined. Internal rating systems are generally designed to measure the borrower ratings for PD and facility rating for LGD in two steps. The PD measurement is called the obligor rating. The LGD measurement is called the facility rating or recovery rating. In this chapter, we have considered the measurement of PD and LGD only. We have dealt with impact of the EAD and maturity in the chapter, "Credit Risk Portfolio Management".

An internal rating system helps FIs manage and control credit risks they incur through lending and other operations by grouping and managing the creditworthiness of borrowers and the quality of credit transactions.

A good internal credit rating system is a key to building sound risk management practices and usually consists of the scale the institution uses to characterise credit risk as well as the credit methodologies, policies and procedures used to assess risk and design of overall scores.

Building a well-functioning internal rating system to measure PD and LGD is an important first step in measuring credit risk. An institution cannot make informed business decisions if it does not measure risks. Risk measurement is also required in pricing decisions. Regulators require credit rating and other measurement systems to be adopted by the banks to sensitise the banks to the quantum of risk by linking risks to the capital adequacy.

Effective internal rating systems also provide early warnings of more serious credit problems for both individual exposures and the quality of the overall portfolio and also identify credit trends. Based on the information generated by a credit rating system, banks can make strategic portfolio decisions about their future participation in various industry sectors.

Rating trends

The Basel Accord has recommended building internal rating systems. The aim is to sensitise the banks towards credit risk and build a level playing field. With the wider adoption of internal rating systems, the following trends are observed.

❑ Ratings are becoming transformed from comparative measurements to absolute measurement. Now a two-dimensional loss probability is attached to the rating. In other words, ratings are measured in terms of default rate and recovery rating.

❑ The number of rating grades has increased to the number recommended by Basel.

❑ Credit rating models and expert systems are becoming developed.

❑ Benchmarking of internal ratings with public ratings is occurring. Banks are using public ratings data to build and validate their internal models.

The internal rating framework prescribed by the Accord has had an impact upon even the rating frameworks of the external rating agencies. Some of these impacts are clearly visible. Rating agencies have started challenging, reinterpreting and supporting their various assumptions used in the external rating. This will further develop the risk management practice. Some examples of these impacts include the following.

❑ Rating agencies have started recovery ratings during the past three years. S&P's started this framework three years ago and Fitch started recently in 2005. This is equivalent to the introduction of two-dimensional facility ratings into the external ratings. The loan rating, a precursor form of recovery rating, has been in existence for more than a decade now.

❑ Rating agencies have started synchronising the rating measures and methods across risk segments or assets class or industries.

❑ Rating agencies are leveraging the Basel II Accord in rating banks.

❑ One of the most important changes already incorporated into the Basel II Accord is to estimate capital charges for all asset classes based on UL only and not on both UL and EL. The historical data and rating migration experience of rating agencies have helped in fine-tuning the capital charges or risk weights for various risk segments or asset classes.

External and internal ratings – an incompatible architecture

External and internal ratings are fundamentally different and there is no mechanical formula to convert one rating into another. The only way to map both is on the basis of the historical loss experience. Banks rarely have such a database whereas rating agencies do. The mapping of external rating to internal ratings can result in the following problems.

❑ Internal rating is a measurement of risk drivers and placing loans into credit grades. Grades do not necessarily represent the credit risk unless grades are calibrated and validated. Without the ordinal measurement of default risk through calibration and validation, it may result in placing loans of different PDs into the same credit grade and loans with similar PDs into different grades.

❑ We know that for the same borrower the default rate of publicly issued bonds differs systematically from loan default rates.

❑ Banks follow PIT and agencies' TTC rating philosophies. This impacts on the PD measurement at the absolute level. Such incompatibility can confuse attempts to tune rating criteria and for banks can seriously distort the measurement of business line profitability, loan loss reserves and capital allocation.

❑ As major agencies rate borrowers with the expectation that the rating will be stable through normal economic and industry cycles, only those borrowers that perform much worse than expected during a cyclical downturn will be downgraded (will "migrate" to riskier grades). In contrast, rating systems that focus on the borrower's current condition (virtually all bank systems) are likely to feature increased migration as cycles progress but, in principle, should exhibit somewhat less cyclical variation in default rates for each individual grade.

❑ An external database used in the internal rating system can be used in various ways. For example, FIs with limited accumulation of data on firms' financial conditions and defaults may use an external database to supplement their data. Each bank uses a different type of database. The database used impacts upon the rating architecture and risk measurement. In particular, characteristics of firms in the sample data that greatly influence default frequencies, such as the firm size, industry and region, should be similar to the user's portfolio.

❑ Even external rating models may be used in some cases. For example, external models designed by rating agencies that have a large set of historical data may be used to assign grades to large firms with good standing for which only a few default samples exist. FIs, in doing so, should have a sufficient understanding of the rating philosophy underlying the model (eg, PIT or TTC ratings) and the design of the model used. It is also necessary that they validate the models themselves rather than depend only on reports made by the model vendor. The same applies to FIs that develop rating models by outsourcing the procedures to an outside agency.

A general framework for an internal rating system: Banks typically produce ratings only for business and institutional loans and counterparties, not for consumer loans or other assets. Consumer and other small loans are generally rated at a portfolio level.

The internal rating systems of a bank are closely determined by the business environment, industry, historical data, exposure size and cost benefit for establishing internal rating at transaction level.

The internal rating systems of banks differ due to the following reasons:

❑ the credit culture and historical experience of each bank is different;
❑ banks give different weight to qualitative and quantitative factors;
❑ Each transaction has a different degree of opaqueness of risks and complexity;
❑ LGD is very difficult to quantify and this difficulty varies with the country, industry, product and seniority of the claim;
❑ there are various approaches to measure the credit rating, risk and loss.

Ratings should be applied to all credit risks consistently. However, depending on the size and nature of transactions, it may be too costly to assign grades to all transactions. Each case should be treated as an exception. However, banks should evolve policies to deal with such exceptions and consistently apply such a policy. All major borrowers and transactions should be subjected to ratings. Small sized loans could be managed by building homogeneous portfolios by segmenting exposures on the basis of product type and characteristics of borrower. Assurance on homogeneity can be obtained from the observed stability default rates and risk drivers. It is important to consider the following points when establishing an internal rating system: the architecture of the internal rating system, the internal rating process, the calibration risk drivers to risk rating and the active use and validation of the rating system.

In the following paragraphs we have explained the architecture of an internal rating system and the underwriting credit process.

Architecture of internal rating systems

The establishment and management of internal rating systems is an expensive proposition. Each institution needs to determine the cost relative to the return and come up with its own design. The asset size of the portfolio determines the need for an internal rating system. There is no single answer to the question as to what is the appropriate architecture of the internal rating system. In addition to asset size, the architecture is also driven by the characteristics of their loan portfolios, operations and the objectives of the rating system. According to the Basel Accord, the internal rating system is required to measure four risk components: PD, LGD, EAD and maturity. According to the Basel Accord, the banks may use the supervisory provided estimates for some of the risk components instead of their own internal risk estimates. Furthermore, it is not necessary that the same risk component be estimated internally for all the risk classes. For example, within corporate assets class, specialised lending (SL) may be using the supervisory estimates. This means an internal rating system need not be set up for the SL or need not estimate all components. Furthermore, the level of validation is also not uniform for the components and risk classes.

Therefore, some banks obviously need to start with a simpler approach.

Number of rating grades

The number of rating grades and the spreading of borrowers in each of the grades is a sign of the maturity of credit risk management in a bank. It is recommended that the bank should have the number of grades according to the variation in the creditworthiness. An appropriate trade-off is also required if the number of borrowers in each grade is too small. With the help of external data, grades may be combined if there are small numbers of borrowers in each grade. However, homogeneity of the grade needs to be maintained. Within each grade, borrowers should be normally distributed.

Deciding how many grades are appropriate is a very difficult question to answer. Experts recommend up to 15–20 grades. The number of grades differs for non-impaired assets and impaired assets. Basel recommends a minimum of seven grades for non-impaired assets. According to the survey conducted by the Basle Task Force, the average number of grades that the sample banks had were 10. Three quarters of the sample had 2–14 grades while a quarter of the sample had 15–24 grades. This included the auxiliary grades indicated by a "+" or "−".

It is observed that over a period of time the cyclical and business conditions cause a shift in the borrowers to the grades in one direction and results in a concentration of borrowers in a few grades. In such cases, the rating definition and number of grades need to be reconsidered. This change should not be frequent since this goes against the stability and sustainability of grades.

Optimal structure of the rating system
It is therefore desirable to establish the number of grades and their definitions in a way that minimises the sum of the variance of default probability and the estimation error of default probability (see Figure 10).

The other problem is that of concentration of exposures in one or two grades. In a survey of credit risk rating at large US banks, 16 large institutions out of 50 institutions surveyed assign half or more of their rated loans to a single risk grade.

Pooling exposures
It may be costly to assign ratings to each transaction. In the case when a rating is not assigned at the transaction level, the rating or

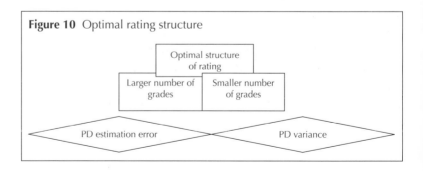

Figure 10 Optimal rating structure

grades should definitely be assigned at the pool or portfolio level. The bank should establish criteria to decide. In general, pooling or segmentation is done on the following three criteria.

❑ *Risk segment*: Basel has identified the risk segments and sub-segments and has also defined the criteria for classifying an exposure to a particular risk segment.
❑ *Borrower characteristics*: Borrower characteristics of each sub-segment may be different. For retail, the recommended characteristics are loan to value, age from origination, geography, origination channel, etc.
❑ *Risk profiles*: These may be based on the credit rating, loss rates or the default rate measures. They can also be based on the measures of past due days, credit scoring, statistical models, expert models, etc. For defaulted assets the segmentation should be based on the risk profiles based on loss or recovery characteristics.

For reliable statistical characteristics, the risk profile of the transaction should be sufficiently homogeneous. The homogeneity is supported by the degree of granularity and the distribution of exposures across segments. Granularity refers to how finely the portfolio is segmented into differentiated risk pools.

The concentration of exposures in a segment (or segments) does not, by itself, reflect a deficiency in the segmentation system. The segmentation method produces reliable estimates of internal-ratings-based risk parameters. For segmentation, the bank has to perform the following tasks:

❑ delineate the segmentation criteria;
❑ demonstrate that there is little risk differentiation among the exposures within the segment;
❑ review their segmentation system at least annually and have clear policies to define the criteria for modifying the system.

The credit rating process described in this chapter or the risk measurement process described in other chapters is a risk segmentation process.

Facility rating
There are two types of facility ratings: one dimensional and two dimensional (see Table 13).

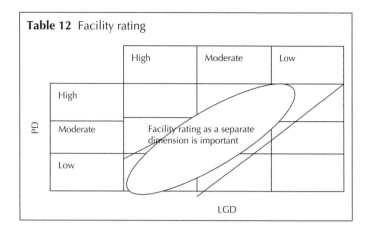

Table 12 Facility rating

Table 13 One- and two-dimensional facility rating

One-dimensional facility rating	Two-dimensional facility rating
Borrower or issuer rating is modified Distinction in credit quality due to recovery or facility is very difficult to indicate	Borrower and facility or recovery rating is indicated separately. This enables distinction of credit quality due to recovery differentiation

One-dimensional facility rating: In one-dimensional rating, the borrower rating is adjusted to reflect the characteristics of the loan transaction.

High PD and high LGD are very risky assets. Low PD and low LGD are AAA type highest-grade assets. The problem is to identify the debt where the PD is high but the LGD is low so that ultimately the credit risk is low (see Table 12).

For the past decade or so the external rating agencies have been providing such ratings as bank loan ratings. Where the recovery or facility is factored in addition to the default risk, this is equivalent to one-dimensional facility ratings.

The one-dimensional rating may be working satisfactorily for the investment grades, where the default risk is very low (so the recovery rating does not matter). However, this may not correctly reflect the credit losses for the regulatory grades or speculative grades.

External ratings cited in Table 10 are the example of one-dimensional rating.

Two-dimensional facility rating: Under the two-dimensional ratings, a borrower's capacity to meet their debt obligation is considered separately from the facility characteristics that influence loss severity. By differentiating the two measures (which are definitely required for the problem assets), risk is measured appropriately and investors are helped to avoid the recovery loss tails. External rating agencies have also started recovery rating in the past three years. This is a refinement of the bank loan rating and is equivalent to two-dimensional facility ratings. The Fed has also recommended two-dimensional ratings for the problem assets.

The recovery or the facility rating emanates from the following sources:

❑ asset quality of the collateral;
❑ loan structure;
❑ debt cushion;
❑ value of the company after default.

For this rating, the post default simulation is worked out and collateral value is stressed and debt a cushion is analysed under the stressed conditions. Asset liquidity provides the difference in the collateral value. Priority claims are subtracted and legal conditions and covenants are analysed.

For example, the Interagency has issued a proposal (2005) on the classification of commercial credit exposures. This framework is based on the two-dimensional facility rating. The proposal divides

Table 14 Two dimensional facility rating

Pass	Marginal	Weak	Default
Pass	Remote risk of loss grade for recovery will convert the entire exposure into pass grade		
For pass grades recovery ratings are not important	Marginal rated assets are criticised assets	Low loss rating convert the exposure into criticised assets	Default rated assets are classified assets with the loss rates of recovery rating
		Weak rated assets are classified assets with the loss rates of recovery rating	

the problem loans into three categories of borrowers ratings and four categories of recovery categories.

❑ *Borrowers ratings*:

　❑ marginal – borrowers exhibiting negative financial trends due to company-specific or systemic conditions;
　❑ weak – exhibiting well-defined credit weaknesses that jeopardise their continued performance;
　❑ default – as per the definition of default.

❑ *Facility rating*:

　❑ remote risk of loss – 0% loss;
　❑ low – ≤ 5% loss;
　❑ moderate – >5% and ≤ 30% loss;
　❑ high – >30% of recorded investment.

One of the factors hindering the spread of use of facility rating is "pooled" collateral and revolving guarantee where each collateral or guarantee is not linked to each transaction but to each borrower. This makes it difficult to assess the exposure of one transaction individually and separately, which is essential for facility ratings.

A two-dimensional system allows banks a far more sophisticated range of responses to risks uncovered on new and existing facilities, and for these responses to apply across a range of lending types. This in turn allows for strategic decisions to be made concerning where to target growth – representing multiple improvements on traditional risk rating systems.

The one-dimensional facility rating approach tended to divide up general and asset-based lending. General lending focuses primarily on default risk and asset-based lending emphasises collateral and recovery (or loss) in the event of default, while two-dimensional facility ratings yield comparable results for all lending.

Where traditional approaches only generate nominal (ie, undiscounted) loss estimates, facility ratings are generally set to estimate real economic losses. Traditional systems build a retrospective loss picture, which is of limited use when it comes to unrated borrowers that have never defaulted in the past. Facility ratings endeavour to identify recovery drivers that can be considered when originating new facilities. Traditional approaches have relied on general understandings – that secured debt is preferable to unsecured, as is more collateral and stronger covenants – while facility ratings actually quantify the degree to which recovery value is improved by these respective factors.

The industry is moving away from the traditional loan-to-value ratio to the facility coverage ratio. The facility coverage ratio is calculated by the following method:

❏ estimate the current and potential exposure including future draws due to revolving credit;
❏ estimate recoveries: recovery in the event of default can come from three sources – realisation of collateral for secured exposures, liquidation or reorganisation of the obligor, or sale of defaulted or distressed securities.

The facility coverage ratio is widely used in bank loan ratings.

Internal rating architecture best practices

Exponential relationship between the rating grade and default rates: No simple and direct relationship (say, linear or log-linear or any other) between rating notches and PD can be defined. However, empirical studies using studies of S&P's and Moody's have found an exponential relationship between PD and rating notch. Therefore, the best practice is to ensure that the relationship among the notches is exponential.

Different rating systems for different portfolios: In general, the credit rating systems are similar but vary considerably in detail. These systems are generally recognised as being reasonably successful at distinguishing the relative risks of different borrowers. There are significant differences across business lines or portfolios.

Cost–benefit analysis of the internal rating system: Establishing and managing an internal rating system is very expensive. Setting up of a rating system should be justified based on the quantum of asset size for a given sub-portfolio. Banks opt for loan syndication, rating through external agencies or other such methods if they do not have the required portfolio size to establish a rating system.

Correlation and maturity: Risk factors, such as loss volatility, or the correlation of risk factors, are generally not taken into account in assigning ratings. Maturity is not explicitly cited as a consideration in the assignment of ratings. However, maturity is generally considered during credit approval.

Internal rating process

Internal rating is a process to distil the information related to the borrowers to measure and compare the risks. The basic steps and assessment criteria are broadly similar to what we have discussed in the generic rating methodology. The internal rating process is similar to the rating process followed by the external agencies. Traditionally, the bank rating architecture differs from the external ratings. However, with the adoption of the Basel Accord, this difference is reducing. The following factors determine whether the borrower is to be rated.

❑ Ratings are applied to those types of loans for which underwriting requires large elements of subjective analysis.
❑ Quantum of the loan: Rating is a costly process, and only large exposures or customers with larger balance sheets are rated.
❑ Quality of the obligor's financial statements: Individual and small and medium-sized enterprises do not have a sufficient quality of financial statement and their assets cannot be ascertained with certainty. The corporates, public sector and regulated institutions generally have higher quality financial statements.
❑ Publicly traded companies have better information disclosure and better quality accounts.

The number of customers to be rated is always small in almost all the banks. Is it possible to rate internally all types of borrowers? The answer is no. This is due to paucity of data and the high cost of rating. Banks sometimes may not even have 100 customers for SL, project finance, asset-based lending and commercial real estate loans. Therefore, Basel has rightly allowed them to rely on the supervisory slotting methodology ie, provided to measure loan risk.

Types of underwriting process

In general there are three types of underwriting processes that are associated with credit rating and loan approval. The credit rating process determines the obligor grade, the loan approval process determines the facility grade and credit audit reviews the final grade and terms for approved facilities.

Credit rating process: Rating a borrower is a process to gather information about their quantitative and qualitative characteristics, that compares them with the standards for each grade and then weighs them by choosing a borrower grade. The comparison or relative measurement is performed with respect to the characteristics of different grades and also with respect to the previously rated loans with characteristics close to those of the loan being rated, and then the rating is set to the grade already assigned to such borrowers. As a part of the rating process, the rater considers both the risk posed by the borrower and aspects of the facility's structure.

Typically, a credit analyst assigns or reaffirms credit ratings as a part of the credit underwriting or as an integral part of the credit approval decision. Ideally, a credit rating process should be totally independent of the loan approval process including a separate approving organisation. In practice, these two processes are intertwined very closely. The rating grade for an obligor determines the limits, covenants and structure of the transaction. Internal rating is generally a four-step process.

❑ The first step is to assess or measure the financial risks using credit scores or a quantitative measurement of risk. This is done generally through the credit scoring model. The quantitative risk models should also consider the impact of the business risks on the financial risks.
❑ The second step is to assess the business risks as explained in the rating process.
❑ The third step is to assess the risk mitigation from collateral seniority and values, credit support and credit enhancement. This step is generally a part of the loan approval process.
❑ Country risk is measured separately for cross border or foreign currency debt. For others, it is measured as a part of the macroeconomic factors and part of business risk.

Rating is generally produced for the following scenarios:

❑ commercial or institutional loans;
❑ large loans to households and individuals;
❑ commercial lease financing;
❑ commercial real estate;
❑ foreign commercial and sovereign entities;

❑ loans and other facilities to the foreign institutions;

❑ loans made by private banking units.

Rating/loan approval process: The rating assignment influences the loan approval process. Underwriting limits and loan approval requirements depend on the rating grade. The rating grade is also confirmed as a part of the loan approval process and the entire approval process must meet the requirements embedded in the bank's credit policies. The credit rating committee approval and loan committee approval will be different if the two processes are different. The number and level of approvers needed for approval typically depends on the size and the proposed risk rating of the transaction: In general, less risky loans require fewer and perhaps lower-level signatures. In addition, signature requirements may vary according to the line of business involved and the type of credit being approved. After approval, the individual that assigned the initial grade is generally responsible for monitoring the loan and for changing the grade promptly as the condition of the borrower changes. Exposures falling into the regulatory grades are an exception at some institutions, where monitoring and grading of such loans becomes the responsibility of a separate unit, such as a workout or loan review unit.

Rating review processes: Once ratings are assigned, there are three types of reviews.

(1) *Continuous monitoring of ratings*: Ratings are monitored broadly against the external events, which helps in generating early warning signals.

(2) *Periodic review of the ratings*: This is in addition to the review at the time of renewal of the facility, which follows the credit rating process. In general, all exposures within an industry are reviewed at the same time by an industry expert. An industry-wide review helps in identifying the inconsistency in rating of borrowers in the same industry. Quarterly, half yearly or yearly reviews are a part of the credit facility renewal process. There are generally two levels of reviews – a review by the relationship manager and a review by the credit committee or credit analyst. All facilities for a group are reviewed at the same time.

(3) *Review by credit audit department*: The rating practices and benchmarks for rating assignment vary from bank to bank. The rating

audit process is important for maintaining the integrity of the rating process. The rating audit department generally reviews only a sample of loans approved. It also examines a sample of loans for each line of practice or sometimes reviews the entire portfolio beyond a certain exposure. The sample is selected on the perception of the riskiness, defaulted or large exposure (say covering the top 30% of the line of business exposure or industry exposure). Review by the credit audit department generally does not require that all ratings produced by the line or credit staff be identical to the ratings assigned by the credit audit department. In general, a two notch difference, either positive or negative, is allowed. The difference of opinion is more when the scale is finer (a higher number of grades). The credit audit department focuses more towards the problem assets and exposure likely to attract the attention of regulators. In some banks the credit audit department assesses the default, watch or regulatory grades. The credit audit department is the final authority to set the credit rating.

Typically, the credit audit department examines each business unit's underwriting practices, and its adherence to administrative and credit policies, on a one- to three-year cycle. As a part of their job, the credit audit department maintains consistency and discipline in the overall rating process. Therefore, when an error is detected, not only is the error corrected but the root cause of the error is identified and corrective action taken. In addition, the credit audit department also has a consultative role. If the relationship manager and credit analysts do not agree on the rating, the credit audit department guides interpretation of the rating definition, standards and their application. In some of the banks, instead of a credit audit, an industry focused review of all exposure belonging to that industry is conducted by the industry experts. This brings more expertise and insights to the ratings and helps in the detection of discrepancies.

Actions triggered from reassessment of ratings: The rating review process may result in reassignment of ratings:

❏ reassignment of a loan to watch or regulatory grades typically triggers a process of quarterly (or even monthly) reporting and formal reviews of the loan;

❑ reassignment to the lower grades impacts the profitability owing to increase in the loan loss provisioning and stopping of interest accrual to the borrower;

❑ reassignment to the default grade triggers propagation of default and recovery of collateral;

❑ reassignment to higher grades impacts profitability favourably;

❑ reassignment owing to restructuring mandates one-time partial write-off.

Calibrating risk drivers to risk ratings: We have explained and dealt with the calibration of risk drivers in the separate chapters.

REFERENCES

Cantor, R. and C. Mann, 2003, "Measuring the Performance of Corporate Bond Ratings", Moody's Special Comment.

Carey, M. and M. Hrycay, 2001, "Parameterising Credit Risk Models with Credit Rating Data", Journal of Banking and Finance.

Interagency Proposal on the Classification of Commercial Credit Exposures Federal Reserve SYSTEM Docket – OP 1227.

Credit Risk Mitigation

INTRODUCTION

One of the primary differences between the Basel Accord I and II has been the recognition of risk mitigation techniques. Paragraph 109 of the Accord has identified the risk mitigation techniques as follows:

> Banks use a number of techniques to mitigate the credit risks to which they are exposed. For example, exposures may be collateralised by first priority claims, in whole or in part with cash or securities, a loan exposure may be guaranteed by a third party, or a bank may buy a credit derivative to offset various forms of credit risk. Additionally banks may agree to net loans owed to them against deposits from the same counterparty.

Therefore, credit risk is mitigated, reduced or transferred using various techniques and instruments:

❏ credit-quality enhancement through guarantees, letters of credit, liquidity support;
❏ reduction in credit risk through collateralisation and netting;
❏ transfer of credit risk through credit derivatives and credit insurance; and
❏ sale of credit risk through loan trading, loan syndication and whole-loan sale, and single-name credit sale.

Credit risk mitigation instruments can be used as a hedge for the underlying credit risk or can be sold and purchased on a stand-alone basis (without underlying). In this chapter, the discussion will focus on the hedging instruments. Instruments purchased/sold otherwise should be dealt with like any other exposure. The risk-measurement techniques discussed here are useful for both as hedging instruments and as independent exposures.

Typical aims of using credit risk mitigating techniques include:

❑ to reduce the credit risk – save economic and/or regulatory capital;
❑ to reduce the large exposures and free up the credit lines with the existing counterparties; and
❑ generate the funds for further investments.

Credit risk mitigating instruments are financial instruments. They are also exposed to credit, market and operational risks like the underlying instruments. However, the quantum of each of the risk may be different. The approach to measurement and management of the risks in the risk mitigating instrument may not be same. Firms dealing in instruments such as credit derivatives (CDs) may manage the CD portfolio from the market-risk perspective and generally focus on the daily changes in the market values, value-at-risk (VAR) and stress tests due to changes in the interest rates and spreads. Firms measure credit risk only if they intend to hold the assets for a longer period. In this chapter, the approach is to evaluate the credit risk of the underlying and the credit risk mitigated by risk mitigating techniques.

With the faster developments in the financial engineering and financial markets, it is very difficult to distinguish between the risk mitigation techniques or instruments. In general, credit risk mitigation techniques can be analysed on the dimensions of credit support and credit structures; funded and non-funded. Since we are talking of financial instruments, there can not be any financial instrument which does not have funding. The question to be analysed is: is it upfront and is it from risk purchaser to risk seller or *vice versa*. A widely accepted description of "funded risk mitigation" has been that upfront funds flow from purchaser to seller.

The focus of this chapter is to analyse, understand and measure the effectiveness of the risk mitigation instruments, techniques and contracts. The risk mitigation instruments and techniques typically establish a credit-protection relationship with the provider of risk mitigation. The protection provider may be the lender itself (that is, collateral, netting and guarantees) or a third party (credit risk

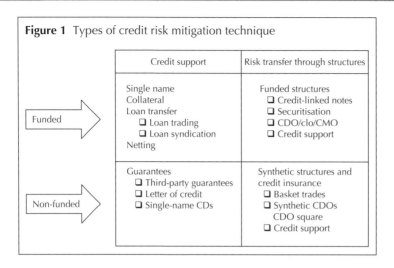

Figure 1 Types of credit risk mitigation technique

	Credit support	Risk transfer through structures
Funded	Single name Collateral Loan transfer ❑ Loan trading ❑ Loan syndication Netting	Funded structures ❑ Credit-linked notes ❑ Securitisation ❑ CDO/clo/CMO ❑ Credit support
Non-funded	Guarantees ❑ Third-party guarantees ❑ Letter of credit ❑ Single-name CDs	Synthetic structures and credit insurance ❑ Basket trades ❑ Synthetic CDOs CDO square ❑ Credit support

insurance, credit default swap and securitisation). Third-party protection contracts typically suffer from asymmetric information, principal–agent problems and incomplete contracts. This creates problems of moral hazard and adverse selection. Later, we discuss problems of the effectiveness of risk mitigation techniques.

Typically, the following factors are analysed in order to understand effectiveness of each risk mitigation technique used.

❑ *Unbundling credit risk*: Risk mitigation typically does not cover the entire credit risk. It covers only some of the credit risk components. Credit risk components may also not be mitigated in full (see Figure 1). The adopted risk mitigation technique itself may generate credit and other types of risks. Therefore, the credit risk of the underlying and risk mitigation techniques needs to be unbundled to recognise, measure and manage the residuary risk in the form of basis risk, legal risk, counterparty risk and operational risk.

❑ *Documentation and legal contract*: Risk mitigation is built on the credit and legal structures. The contractual documents and standardisation of contractual terms play an important role in risk mitigation.

❑ *Valuation of credit risks and financial assets*:

 ❑ cashflow (both underlying and risk mitigation) during the entire life cycle of the credit risk mitigation technique used;

❑ definition of credit events, and triggers;
❑ potential exposure and counterparty risks; and
❑ valuation of loss or claim and settlement.

ISSUES IN THE EFFECTIVENESS OF RISK MITIGATION TECHNIQUES

Structure innovations, nonstandard contracts and partial transfer of risk have created a series of issues in the effectiveness of risk mitigation, which we now turn to.

Risk mitigation may not reduce the risks to the desired extent

Risk mitigation through structures may not lead to the desired reduction in the risks (see Figure 2). Risk mitigation through structures has two important constraints. First, full risk is rarely transferred for reasons of moral hazard and the inability to measure the risk or complete the unbundling of the credit risk mitigated from the credit risk retained. Some risk is always retained. The risk transferred through structures and risk retained rarely add up to original risk due to residual risk, legal risk, counterparty risk and operational risk (all these risks are nonexistent otherwise).

Secondly, even if the entire risk is transferred (which is rarely possible practically), the impact depends upon the correlation of

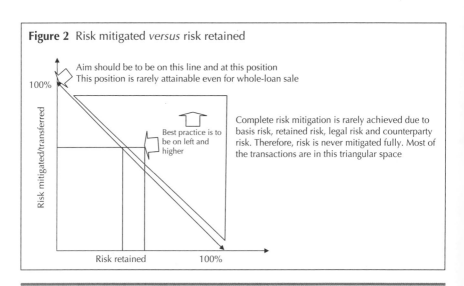

Figure 2 Risk mitigated *versus* risk retained

Aim should be to be on this line and at this position
This position is rarely attainable even for whole-loan sale

100%

Risk mitigated/transferred

Best practice is to be on left and higher

Complete risk mitigation is rarely achieved due to basis risk, retained risk, legal risk and counterparty risk. Therefore, risk is never mitigated fully. Most of the transactions are in this triangular space

Risk retained   100%

the credit risk of the assets transferred with the credit risk of other (retained) portfolios of the bank. It means that, for the same assets and similar risk-transfer contract, the actual risk transferred varies across the transferring banks. The marginal risk contribution (MRC) concept explained in the chapter on "Credit Risk Portfolio Models" is also applicable here.

Full risk mitigation is very rare: due to various constraints credit risk can not be transferred fully. Also, as we have seen, the risk retained and risk transferred do not add up to 100%. When the risk is transferred, basis, legal, counterparty and operational risk get added.

Basis risk

❑ The event triggering the default should be relevant from the underlying or hedged-assets perspective. Events for hedged assets in the trading book and hedged assets in the banking book are different. In the trading book, mark-to-market is relevant and, for the banking book, assets restructuring, bankruptcy and default are more relevant.

❑ Reference obligations and entities: Events triggering payment are tied to reference obligations and entities. Reference obligations and entities are generally different from the bank's portfolio. To reduce the basis risk, the all the obligations and entities in the underlying portfolio should be part of the reference obligations and entities. The next best is if obligations are not same atleast entities should be same, still the next best is to have in the reference portfolio entities belonging to the group of the entities or the guarantor of the underlying portfolio.

❑ Maturity mismatch – protection purchased expires before the asset's final repayment date.

❑ The bank's internal definition of default and the contracted credit event triggering the contingent payment differ. The contracted credit event definition should be stricter.

❑ Currency mismatch – the currency of the hedged obligation and that of the credit derivative contract are different. In case of mismatch, there should be a currency hedge to support the currency mismatch.

❑ To make a cash settlement, the contract should describe a price discovery mechanism.

Legal risk

❏ Restructuring events were the primary source of disputes. In the past few years, the International Swaps and Derivatives Association (ISDA) documentation has standardised restructuring events. Banks have been successfully using ISDA documents to avoid the disputes. For example, when WorldCom filed for bankruptcy in July 2002, there were around 600 credit default swap (CDS) contracts with a notional sum of approximately US$7 billion on WorldCom as a reference entity. There were around 4,000 CDS contracts for an approximately US$10 billion notional amount on Parmalat SPA as a reference entity, when it defaulted in 2003. In both these credit events, contracts were settled without settlement problems, disputes or litigation.

Counterparty risk

❏ This is the risk that the protection seller is not able or willing to make contractual payments. This is a major issue if reference assets and counterparty credit quality are correlated. Counterparty risk is typically managed through the posting of collateral and not through adjustment of the price of protection.

Operational risk

❏ All types of credit risk transfers still have limited straight-through processing, and processes may not be well established. With growing volumes, CDs are exposed to a larger quantum of operational risks. Banks have started managing operational risks on the credit derivative desks through risk-control self-assessment.

Information asymmetry

Information asymmetry creates problem of adverse selection and moral hazard. It leads to adverse selection if, in spite of doing its best, a bank cannot get the information and reduce the risk. It is a moral-hazard problem if the bank has the information but it is not sharing or does not bother to obtain information. Typically, adverse selection is against a lending bank and moral hazard is against a protection seller. Since adverse selection is very difficult to manage and measure, according to Duffee and Zhou (2001), the issue is still unresolved whether use of credit derivative is beneficial or harmful to the bank purchasing the protection. They have also argued that the asymmetry of information is not same during the entire life of the loan. Generally, during the repayment period or at the time of maturity of loans, better information is available. Therefore, to manage adverse selection, the lending bank may purchase the credit protection during the earlier period and take the credit risk during the last years of loan. According to Gorton and Pennacchi (1995), the moral hazard problem can be reduced if the purchaser of the protection holds some fraction of risk.

At the heart of adverse selection and moral hazard is the conflicting interest between risk purchaser and risk seller. The risk seller may be less interested in the prevention or control of losses that it can pass on to the risk purchaser. The industry has realised this conflict and has been devising various ways to reduce the conflicting interests.

The Committee on the Global Financial System (2003) has identified the principal–agent problem, the incomplete-contracting problem and the information-asymmetry problem as issues for ineffective risk mitigation techniques.

Credit support reduces effectiveness of risk mitigation
Credit support can be indirect or direct. Direct credit support enhances the credit quality. Credit support is provided through posting assets or rights over assets or providing liquidity, or indirectly through guarantees or letters of credit or through representations from a better creditworthy guarantor. Credit support is also provided through credit structure.

The following credit support clauses limit transfer of risk in a structured arrangement.

Early-default clauses: Early-default clauses typically give the purchaser of a loan a right to return the loan to the seller if the loan becomes 30 or more days delinquent within a stated period (say six months) after the transfer.

Clean-up calls: A clean-up call is an option to investors to get paid for their remaining investment in a securitisation before all the transferred loans have been paid. Clean-up-call support may be provided by the originator, servicer or third party. Generally, clean-up calls are treated as recourse to the originator (for the lower quality loan which is yet to be repaid) and provide a mechanism to substitute the lower quality loans still outstanding.

Liquidity support: There are two types of liquidity support – support for the assets that are less than and those that are more than 90 days overdue. Liquidity support for the assets overdue for less than 90 days is also called funding of a cash-collateral account. If there is a shortfall in loan collections in any period that prevents asset-backed note holders from being paid, the cash-collateral account may be drawn down for the payments due to investors on the asset-backed notes issued. Losses that are overdue for 90 days or more may also be absorbed by the originator to the extent guaranteed by first-loss clauses or by the equity and subordinated tranche.

Retained subordinated interest: The sub-ordinated interest is retained by the originator to protect the investors from the losses. It is represents additional loss to be borne by the originator over above the pro rata loss-share of the assets retained by the originator. The retained interest is applicable to securitisation, credit derivative or other structure built to transfer the credit risk. This also includes repurchase agreement on the sold assets.

Recourse: Risk retained by the bank in form or substance, directly or indirectly associated with an asset it has transferred.

Residual risk: The credit risk to which a bank is exposed due to credit-support or credit-enhancement techniques. The exposure may be direct or indirect. The residual risk is the credit risk in excess of the pro rata share of the bank's claim on the asset.

Early amortisation: Structures generally provide for early start of the amortisation phase if the deal performs poorly. Generally, structures provide for an over-collateralisation test and interest-coverage test. If either ratio falls below a specified threshold, early amortisation is triggered. Early amortisation includes the waiver of contractual charges to be paid to the sponsor for servicing the hedged assets and/or winding up the structure. Winding up of the structure reduces the time period for which investors are exposed to the credit risk.

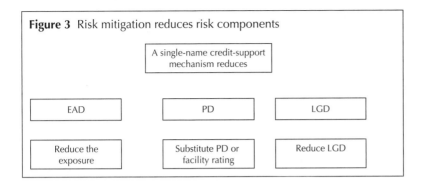

Figure 3 Risk mitigation reduces risk components

Over-collateralisation: This is a form of credit enhancement covering a predetermined amount of potential credit losses. Over-collateralisation means assets available for satisfaction of the claim exceed the face value of the securities.

Risk mitigation techniques reduce risk components. This is illustrated by reduction in the capital computation allowed by the Basel II Accord.

Optimisation of risk mitigating technique to risk components is attempted because measure of risk component varies with risk segment/asset class. Banks optimise risk mitigation allocation within the legally enforceable contracts with obligors; in other words, banks can allocate the collaterals so as to reduce capital to the maximum extent possible. However, no double counting is permitted. No double counting means, if a credit risk mitigant has been considered for reduction in any of the risk components, it cannot be considered again. Optimisation of risk mitigation across the various facilities for the same obligor or across the obligors within a group has emerged as a major requirement for collateral management software.

Correlation of risk mitigating instruments with the underlying

Quantum of correlation has emerged as a major area of concern in risk mitigation. For example, collaterals and guarantor should not have high positive correlation with the obligor. Correlation measurement is a critical element in every credit structure. And we know that correlation is the most difficult risk component to be measured. Financial industry has been developing correlation-measurement techniques. However, many assumptions are still made to measure correlation. The correlation trading market is also evolving and hopefully development in this market will improve the correlation measurement techniques.

LOAN SALE

One of the simplest techniques of risk mitigation for defaultable assets is to sell off those assets. Loans and bonds are the two broad categories of defaultable corporate asset. The difference between the two has been the existence of the secondary market. The secondary market for bonds is well established, well developed and mature; the secondary market for loans has been developing over the past decade or so.

Let us study the typical attributes of loans that have hindered and helped in the development of secondary loan market.

❑ Loans typically are of a lower rating grade, so making them saleable typically requires collaterals. Bonds are typically investment-grade and unsecured. Therefore, lenders typically have a better legal structure, nature and quality of collateral for loans.

❑ No doubt loans are supported by collaterals. However, collaterals are relevant only in effective bankruptcy regimes. A secondary market for loans still does not exist in many countries. The primary reason for the underdevelopment of the loan market has been the nonexistence of a bankruptcy regime. Countries such as the USA with a well-developed bankruptcy regime have developed a liquid loan-sale market.

❑ Covenant structure – covenants for loans are stricter than for bonds.

❑ The lenders in the loan markets have advantage over the investors in the bond markets by way of being privy to the private information about the obligor.

❑ The loan-transfer or loan sale is constrained by the relationship with the customer. Loan-transfer structures are evolving in such a way that the lead manager retains the relationship while a part of the loan is syndicated or sold in the secondary market. Figure 4 shows sophistication in the loan sale for corporate loans.

❑ Secondary loan markets are now being extended for trading in the retail and SME loans through whole-loan transfers, securitisation

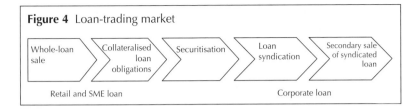

Figure 4 Loan-trading market

Whole-loan sale → Collateralised loan obligations → Securitisation → Loan syndication → Secondary sale of syndicated loan

Retail and SME loan   Corporate loan

of credit card receivables and housing mortgage receivables through collateralised loan obligations.

Whole-loan sale

This is a mechanism for transferring risk associated with loan origination. Generally, residential mortgage loans and auto loans are sold to third parties in a whole-loan sale transaction. The actual amount of risk transferred depends upon the correlation of the assets transferred with the remaining portfolio of the bank, and the sale contract and retained risks or guarantees under the contract. It is generally assumed that in a whole-loan sale a greater part of the risk is transferred. However, it may be noted that the whole-loan sale may impact on the business model of the seller. Here are some of the factors to be analysed to assess the extent of risk transferred.

❑ *Arm's length transaction with third party*: This is assessed from the sale documents and pricing of the transaction. There should not be any obligation on the seller of the loans to repurchase loans and purchase should not have any recourse to sell back loans or part of the loans.
❑ *The remedy provided in the contract for breach of any representations and warranties by the seller is limited to the repurchase and substitution of the loans*: The intention of this provision should be explicitly to limit the seller's support, since the seller may remain exposed to the underlying performance of the loans if it is forced to replace a significant number of loans. A key determining factor against the claim for breach of representations and warranties is the strength of the seller's operating and documentation procedures. With strong loan processes and controls in place, the seller will be better able to reduce breaches of representations and warranties so as to minimise the necessity for repurchase or substitution of the loans.

Even if both these conditions are met, a potential offset to the arm's length nature of a whole-loan sale is the implied support. As is the case with securitisation, the true sale character of a whole-loan sale can be undermined by a seller if it supports its underperforming sold loans. Practically, although the loans are sold on a non-recourse basis, a seller may feel some pressure to prevent investors from taking losses in order to preserve its reputation, the price at which it may be able to access funding in the future, and, ultimately, market access itself.

Loan syndication

Syndicated loans are credits granted by a group of banks to a borrower on a single-loan-agreement contract but a separate claim on the borrower. Syndicated loans combine features of relationship lending and publicly traded debt. It is a mechanism for sharing the credit risk among various financial institutions. Since it is a relationship loan, it presents a smaller burden of disclosure and marketing upon the borrower compared with bond issuance. With standardisation of documents, contracts and loan-trading practices, the syndicated loans are also traded in the secondary market.

Unbundling of credit risk

Based on the seniority of the claim, the lending banks are divided into two groups: senior syndicate banks and others. The senior syndicate members typically work as mandated arrangers, lead managers or agents. Senior banks are appointed by the borrower to bring together the syndicate of banks to lend money at the terms specified by the loan. The syndicate is formed around the arrangers or the senior members. Senior banks are, typically, the banks with which the borrower has relationship. Senior banks contribute a portion of the loan. The other members' roles will vary according to the size, complexity and pricing of the loan as well as the willingness of the borrower to increase the relationship with other banks.

Syndicated credits lie somewhere between relationship loans and disintermediated debt. There are several reasons for loan syndication – from fee earning to avoiding excessive single-name exposure, to comply with regulatory limits on risk concentration etc.

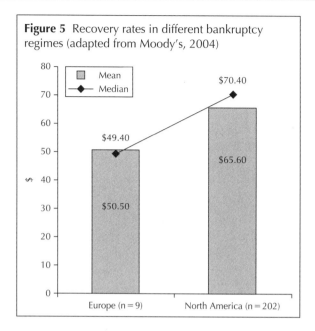

Figure 5 Recovery rates in different bankruptcy regimes (adapted from Moody's, 2004)

Better collaterals are provided in syndicated loans. Recovery rating or LGD plays a determining role in the risk analysis. Better collaterals lead to better recovery ratings and improve the overall rating. Borrowers with investment-grade ratings are unlikely to offer the collateral needed for a syndicated loan transaction, so investment-grade companies rarely borrow on syndicated loans. Therefore, the syndicated-loan market is more developed in the legal jurisdiction with a mature bankruptcy process. The recovery rate is determined by the capital structure of the borrower and efficiency in the bankruptcy process.

The starting point for rating syndicated loans is the obligor ratings. The methodology is broadly similar to that explained in the chapter "Credit Rating". The next step is an analysis of collateral and the legal contract. The recovery rating generally determines the rating for issue.

Collaterals provided for syndicated loans are generally not financial assets. Collaterals are generally physical or other business assets. Broadly there are two types of collateral: those whose value is impacted by the bankruptcy of the business and those whose value not impacted to a great extent. The value of inventory,

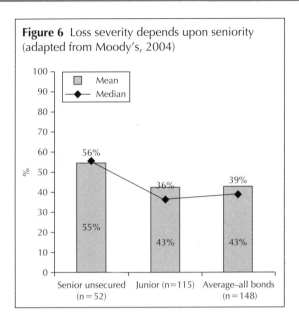

Figure 6 Loss severity depends upon seniority (adapted from Moody's, 2004)

machines or the shares of the enterprise becomes impacted by bankruptcy. Value of accounts receivable, real estate and vehicles is independent of the bankruptcy. Recovery and collateral are relevant only after default. Therefore, collateral value under the default scenario is relevant. In addition to the collateral value, collateral seniority is also relevant.

Both the default rate and recovery rates are significantly lower for loans compared with bonds. This has been empirically tested by rating agencies.

Documentation

Syndicated credits are increasingly traded on secondary markets. The standardisation of documentation for loan trading, initiated by professional bodies such as the Loan Market Association (in Europe) and the Asia Pacific Loan Market Association, has contributed to improved liquidity in these markets.

A measure of the tradability of loans on the secondary market is the prevalence of transferability clauses, which allow the transfer of the claim to another creditor. The US market has generated the highest share of transferable loans (25% of total loans between 1993 and 2003), followed by the European marketplace (10%).

The secondary market is commonly perceived as consisting of three segments: par/near par, leveraged (or high-yield) and distressed. Most of the liquidity can be found in the distressed segment.

Pricing syndicated loans
Funded
❏ Receive coupon and principal until default/maturity.
❏ Coupon = Libor + spread + facility fee + utilisation fee.
❏ Receive recovery at default.
❏ Spread may be rating-dependent.

Unfunded
❏ Receive fee until default/maturity.
❏ Total fee = commitment fee + facility fee.
❏ Pay loss amount on default = draw − recovery amount.

Credit line
❏ Funded and unfunded amounts are time-dependent.

COLLATERAL
Collaterals are the assets transferred to the lender or charged with the lender's right to recover its exposure by selling the collaterals. Collaterals provide an additional source of cashflow for the lender. Collateral value, volatility in values, liquidity, and recovery right on the collateral are the key factors in credit risk mitigation through the use of collateral. Collateral will be able to reduce the credit risk only if it satisfies the following:

❏ reliability in valuation – collateral can be revalued reliably preferably on the basis of periodically published market guide prices;
❏ lower volatility – collaterals maintain their value over time and have volatility comparable to equities, bonds or gold;
❏ legal certainty – the documentation must be binding on all parties and must be legally enforceable in all relevant jurisdictions; and
❏ liquidity – the liquidation cost of collateral is reasonable.

Unbundling credit risk
Valuation and liquidity are far better in financial assets than the physical assets. Within the financial assets, volatility in superior grades is

smaller than speculative grades. To get better capital measurement, the Basel Accord recognises only the higher-grade financial assets as eligible collaterals for capital computation. Other collaterals (physical assets and lower-grade financial assets) also provide risk mitigation; however, such mitigation is comparatively less effective.

Collateral has a twofold impact. As an important driver of facility and recovery rating, collateral improves the final risk rating and quality of an obligation before default, and extent of impairment and recovery rating after default. This results in a reduction in the risk premium demanded, loan-loss provisioning (impairment is net of collaterals) and regulatory capital. The regulatory capital for the past due assets (defaulted assets) is required only for the unsecured portion of the assets.

Collateral management is very much a team effort within an organisation, and it is important that all members of the team understand the information flow.

There are two ways to view collaterals.

❑ To hedge credit exposure or potential credit exposure, collaterals are posted by counterparty or by a third party on behalf of the counterparty.
❑ Collateralisation is a credit-enhancement technique and a means of mitigating credit risk. Credit-enhancement techniques range from simple, transaction-specific risk-reducing measures and extend to sophisticated portfolio management techniques (such as the use of credit derivatives).

Collateral valuation

Collateral and the underlying transaction should not be highly positively correlated. Every collateral value is volatile. Volatility in financial collaterals can be measured with precision. Volatility measurement is more reliable in liquid markets. For these reasons, Basel recognises only higher-quality financial collaterals. Basel has recommended a 99th-percentile, one-tailed confidence interval for volatility or value measurement. It has prescribed a framework to measure collateral value, which is represented as Figure 7.

Standards for the holding period for each type of assets should be supported by the liquidity for the asset. Typically 5, 10 or 20 days' holding period is used.

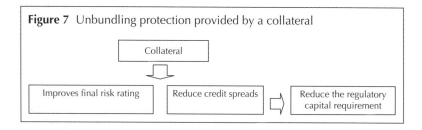

Figure 7 Unbundling protection provided by a collateral

The best-practice historical data are for a minimum of one year and the data should be refreshed at least once in three months. Haircuts are applied for each type of the collateral volatility.

Both price and currency volatilities are price-related. The standard-deviation framework (VAR models) at a certain confidence level is used to measure the haircut. Using VAR models and assuming typical volatilities for assets, the Accord has computed standard haircuts.

The framework for semi-financial and physical assets should also be similar. However, as one moves from financial to semi-financial to physical collaterals, it is more difficult to measure volatility, and volatility measurement is based on approximations and assumptions.

Non-recognised collaterals do not get the same treatment as recognised collaterals; nevertheless, they reduce the credit risk. The Accord recognises their contribution to risk reduction and has considered them even for capital requirement for past due loans (Paragraph 90 of the Accord) and reduced LGD estimation (Paragraph 287 of the Accord).

Netting as a form of collateral

For credit risk mitigation to be effective, the bank should have a legal contract to net the mitigant provided against the exposure. Mitigation in the form of collateral is driven by the charge documentation; mitigation in the form of other deposits and payables should have an on balance sheet netting contract; for guarantees, collateral posted and credit derivatives generally, ISDA documents with netting clauses are used.

For the positions to be netted, both should ideally have the same volatility (should be in the same book). Positions with varying degrees of volatility (positions in a banking book have different volatility as compared to positions in a trading book) can be netted if their volatility is brought to same level by daily mark to market.

Any legally enforceable form of bilateral netting (that is, netting with a single counterparty) to be eligible for netting may require the following conditions to be satisfied.

A valid contract
❏ A contract to create single legal obligation.
❏ Legal opinion supports the validity of netting contract.
❏ The bank has a policy and procedure to review continuously the existing netting contracts based on the legal developments.
❏ The netting agreement is valid under the relevant jurisdiction.

Event of default definition and triggers
❏ Clear definition of event of default.
❏ Computation of the net payable or receivable on the contract.
❏ Rights of the non-defaulting party to terminate and close out the contract.
❏ Liquidation and set-off of collateral.
❏ Computation of value of the mitigation using the methodology as applicable to the mitigant.

CREDIT DERIVATIVES
CD are the financial contracts for exchanging payments; in credit derivatives at least one leg of the payment is linked to the performance of a specified underlying credit-sensitive asset. CD provide a mechanism to "unbundle" credit risk from the credit instruments. CD enable trading in the credit risk separately. There are two parties in a CD: the protection seller and the protection buyer. The protection seller receives premiums or interest-related payments from the protection buyer in return for assuming the credit risk on an asset or a group of assets. The protection buyer a receives credit-event payment from the protection seller. Credit event is a scenario or condition agreed between the contracting parties; credit event triggers the credit-event payment from a protection seller to the protection buyer. A credit event typically

❏ triggers payout by the protection seller; and
❏ determines the deliverable obligation in the event of the payout.

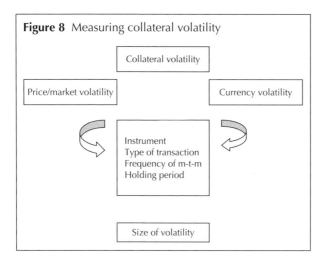

Figure 8 Measuring collateral volatility

CD contracts are linked or referenced to the reference asset. Underlying assets refer to the assets on which credit risk protection is bought. To reduce the basis risk, the reference assets and the underlying assets should be same. Many times both are different.

Compared with loan trading and loan selling, credit derivative transactions are confidential. A reference entity whose credit risk is transferred is neither a party to a credit derivative transaction nor is even aware of it. This confidentiality enables risk managers to isolate and transfer credit risk discreetly, without affecting business relationships. Typically, credit protection through a CD is purchased for existing loan. In a typical credit derivative transaction, the protection purchaser (the creditor's bank), for a fee, transfers some or all of a credit risk to the protection seller. Typically there are three types of CD, as depicted in Figure 8.

Structural innovation, as arrangers look for different ways to place different parts of the capital structure to different types of investor, makes it extremely difficult to distinguish between the synthetic and cash credit derivatives.

According to one estimate, single-name credit default swaps represent about 60% of the total volume of credit derivatives; credit derivative index products represent about 25%; credit default options, first-to-default baskets, synthetic collateralised debt obligations (CDOs) and tranched credit products account for the remaining 15% of the credit derivatives market.

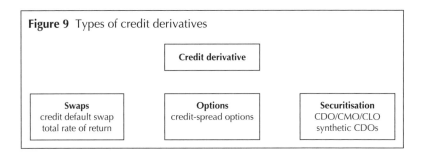

Figure 9 Types of credit derivatives

	Credit derivative	
Swaps credit default swap total rate of return	**Options** credit-spread options	**Securitisation** CDO/CMO/CLO synthetic CDOs

Most liquid tenor is 5–10 years and accounts for 70–75% of the volume. Remaining volumes are equally divided between less than 5 and more than 10 years' tenor. As we have seen in the section on effectiveness of risk mitigation, CDs reduce the credit risk but simultaneously increases residual risks such as basis, legal, operational, liquidity and market risks.

In the remaining part of this chapter we shall discuss credit derivative swaps which are the most important CD instrument.

Documentation and legal contract

Typically, ISDA documentations, standards and definitions are used for CD contracts. Definitions of credit events and valuation of the contingent payments are the most important elements in a CD contract.

Credit event

The contractual *credit event* definition determines the extent of risk transference from the protection buyer to the protection seller. To ensure that credit event occurring on the underlying asset triggers the event payment under the terms of the credit derivative, the set of credit events should contain as wide a range of triggers as possible. If the set of credit events or the set of credit triggers are restrictive, the credit derivative hedge transfers insufficient risk.

The other important definition is the size and nature of any *materiality threshold*. Materiality threshold requires a given level of loss to occur before a payment on credit derivatives is triggered. If these thresholds are set too high, a significant loss will be incurred on the underlying asset without a credit-event payment being received. CD contracts generally cover the following events:

❑ failure to pay;
❑ write-down or principal reduction;
❑ ratings downgrade;
❑ bankruptcy; and
❑ restructuring.

In addition CDs also cover the following events:

❑ obligation acceleration;
❑ obligation default; and
❑ repudiation/moratorium.

Finality of credit events

Contractual terms should define the finality of the credit event. The acceptable finality of credit event actually comes from:

❑ a well-known news source;
❑ corporate filing; or
❑ a court document.

Restructuring event

It is quite possible that a restructuring event may be taken as a credit event according to the contractual definition but may not diminish the financial obligation. Restructuring is the toughest event to handle – it is very difficult to define restructuring event in the contract and it is also difficult to define it in the practice. This is because restructuring is a soft event, where loss to the owner of defaultable assets is very difficult to measure; and restructuring builds a complex schedule of maturity and repayment of the debt.

Sometimes restructuring creates an opportunity to make profits for the protection buyer. Restructuring is driven by the contract terms between the seller and buyer of the protection. In parallel with the rapid growth of the CDS market, the menu of contractual terms available to the parties to a CDS contract has expanded. In the past five years or so, the ISDA has attempted to address the challenges of restructuring by publishing four types of restructuring terms.

❑ There are two extremes – *full restructuring* and *non-restructuring*. The non-restructuring is gaining acceptance among participants. Some of the CDS indexes of North America are traded under non-restructuring clause.

CREDIT RISK MANAGEMENT AND BASEL II

❏ Between the two extremes there are two types of acceptable restructuring. These are called *modified restructuring*. Both types of modified restructuring limit the extension of the maturity period of the contract. After the termination date of the contract the deliverable obligations are restricted to 30 months for the modified restructuring and 60 months for the modified-modified restructuring.

Valuation of a contingent payment

If a credit event occurs, there are two types of settlement: physical and cash. In a physical settlement for the credit loss, the protection buyer transfers the defaulted obligation to the protection seller and the protection seller pays the par value of the debt to the purchaser of the protection. For a cash-settled credit derivative, to measure the credit losses, the three most important factors to help measuring the losses of protection purchasers are the following.

❏ The time horizon within which losses are to be valued – immediately after the date of default, lesser information is available in the market. Only after some days of the default, a larger set of information about the obligor is available. The contract should define the time period for which prices and other information will be considered to measure credit losses.
❏ Recovery assumption – this varies according to the seniority and covenants. Recovery assumptions also vary across the bankruptcy jurisdiction.
❏ The referenced obligation and the obligations having same seniority with the reference obligation drive the credit events and not the underlying obligation.

Unbundling credit risk

Credit default swaps provide default insurance for defaultable securities. The CDS protection seller pays the contingent amount to the protection purchaser at the time of a credit event. The contingent amount is generally the difference between the face value and market value of the bond. In return for the protection, the purchaser pays the premium.

Credit risk should be unbundled effectively to transfer the risk of loss due to the changes in the credit quality, including default.

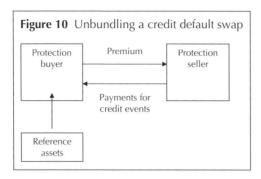

Figure 10 Unbundling a credit default swap

However, the skills and tools available to the purchaser and seller may not be very effective in unbundling.

Valuation of credit default swap
For analysing the pricing of credit default swaps, the cashflow can be divided into two: cashflow as on the date of the origination and cashflow during the life of a CDS, which includes cashflow due to credit events. As on the date of the origination, the aim of the CDS valuation is to measure the risk premium for the credit risk protection promised in the CDS contract. This is to measure the CDS spread, or the CDS premium.

During the life of a CDS, one or more of the following change:

❑ credit quality of the underlying;
❑ risk-free interest rates; and
❑ premium demanded (credit spreads) for a given credit-quality change.

The promised premium to the protection seller remains constant. This changes the CDS value. To model CDS valuation, the Darrel Duffie model makes the following assumptions.

❑ There exists a default-free floating rate note (FRN).
❑ For a default-free FRN, there is no embedded interest-rate swap.
❑ To get efficiencies of the financial market following assumptions are made by the model:
  ❑ no transaction costs, no taxes; and
  ❑ it is cost-free to the short underlying FRN.

Purchasing a protection through a CDS of a notional amount of US$100 is equivalent to purchasing a default-free FRN of US$100

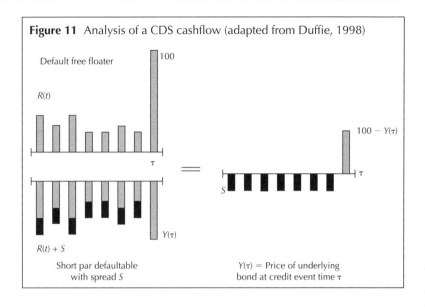

Figure 11 Analysis of a CDS cashflow (adapted from Duffie, 1998)

with a risk-free floating interest rate R(t) and US$100 repayable after the period T from the protection seller and selling a default-able FRN to the protection seller. The risk seller receives the coupon of R(t) + S (the risk spread) and in case of default the difference between the face value and market value is paid by the protection seller. Duffie extends the reduced-form model to model credit events as a hazard rate in terms of the Poisson process.

Trading in CDs

CDs are working as instruments for change. For the banking industry, CDs are not working as an instrument to transfer the credit risk or to escape the ills of business/economic cycle since a substantial portion of substandard and equity tranches are held by the banks only and not by the insurance company, as is generally believed.

❑ Banks are retaining first-loss piece. This piece is substantial.
❑ The most actively traded names in the credit derivatives market are investment-grade.
❑ The aggregate amount of risk transfer that has occurred is small relative to banks' and insurers' overall exposures and relative to the notional size of the market.

In addition to using credit default swap (CDS) as a hedge, banks are trading in CDS like any bond. Many banks apply the same VAR framework to CDS as is applied to other traded instruments. That is, there is a focus on the market values of CDS, their volatility and their correlations with other contracts. VAR-based measurement may be good for hedging credit spread or nonlinear movements in prices but not the credit risk for a name. The problem with the VAR-based approach is that swap spreads on a particular name can remain relatively stable for a long time and then change dramatically if the creditworthiness deteriorates. Unlike other derivative contracts, where the notional amount is not a measurement for the counterparty risk, in CDS the entire notional amount is exposed to the credit risk.

CDS spreads provide a market-based mechanism to measure credit risk on a continuous basis. This helps in building indicators for pre-warning and surveillance, which is otherwise not possible for credit risk. However, it may be noted that a CDS spread is not always a true measure of credit risk. Market liquidity plays a major role in the CDS spread in addition to the credit stress. The best practice is to use CDS spread information along with the fundamental credit analysis.

Credit derivative index

JP Morgan in 2001 launched the first index on credit default swaps: the High Yield Debt Index (HYDI). In 2004, index families were established: Dow Jones CDX index family for US and the Dow Jones iTraxx index family for Europe and Asia. Within each index family there are separate indexes for investment-grade and for high-yield-grade.

For example Dow Jones CDX.NA.IG Index (the "IG Index") is composed of 125 investment-grade entities based in North America, distributed among five subsectors. Dow Jones CDX.NA.HY Index (the "HY Index") is composed of 100 non-investment-grade entities based in North America. Each index reflects the performance of a basket of single-name credit default swaps. The CDS indexes are not perpetual. Each index has a fixed composition and fixed maturities. The new indexes are launched periodically, say twice a year. The index continues to trade till maturity.

❑ CDS indexes help unbundle the credit risk and trade in pure credit risk.

❏ CDS indexes provide an efficient mechanism to purchase credit protection. The protection purchased can be on specific credits or a broad index of credits.
❏ CDS indexes provide liquidity for the transfer of credit risk when loan markets may have liquidity problem.
❏ CDS indexes provide freedom to tailor the bonds and investments.

Investors cite liquidity as the key reason for the popularity of CDS indexes, since now CDS indexes have got wider acceptability among dealers and investors, even when the single-name CDS market is less liquid.

CDS indexes are a useful tool for traders in credit risk for taking or hedging macro exposures. Sector-level subindexes are used for taking relative-value positions on the outperformance or underperformance of a given sector, similar to the equity markets.

Since CDS indexes enable trading of credit risk, this is an appropriate point to reiterate that, unlike the market risk, where long and short positions may hedge each other, the difference between protection bought and protection sold is not the way to hedge credit risk. The only way to be completely hedged in credit risk is to be long and short on the same reference entity with minimal basis risk with creditworthy counterparties.

Information systems for credit derivatives

With the increased volume, straight-through processing (STP) is required to automate the credit derivative life cycle, which includes pre-trade, trade and post-trade changes in the trade characteristics, servicing and settlement.

Generally, Excel based models are used for valuation of CDSs. In the past couple of years, information systems for the entire life cycles of CDSs have developed. This has become possible thanks to XML technologies and standardisation through ISDA documentation and (Financial Products Markup Language) FpML messaging standards.

Guarantees

In order to provide effective credit risk protection, the guarantees and credit derivatives must have following characteristics.

❏ They must be clearly defined and incontrovertible. A guarantee or credit derivative must represent a direct claim on the protection provider and must be explicitly referenced to specific exposures or a pool of exposures, and the event, so that the extent of the cover is clearly defined and incontrovertible.

❑ Clear definition of the event helps in triggering the protection. Most of the disputes concern restructuring.

❑ The bank must have the right to receive any such payments from the guarantor without first having to take legal action in order to pursue the counterparty for payment.

❑ The contract must be explicitly documented and assumed by the protection provider.

Securitisation

Securitisation is a process through which a variety of financial assets are packaged into securities that are then sold to investors. The securities issued to the investors supported or backed by the assets and are known as *asset-backed securities* (ABSs). *Collateralised debt obligation* (CDO) is a generic name for asset-backed securities. They are backed by residential mortgages, bank loans, loans to SMEs, credit-card receivables, commercial mortgage and so on. The cashflows generated by the underlying assets are used to pay principal and interest on the securities in addition to transaction expenses.

In addition to being a tool of risk mitigation, securitisation is a type of secured financing. However, it differs from a typical secured lending – in the securitisation transaction borrowers are isolated from investors by the presence of a special-purpose vehicle (SPV) built for securitisation.

One of the important ingredients in developing efficient, deep and liquid markets for securitisation is the standardisation of products and the credit underwriting practices of the underlying exposures. The required standardisation has been achieved for mortgage loans in the US and many other markets. Similar standardisation initiatives are under way for auto financing loans. If a whole loan is sold on a regular basis to the investors using an SPV, the seller becomes more of an originator and servicer of loans rather than a holder of the loans that it originates. The business model, capital structure, risk profile and earnings volatility are significantly different for this originator-cum-servicer.

For a securitisation to be an effective tool for risk mitigation the following three conditions must be satisfied.

❑ *Have a clean break between the originating bank and the securitised assets – economically and legally*: For capital relief, the Basel Accord imposes a number of restrictions on the originating banks to achieve a clean break between the originating bank and the securitisation transaction. This includes transfer of assets, limited roles for the originating bank and separation of the seller of the assets legally and economically from the securitised assets. The very same conditions are applicable for effective transfer of risk. According to Paragraph 13 of the consultative document on assets securitisation (2001), the transferred assets should be legally isolated from the transferor and the separation should be effective even under bankruptcy and receivership. Further, the clean-up call should represent a relatively small percentage of securitised assets. In addition, according to Paragraph 554 of the Accord, significant credit risk should be transferred to the third party, and the transferor should not maintain a direct or indirect control over the transferred assets.

❑ *Do not provide the credit support*: In a securitised deal, the following types of credit support, if provided by the originating bank, severely restrict the risk mitigation:

❑ the first and second loss position;
❑ liquidity support;
❑ clean-up call;
❑ early amortisation; and
❑ increase in the yield payable to parties other than the originating bank on the tranches in response to a deterioration in the credit quality.

❑ *Have the tranches rated*: Tranches to be invested by banks should be rated. Otherwise a look-through principle may be applied for the senior tranche.

Unbundling credit risk

The purpose of securitisation is to repackage the credit risk in the underlying and sell the risk as a part of the investment tranches. Risk transfer is determined by risk contracts, risk ratings of the tranches sold and the tranches retained. Broadly, there are

Table 1 Types of tranche

Senior tranche	A tranche in a CDO having the lowest risk weight or the highest rating
Mezzanine tranche	Tranches in CDO having risk between the senior and equity
Equity	A tranche with highest risk weight or the lowest debt rating

Table 2 Quantum of risk transferred – retained *versus* sold risk

Quality of loan assets retained sold	High	Medium	Low
High	Quantum of risk transferred is high	Quantum of risk transferred is medium	Quantum of risk transferred is low
Medium	Quantum of risk transferred is medium	Quantum of risk transferred is high	Quantum of risk transferred is high
Low	Quantum of risk transferred is high	Quantum of risk transferred is high	Quantum of risk transferred is highest

three types of tranche (see Table 1), depending upon the risk rating.

Risk transference under securitisation is a continuum from the least risk transferred to the most risk transferred (see Table 2). Securitisation transactions, where the first-loss piece and subordinated tranches are retained by the seller, would fall in to the "least risk transferred" category while sales of low-quality assets on a servicing-released basis would fall into the "high risk transferred" category.

Even when a loan is securitised, the originator primarily remains exposed to credit and prepayment risks by retaining an interest in the securitised assets through the first-loss piece (or equity piece) and subordinated tranches in the transaction.

In most of the securitisation structures, the credit risk of a reference portfolio is divided into tranches with different seniority. Mezzanine and equity are junior tranche to senior tranche and equity is junior to mezzanine. The senior tranches are more creditworthy than the mezzanine tranches, which in turn are more creditworthy than the equity tranches because they have a higher priority in receiving cashflows from the reference portfolio and have lower credit risk. The quantum of each measure of risk is shown in the Table 3.

Table 3 Distribution of risks in CDO tranches *versus* the risk in underlying exposures (adapted from BIS, 2005, the Joint Forum paper, "Credit Risk Transfer")

Tranche	Weightage (%)	Spread sensitivity	Expected loss	Unexpected loss	Risk weights (%)
Equity	0–3	15 times	17 times	13 times	1250
Mezzanine	3–10	7 times	5 times	6 times	100–200
Senior	10–100	0.3 times	0.1 times	0.3 times	20

CDO losses are assigned sequentially to the protection seller. The sequence of protection sold starts from the most junior tranche holder and so on. In a CDO, as defaults occur, losses are assigned to the *protection sellers* sequentially. The portfolio loss level at which a specific tranche begins to experience losses is referred to as the tranche's *attachment point*, and the level at which the tranche no longer experiences losses as the *detachment point*.

Table 4 Attachment and detachment point of tranche

Tranche	Marginal losses	Attachment point	Detachment point
Equity	The first 3% of securitised assets	0% loss	3% loss
Mezzanine	The next 3% of the securitised assets	3%	6%
Senior mezzanine	The next 4% of the securitised assets	6%	10%
Senior	The next 12% of the securitised assets	10%	22%

In general, it is difficult to assess the risk that underperforming loans may be supported in the future other than perhaps to observe the seller's pattern of behaviour in any given transaction. If the originator has supported even one existing transaction, its ability to realise any risk transference from a whole-loan sale is limited.

For effective risk mitigation, the junior and subordinated tranches should be transferred. Some of the methods for transferring risks in the subordinated tranches are as follows.

❑ Purchasing insurance and reinsurance for the credit risk – This will reduce the size and existence of subordinated tranches.

❑ There are investors for subordinated tranches and some of the originators are transferring risks through net interest margin (NIM) bonds to the third parties.

❑ Credit derivatives and other synthetic structures and now combined with the securitisation structures to reduce the credit risk.

The Basel Accord aims to align the capital charges to the underlying economic risk of the exposures. The transference of a portion of the risk may result in the release of the capital initially allocated to support the risk. This effect is known as *capital relief*. Capital relief is one of the significant drivers for pricing in the securitised transaction.

❑ In general, lesser capital is required for rated securitised assets as compared with unrated securitised assets.

❑ Within unrated securitised assets, the Accord encourages holding the underlying assets instead of the unrated securitised assets.

❑ The Accord encourages securitisation of commercial mortgages and credit-card receivables. For CDOs, the Accord is neutral on the question of securitisation *versus* the underlying pool of corporate bonds. The Accord is slightly in favour of the securitisation of residential mortgage-backed securities.

❑ The Accord strongly discourages exposures in sub-investment-grade tranches.

The Basel Accord provides capital relief for the assets securitised. Whether securitisation really reduces the capital is always a question for the following reasons.

❑ The first reason is obviously the quantum of risk transferred.

❑ Impact on the retained portfolio: Selling of loan reduces the diversification of the retained portfolio or it reduces the average credit quality of the retained portfolio. The retained risk and the quality of the loan assets transferred determine the amount of risk transferred.

❑ Risk-based service-related compensation exposes a bank to the upside and downside of the loan performance, so affects the measure of the risk actually transferred.

Granularity of the underlying determines the risk in tranches
Granularity and the risk characteristics of the underlying pool determine the risk in a securitised tranche. Granularity plays a very important role in the distribution of risk across the tranches. According to Paragraph 20 of Basel Committee on Banking Supervision (2002),

> … non-granular pools tend to shift greater amounts of systematic risk to more-senior tranches compared with otherwise identical securitisations of highly granular pools. This arises because the less granular pool will tend to exhibit greater probabilities of experiencing relatively high loss rates.

Tranche rating is the first step towards risk transfer
Banks have two types of role in a securitised transaction – as an originating bank and as an investor. For the originating bank, securitisation is a risk mitigation technique and for the investing bank it is like any other investment, driven by credit risk/rating measurement for the investment. For the originating bank, the effectiveness of the risk transfer, credit support provided, loss position and investment/retention of lower tranches determine the effective risk mitigation. For both originating and investing types of bank, investment in unrated tranches is discouraged, since most of the equity and junior tranches are unrated.

Basel has prescribed look-through principles for unrated senior tranches while discouraging investment in any other unrated tranches. This is based on the basic principle of risk management – you cannot manage something you cannot measure.

Documentation and legal contract

Securitisation is a mechanism for repacking and redistributing the credit risk through structured finance instruments. Structured finance instruments can be defined using three distinct characteristics:

❑ pooling of defaultable assets;
❑ credit risk on the assets is transferred from originator to stand-alone special-purpose vehicle; and
❑ liabilities are tranched. Tranching is a process of creating one or more classes of security. Rating of some of the created securities

Figure 12 Securitisation

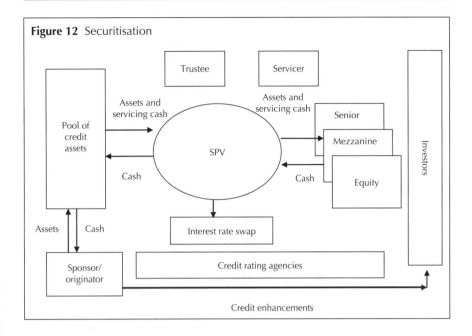

is higher than the average rating of the underlying collateral asset. Often the rated securities are generated from the pool of unrated assets.

In addition to the contract between investors, risk seller, protection provider and lender, the covenants and contracts with managers and trustees managing the structures have also become important since a lot of discretion is given to the structure managers and trustees, and many times their compensation is linked to the performance of the structure.

An SPV is a legal entity created by sponsor or originator by transferring assets to it. The SPV can be a limited-liability partnership, company or a trust.

Sponsors or originators are the banks interested in mitigating their credit risk. *Sponsor* denotes the entity that places the portfolio in an SPV for issue of notes. Generally, SPVs are structured in such a way that they cannot go bankrupt. This reduces the premium to be paid for bankruptcy for the debt issued by the SPV. To avoid the adverse selection, moral hazard and future litigation over the quality of the assets, the debt issued by the SPV is generally rated by the credit rating agencies. This also enables the SPV to create investment-grade – and some institutions cannot invest in any security lower than investment-grade.

At the same time as, or prior to, setting up an SPV comes the selection of the assets to be transferred. In order to make investors buy the debt of the SPV, various types of credit risk and liquidity support are also provided by the originator, or support is purchased from third parties.

Typically, a master-trust structure is used to get the flexibility in structuring tranches with different repayment terms and characteristics, and provide the ongoing ability to transfer assets and ability to offer multiple series. It also helps in mitigating the temporary cashflow mismatches due to heterogeneous collateral assets and liabilities.

There are three types of risk in the underlying portfolio and each of the tranches: risk of expected loss, unexpected loss and sensitivity of the credit spreads of the tranche to the changes in the credit spreads of the underlying. These can be divided into additive and non-additive.

As shown in Table 3, each tranche has a different quantum of the risks.

Contractual provisions known as "triggers" give additional protection to senior tranches. If defaults cause the reference portfolio's par value or interest proceeds to decline below a trigger level, cashflows from the reference portfolio are diverted to pay down the principal balance of senior tranches before more junior tranches can receive interest and principal payments.

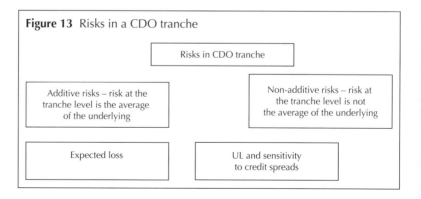

Figure 13 Risks in a CDO tranche

Risks in CDO tranche

Additive risks – risk at the tranche level is the average of the underlying

Non-additive risks – risk at the tranche level is not the average of the underlying

Expected loss

UL and sensitivity to credit spreads

The structures are of two types:

❏ static structures – composition of the underlying reference portfolio does not change during the life of the transaction; and
❏ dynamic structures – the underlying reference assets keep
changing.

The set of contractual terms governing the payout of interest and
principal payments on the reference portfolio is called a *waterfall*.
As an example, take a hypothetical cash arbitrage CDO referencing
a portfolio of high-yield bonds and structured as follows:

Pass-through structures: Under the pass-through structures the payment flows to the tranche owners are closely tied to the cash
inflows from the underlying exposures. Interest is generally paid
monthly; repayment of principal to the tranche holder is a closely
linked to the amortisation schedule and the capital repayment rate
of the underlying collateral.

Pay-through structures isolate the tranche holders from the fluctuations in the cashflow from the underlying exposures and typically
follows the promised cashflow waterfall.

The cashflow to the tranche owners is ensured through credit
support, credit enhancement, liquidity support, early amortisation
triggers, first-loss position, etc.

Synthetic securitisation does not intend to raise cash by transferring loans, but instead merely transfers the credit risk. It replicates
the economic risk-transfer characteristics of a securitisation transaction without having to remove the assets from the originating

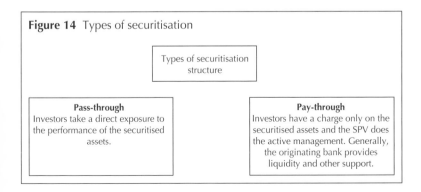

Figure 14 Types of securitisation

bank's balance sheet. It is more flexible than the traditional securitisation.

Pricing synthetic CDOs

Copula models provide a mechanism to decouple the stand alone default probabilities from the correlation structures. Since correlation is an important element in securitisation of portfolios of assets; copula models have gained wider acceptability and use.

A normal copula function provides the capability of computing joint default probability. Therefore, specification of correlation to a copula function can generate a loss distribution. The existing models and market practices use the single-factor model to generate the correlation structure. The single-factor models have following advantages:

❑ there is a reduced number of parameters and much-reduced computation complexities;
❑ there is generation of closed-form equations;
❑ once the value of the factor is fixed, the individual obligor default becomes conditionally independent, which enables

Table 5 Sources of correlation data

Correlation models	Shades of models	
Assets/structural	Correlation between the time series of asset values derived from the equity market	Correlation in the equity values
Historical default experience	Correlation is forecast sector level	Forecasting a distribution of possible default experiences
Spread changes	Link assets moves to the spread moves. Existing CDO models are not based on spread but on default	Correlating movements in the spreads
Implied correlation from standard tranche price of the larger credit indexes (correlation as an input to the model)	Implied correlations vary across the different tranches on a single portfolio. Implied correlation of first-loss tranche is not applicable to second-loss tranche	Base correlation or detachment correlation is used

the portfolio loss to be sum of independent obligor losses; and

❑ conditional portfolio losses can be convolution of individual loss distribution.

Approximation in a single factor CDO valuation model

Correlation structures are the most important input to a CDO valuation. In the previous paragraphs we have seen the type of correlation structures and shades of their models.

The other important approximation is the averaging spread for the underlying portfolio. Different models take this input at different levels of granularity. Most of the models take average spread for the portfolio and not spread at a name level.

Another important approximation is to compute losses through numeric integration equation instead of adding the convoluted structures. This is done through subdividing the exposures and multiplying the number of obligors to get a limit portfolio (where obligors tends to infinity or exposures tend to zero).

The other assumption is the default time – there are three ways to model default time – on the payment date, at the beginning of the period and at the moment the default occurs. Modelling default at any point in time adds many complexities to the model. Assuming default at beginning or end of the period adds to inaccuracies. Therefore, default is assumed to occur midway between the payment dates.

Modelling of structured finance depends upon the concentration and correlation assumptions. All of them have concentrated collateral pools. Commercial Mortgage Based Securities (CMBS) and CDOs normally lead to a fat-tailed loss distribution. Like all other credit exposures, computation of correlation and the impact of correlation on the credit risk has been a major problem in the pricing of risk mitigation and measurement of the transferred risk for CMBS and CDO, while for Asset Backed Securities (ABS) and Residential Mortgage Based Securities (RMBS), since their granularity level is very high, expected loss plays a major role in rating of tranches. Moody's in their methodology have assumed the cumulative default distribution for ABS/ RMBS to be the lognormal distribution.

Valuation of securitisation

Like any other credit risk measurement model, measurement of risk and price is based on assumptions and approximations. The basic inputs to the model are as follows.

❏ Assessment of probability of default to each obligor within the underlying portfolio.
❏ Assumptions on recovery rates.
❏ Assumptions on correlation. Correlation assumptions are the key to the loss estimations and tranche ratings. The correlation assumptions depend upon the exposure size of the underlying portfolio. For the smaller exposure size, such as SME loans, typically lower intra- (8–12%) and inter-industry (3–5%) correlation is assumed.
❏ Structural-analysis assumptions on the distribution of cashflow among the tranches. This involves:

  ❏ analysis of legal documentation;
  ❏ cashflow analysis and modelling; and
  ❏ rights and obligations of assets managers and servicers.

Therefore, measuring CDO risk involves measuring of each of the risk component. Due to various assumptions, CDO measurement is exposed to significant model risks.

Structural analysis is a significant determinant for the funded CDO tranche rating compared with a synthetic CDO. One criticism of the CDO rating has been the existence of wider differences in the rating approaches by the rating agencies. This can lead to significant differences in the ratings assigned by the agencies. The quantum of differences is different in senior and junior tranches. This may results in cherry picking the rating agencies.

In CDOs, the financial and quantitative analysis and contract analysis play a more important role than the economy, industry and management analysis (since cashflow to serve the debt does not come from the working of the firm but from the contract). Default rates of the underlying exposures and correlation play a major role and we know from the chapter "Credit Risk Portfolio Models" that correlation is the most difficult credit risk component to be measured.

Approaches to CDO pricing
Measurement of correlation is the major determinant of CDO pricing models. Five types of model have emerged.

❑ Measuring the correlated default intensities. This is extension of the reduced form models.
❑ Extension of structural approach to two obligor to many obligors portfolio. The correlations between the assets of the obligors are determined by one or more factors. The credit spread can be modelled and linked to the tranches.
❑ A copula approach to specify directly the dependence structure, though in a somehow ad hoc way.
❑ The Gaussian copula model, introduced to the credit field by Li (2000) has become an industry standard. Theoretical foundations of this approach such as credit-spread dynamics are not widely accepted.
❑ Various authors have also considered tail dependence within default times or default events. This would lead to fat tails in the credit-loss distributions. For this purpose, copulas such as Clayton, Student t, double t and Marshall–Olkin copulas have been proposed.

Credit rating agencies have been using the following three types of correlation methodology.

Concentration analysis
Modelling correlation in underlying exposures is very difficult. To manage the adverse impact of correlation, the following types of limit are prescribed:

❑ based on inter- and intra-industry/portfolio assumptions or measurement, a minimum number of industries or portfolios;
❑ limits on the individual exposures of single obligors in the underlying;
❑ in the cashflow analysis, a measurement of the maximum number of simultaneous defaults that a tranche should be able to withstand; and
❑ weighted-average credit rating or expected loss of the underlying.

Concentration under various scenarios is also analysed.

Analytical method for valuation of securitisation
Moody's have extended their binomial expansion method to correlation analysis of underlying in a CDO. This method builds expected loss distributions from a reference portfolio that is a set of homogeneous and uncorrelated assets. The reference portfolio has same diversity score as the actual portfolio and has total number of assets equal to the diversity score (DS). Since the assets are homogeneous and uncorrelated, the only relevant scenarios are 0 to DS assets defaulting. The probability of each scenario is estimated using binomial distribution. The expected loss distribution is generated for a set of stressed scenarios, for a set of PDs and LGDs. The loss distribution is used to assign expected loss and associated ratings to the cashflow tranches.

Monte Carlo method for valuation of securitisation
The next – to consider a higher number of scenarios (but not all combinations of pairwise defaults – a smaller set of combinations is generated by making assumption a correlation structure) are generated using Monte Carlo simulations. Generally, Merton-type models are used to generate the desired correlation structure. In MC, assets-value correlation is used as a proxy for the default correlation. Correlation inter- and intra-industry and correlation with economic is modelled. The available models assume constant correlation.

Extending the MC method to cashflow division among tranches is extremely difficult; therefore, generally, cashflow is modelled only under stressed conditions.

Case study: Moody's methodology for rating CDOs
This methodology is based on building multiple potential collateral loss scenarios. The portfolio is mimicked through representative assets. For each of the assets, both the common default probability and the common recovery rate is assumed. Correlation between each pair of assets is estimated. The number of defaulted representative assets is an indicator of the level of losses. Loss-probability distribution is generated for various default scenarios. The CDO waterfall structure is combined with the loss distribution to estimate probability-weighted loss for each tranche of CDO.

Model risks

The key risk in a securitised structure is the assumption regarding correlation. Market participants have developed complex models to measure the correlation between the different credit risks. However, each of the models makes many assumptions in order to make a model workable. This has resulted in a large amount of model risk. Residual risks along with model risks result into nonlinear risks.

Case study: Securitisation market meltdown in the US

The US market had an unprecedented wave of downgrades in the CDO sector, and in particular the high yield collateralised bond obligations (HY CBOs) in 2000–2. In addition to the adverse credit environment after the Internet meltdown, the other three most important factors contributing to the surge were:

❑ adverse selection of collateral assets and sidestepping of structural features by collateral managers;
❑ interest-rate risk overhedging; and
❑ insufficient structural protection against credit risks arising from collateral reinvestment.

CDO square

This is a synthetic CDO. It is CDO built on the investment-grade tranche of other CDOs. The assumptions on the assets correlation is a key element. An extension of the Monte Carlo method for CDOs is the look-through model for the CDO square. This allows correlation modelling between CDO tranches based on the correlation assumption for the underlying assets and requires no additional assumptions.

In a securitisation transaction, risk transfer is in proportion to the lower rated tranche sold to the investors. Ideally, the proportionate capital should be saved.

A major problem with the CDO square is likely to be measurement of correlation. The CDO square requires measurement of correlation not only among the underlying assets but also among the underlying tranches. Indeed, a squared complexity.

Correlation trading

The growth of new portfolio products, such as single-tranche CDOs and nth-to-default baskets, has given rise to what is known as *correlation trading*. By delta-hedging a single-tranche CDO, a dealer can eliminate most, if not quite all, of its sensitivity to the individual credit spreads in the reference portfolio. What remains unhedged is the exposure to correlation. When market participants disagree about the correct correlation to apply to a CDO tranche, they will put different values on the tranche. Those who put a lower value on the tranche will sell to those who give the tranche a higher value. Different perceptions of hedging costs, different model structures or model parameters can also lead market participants to disagree about the value of single-tranche CDO.

First-to-default baskets are the instruments for correlation trading. They have a first-loss exposure to the reference portfolio. Unlike CDO tranches, first-to-default baskets typically involve small-reference portfolios.

Gamma risks in CDOs

There is a risk that the tranche and the hedge position will respond differently to a large move in credit spreads or to a default of one of the underlying credits. This leads to widening of spreads and is called *gamma risk* or *convexity*.

It means changes in the credit spreads of the underlying assets change the credit spread on the tranches disproportionately. The change is not necessarily in the same direction. There are two scenarios: changes in the credit spread of one of the assets and changes in the credit spread of all assets. Generally, they have opposite effects on the changes on the tranche spreads. The equity tranche and the senior tranche also have opposite exposures to gamma risk.

REFERENCES

Basel Committee on Banking Supervision, 2001, "Assets Securitisation", consultative document.

Basel Committee on Banking Supervision, 2002, "Second Working Paper on Securitisation".

Basel Committee on Banking Supervision, 2005, "Credit Risk Transfer", Bank for International Settlement.

Committee on the Global Financial System, 2003, "Credit Risk Transfer", Bank for International Settlement.

Duffee, G. R., 2001, and Chunsheng Zhou "Credit Derivatives in Banking: Useful Tools for Managing Risk", web site of Haas School of Business.

Duffie, D., 1998, "Credit Swap Valuation", *Financial Analyst's Journal*.

Gorton, G. B. and G. G. Pennacchi, 1995, "Banking and Loan Sales: Marketing no Marketable Assets", *Journal of Monetary Economics*.

Li, D. X., 2000, On Default Correlation: a Copula Approach, Journal of Fixed Income.

Moody's, 2004, "Recovery Rates on North American Syndicated Bank Loans, 1989–2003", special comments, March 4.

7

Loss Given Default

INTRODUCTION

In the first chapter we examined and understood the importance and need for unbundling credit loss and the credit risk component. LGD is one of the risk components identified by the experts and also by the Basel Accord. As noted in the same chapter "LGD represents the fractional loss due to inability to recover the claim. This is the fraction of exposure which is not recovered".

Since LGD is a fraction of the exposure, is measured through recovery rate which is equal to i-LGD. Both the LGD and RR measure loss severity on default. LGD or RR measurement has various practical difficulties in measurement and modelling. We have identified these difficulties in the section below. Due to these difficulties various approaches for LGD measurement have evolved.

Conceptually, LGD and RR are the same but may be used differently. Some experts opine that LGD is the term generally used in the context of tradable assets and represents loss in the market value of the bond immediately after default. Others believe that RR is the term used for the amount recovered after default for non-traded assets such as bank loans. Bank loans have work-out-RRs. Banks have used the RR concept for many years and have been building and using recovery tables for a very long period of time. Although tables were used to modify/notch the ordinal credit rating they have historically not been used for facility rating as a part of two-dimensional rating, as is now often done. LGD and RR are also useful in EL and UL. The LGD concept has received a greater acceptance in the past five years or so since the Basel Accord exposure drafts started using it. It has also become an area of intensive research. Most of the rating agencies have strengthened their offerings for LGD estimation. However, their offering outside the US still continue to be weak due to the non-availability of loss data outside the US.

There are very few empirical studies available on the LGD and RRs. Typically, only mean values and quintiles are computed and sometimes only factors influencing LGD/recovery are identified and described without computing recovery or LGD.

In this chapter, we have made an attempt to explain LGD/recovery measurement models, factors influencing LGD/RRs and the Basel II Accord treatment for LGD.

MEASUREMENT AND ESTIMATION OF LGD

LGD and RR are required for investment decisioning, facility ratings, pricing, loss provisioning and capital allocation and portfolio credit risk estimation.

Data availability has been a major issue in measuring LGD. Data availability has been such an important issue that little of the work in this field has considered modelling. Most of the models measure central tendencies or are based on regression techniques. In the recent past market information based reduced form models have started modelling RRs as stochastic variables on the assumption that the market price reflects default and RRs (credit loss).

The following paragraphs explain broadly the scale of complexity, data requirements, approaches and building blocks for recovery modelling.

Recovery modelling is difficult

Recovery modelling has many associated problems.

❑ Recovery has far fewer data points than the default rate.
❑ The accounting data regularly collected for loss is not suitable for recovery modelling.
❑ Recovery quantum and process is determined by the bankruptcy legal framework in the particular country.

❑ Many jurisdictions do not have a well-defined bankruptcy framework.
❑ The existing recovery models have low predictive power.

Therefore, we are presented with a vicious circle. The banks do not use recovery models because they have lower predictive power and models are neither built nor improved as banks do not use them. To manage the data and model problems, various approximations and assumptions are made. For example, the following approximations and substitutions are employed in various LGD models.

❑ The most widely used assumption is constant LGD.
❑ Where LGD is treated as a random variable, the following assumptions are made by different models:

  ❑ LGD is assumed to be beta distributed;
  ❑ LGD is assumed to be identically distributed over time and across all the borrowers;
  ❑ zero correlation is assumed among the LGD of different borrowers and hence there is no systemic risk due to LGD volatility;
  ❑ LGDs for the same borrower across time are assumed to be constant.

These assumptions and approximations do not always hold. The shortcomings of the previously mentioned approaches are as follows:

❑ Most of the studies available relate to the US. In general, such studies are extrapolated while the bankruptcy laws, firm structures and industry conditions vary greatly between countries.
❑ For a skewed portfolio, a constant LGD assumption does not hold.
❑ A lack of correlation of LGD between the borrowers does not hold for a portfolio with significant industry concentration.
❑ The data available is generally for a short period only.
❑ For an industry under stress, the assumption that PD and LGD are independent of each other does not hold.

Data sources
The following sources of data are the most commonly used for LGD modelling:

❑ internal historical LGD of a bank – divided according to the assets class/risk segment;

Table 1 Recovery table

	Line of business 1/ product of the bank or industry of the borrower	Line of business 2/ product of the bank or industry of the borrower	Line of business 3/ product of the bank or industry of the borrower	Line of business 4/ product of the bank or industry of the borrower
Senior secured collateral type 1 collateral type 2 ...	60–75	50–60	45–55	40–55
Senior unsecured debt type 1 debt type 2 ...	35–45	35–45	30–40	25–35
Subordinated loan purpose 1 loan purpose 2 ...	10–15	20–30	15–25	10–20

❑ loss data collated by regulators or other similar organisations;
❑ loss data extracted from market based information on price and spreads;
❑ loss data published by rating agencies;
❑ judgmental adjustments in any or all of the above.

Recovery tables

This is the basic, preliminary and most simple measurement of RR. For the past two decades or so, banks have been developing and using tables based on their recovery experience as modified by the judgmental factors for debt type, seniority class, collateral type, loan purpose and line of business segment. Table 1 is an example of this.

For a structured finance style of transaction, there is no issuer rating since the special purpose vehicles have very nominal capital. The issue rating is driven by the instrument type and seniority and is similar to the recovery table (although the RRs may be higher). Since 2005, Fitch has started distress recovery rating to distinguish between issue rating and recovery ratings.

This table is also called a look-up table. Recovery tables generally blend historical RRs (the history only covers a very brief period) with subjective adjustments. The major drawbacks of recovery tables are as follows.

❑ Each cell represents average recovery. Within a given cell, the issuers are not differentiated. According to Fitch, the average RRs are relevant for investment grade and high-speculative grade issuers and not for medium- and low-speculative grade issuers.

❑ Recovery tables typically use the same estimates of recovery irrespective of the time horizon over which default might occur. This means that important considerations such as the point in the credit cycle or the sensitivity of a borrower to the economic environment are ignored.

❑ Recovery tables are updated infrequently.

Ultimate recovery – the internal data-based approach
Every bank suffers defaults and then makes recovery attempts. LGD is determined by the following factors:

❑ loss of principal;
❑ carrying costs of non-performing assets; for example, interest income lost or foregone;
❑ recovery and workout expenses (collections, legal, etc).

Financial institutions follow prudent accounting polices. The periods and accounting heads for booking recovery and losses are determined by accounting policy, which focuses on prudence and not the computation of LGD or the risk policies. As explained in the chapter "Credit Risk Fortification", the accounting prudence and risk management may not always converge. This results in the two approaches for the estimation of LGD: accounting LGD and economic LGD.

Accounting LGD
Charge-off for credit losses is determined by accounting policies and standards, product type, past due days and collateral/risk mitigation available. The accounting charge-off may or may not represent LGD. Accounting credit losses are continuously adjusted for restructuring and recovery. The exposure keeps varying. The purpose of booking accounting losses is to maintain the regulatory required quality of assets.

Existing product processors or the general ledger or management inforamtion systems are generally inadequate to compute/capture LGD from the accounting entries. Elsewhere, we have defined business requirements for a system to compute accounting LGD.

Economic LGD

Economic LGDs are driven by risk prudence and fair value. Economic LGDs include all economic costs and all recoveries, discounted at appropriate discounting rates. Accounting and economic LGD and charge-off differ from each other. This is applicable for middle market firms, mid-tier lending and for loan assets. There are three approaches to discount rates:

❏ discount the stream of recoveries and workout costs by a risk adjusted discount rate;
❏ the risk neutral approach – convert the stream of recoveries and the stream of workout costs and compute the cashflow equivalent and apply the risk-free interest;
❏ a combination of the previous two approaches.

Post-default prices – external market-based information

This is suitable for larger firms, syndicated loans, corporate bonds and preferred stocks. This is a method based on post-default debt prices. For market-traded instruments, two types of information may be available:

❏ price;
❏ risk-neutral discount rates.

Recovery of market value

Market value of the instrument can be observed if it continues to trade after the default. The value observed in the market immediately after default is called the recovery of market value. In some countries, there is an active market for trading default bonds. Trading in defaulted bonds continues for some time (eg, 3 months) after default. The market value is the best estimate for the expected recovery and the market price reflects the expected recovery suitably discounted. This is the most widely used and acceptable measure of RR. The recovery of market value is a measurement for the recovery risk (uncertainties in the recovery), recoveries on discounted principal, missed interest payment and restructuring costs. However, the problem is that outside the US the market for defaulted bonds is either non-existant or does not have the required depth or liquidity.

Market prices are impacted by supply and demand in the distressed debt market. Rating agencies like Moody's publish a bankrupt bond index to capture price variability caused by supply and demand. They have found that presumed imbalances in supply and demand do not drive market prices of defaulted debt and thus market prices are efficiently stated and predictive of future recoveries. However, during an economic distress (when the supply of default securities increases substantially), the demand supply is an important factor.

Loss measured from risk-neutral discount rates
observed in the markets
Risk-neutral discount rates are applicable to the instruments as inferred from the market prices of equity, bonds or default swaps. A risk-neutral measure is the fair expected value of its future pay-offs. This approach typically requires measurement of the PD as a first step and then estimating the LGD to reconcile the market price for bonds, equity and default swaps. There are two approaches to measure loss:

❑ *Recovery of face value*: This is observed from the set of estimated cashflows resulting from the workout and/or collections process, properly discounted and the estimated exposure. An estimation of the recovery of face value is more complex than for the recovery of market value. The availability of recovery rating helps in adopting this approach. There are two types of complexities.

  ❑ The timing of various cashflow components from the distressed asset has an important bearing.
  ❑ Discount rates at which these cashflows should be discounted. The recommended rate is a rate on the assets of the similar risk.

❑ *Recovery of treasury value*: RR is derived from risky (but not defaulted) bond prices using a theoretical asset-pricing model.

Another approach is to consider credit spreads on the non-defaulted risky (eg, corporate) bonds currently traded.

Ultimate recovery *versus* post-default prices

Post-default prices on currently defaulting bonds are observable in the current period, while ultimate recoveries are not. The recovery method to be used varies with the investment objective. Investors interested in current income use post-default measures while investors aiming to earn capital gains use ultimate recovery.

Applicability of recovery measurement techniques

The above techniques are useful at different points of time in the life of an asset (see Figure 1).

Obviously, the definition of default plays a very important role in the RR. Banks can improve RRs both before and after default by developing structured processes for distress management. Identification of distress, along with early and timely enforcement of covenants, provides the bank with an opportunity for renegotiation and a possibility of improving the collateral position.

According to paragraph 470 of the Accord, LGD estimates must be grounded in historical RRs and, when applicable, must not solely be based on the estimated market value of the collateral. Historical RRs are important even for defining economic downturn.

Approaches to measure historical LGD

There are two approaches to measure LGD from historical data:

❑ exposure weighted (for one year), given by the ratio of the total amount lost and the total exposure of the defaulted loans

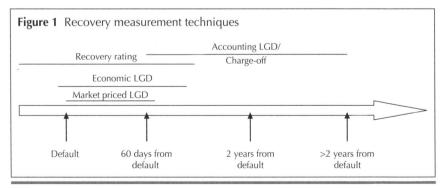

Figure 1 Recovery measurement techniques

Table 2 Measuring historical LGD

Facility	US$10 million		US$1 million	US$5 million
Loss	US$8 million		US$0.3 million	US$2 million
LGD for each facility	0.8		0.3	0.4
Average LGD	(0.8 + 0.3 + 0.4)/3 = 0.5			
Exposure weighted average LGD	(8 + 0.3 + 2)/16 = 0.64375			

$$Exposure\ weighted = \frac{Total\ lost}{Total\ exposure\ of\ defaulted\ loans}$$

❑ the average LGD, given by the sum of the LGDs divided by the number of LGDs

$$Average\ LGD = \frac{\sum LGD}{Number\ of\ LGDs}$$

Table 2 shows the difference between the two approaches.

Time weighting

Time weighing is performed by averaging the above two measures across the years. This smoothes out the high LGD with low LGD and therefore understates the expected LGD. Under both approaches the losses of all years get the same weight. Some experts argue that the recent data is more relevant than the past as it will have been generated during the same phase of the business cycle (eg, the cycle is going down). The losses/data for recent years can be given more weight accordingly. The Accord allows this for the retail segment.

Fixed or constant LGD

The Accord has provided an estimate of average LGD (45%) under the foundation approach for an unsecured corporate loan.

LGD models

In general, there are two types of LGD models:

❑ interpolating and extrapolating the historical LGD through regression techniques or analytical techniques based on the assumptions about LGD distribution; this includes modelling distribution and volatility of recovery.

❑ modelling recovery as a stochastic variable, estimated from market based information.

Before default the exposure is subjected to default risk and after default the EAD is subjected to the recovery risk. There is a risk that actual recovery may not be equal to the estimated recovery. Techniques similar to credit scoring are applied to RRs.

There are two approaches to the modelling of RRs as a stochastic variable. The RR is modelled as a function of assets value – default is declared when the assets value falls below a certain level. In this case, the expected RR is assumed to be constant. The second approach is where lenders allow a firm to continue as long as the expected RR is likely to be higher than the targeted RR. Each class of lenders in various bankruptcy regimes in different phases of the economic cycle will have different expected RR. For a given class of lender in a given bankruptcy regime the targeted RR varies according to the economic cycle.

Since the recovery streams are uncertain and such uncertainty cannot be diversified away, appropriate premiums are applied in computing present value.

The availability of historical recovery data is a major challenge for modelling RRs. The availability of loss data determines the degree of complexity and the choice of methodology for LGD modelling.

Measuring central tendency

This model is used to interpolate and extrapolate historical LGD experience. Central tendencies are measured and various assumptions on central tendencies are made to predict LGD (See Table 3). This is the simplest LGD model.

These measures help describe the variable's shape or distribution. If a variable is not normally distributed, a transformation can typically be devised that will convert it to be normal.

As a part of the measurement of central tendencies, LGD distribution is assumed to be lognormally distributed. One of the benefits of the use of the lognormal distribution is that the collateral value is guaranteed to be positive.

Building regression analysis equation

The regression equation is built to enable the user to predict LGD. The regression equation is built between the LGD risk factors and the

Table 3 Central tendency in recovery

	Mean recovery	Median recovery	Standard deviation	Co-efficient of variation	Count
All	51	45	38	73	950
Senior secured	83	100	25	30	264
Senior unsecured	64	71	31	48	141
Senior subordinated	50	43	34	70	125
Subordinate	28	18	28	100	395
Junior subordinate	12	5	18	143	25

Source: Adopted from "Suddenly Structure Mattered: Insights into Recoveries from Defaulted Debt" Karen Van de Castle, David Keisman and Ruth Yang. Published in Standard & Poor's Credit Week, May 2000.

historical LGD experience. LGD risk factors include various factors such as collateral, firm, industry and macroeconomic factors. For example, LossCalc, a Moody's KMV model, uses regression analysis to map historical LGD on to risk factors in a three-step process.

❑ *Transformation*: In this step, risk factors are transformed into intermediate aggregate level risk indicators. For example, macroeconomic variables are transformed into composite indices.
❑ *Modelling*: A regression equation is developed for the risk factors and risk indicators.
❑ *Mapping*: The model developed is statistically mapped onto the historical LGD.

LossCalc model

Based on empirical studies, the model assumes that defaulted debt values can be approximated to a beta distribution. The beta distributed debt values are converted into normal distributed recovery values.

The modelling phase of the LossCalc methodology involves statistically determining the appropriate weights for risk factors and risk indicators for a linear weighted regression equation.

Measuring and predicting recovery rate from defaultable debt price

This approach is based on the reduced form models and has been explained in detail in previous chapters.

The market price of a defaultable bond represents the following three components:

❑ receipt of coupons prior to default;
❑ promised face value in the absence of default;
❑ value of the recovery payout if the firm defaults, which is determined by RR and the hazard rate.

Therefore, given coupons, the promised face value and the hazard rate, the RR can be estimated from the market price of a defaultable bond.

These models are used to infer the market's expectation of RR implicit in bond prices. These models are also useful for benchmarking RRs in terms of risk-neutral expected RRs over time and across the rating categories.

These models recognise that the actual/physical level of recovery is a random variable, and LGD and expected value can be modelled for valuation purposes.

Relation between probability of default and recovery rates
In the past three decades or so, most of the work in the credit risk field has been on the estimation of PD and portfolio losses. Traditionally, both of these models have assumed recovery to be constant and independent of PD. Furthermore, the research in the past has concentrated more on the systemic risk rather than idiosyncratic risk, such as LGD. This may also be due to the absence of a market for bankrupt bonds outside the US. Only during the last decade has the literature and research on the estimation of recovery risk gained importance. The research in the LGD arena started after the Basel Accord identified LGD as a risk driver. Empirical evidence has now established the following facts about RRs:

❑ economic downturn has an impact on the RRs;
❑ collateral values and RRs are volatile, have high standard deviation and fat tails;
❑ RRs have an inverse relationship with default rates;
❑ recovery data is scarce and is therefore modelled with simplifying assumptions, such as constant and single RRs.

Table 4 summarises the various assumptions on RRs and the relationship between RR and default rate.

Table 4 Relationship between RR and default rate

Inversely related to PD	Endogenous LGD inversely related to PD	Stochastic LGD inversely related to PD
Independent of PD	Constant LGD, independent of PD	Stochastic LGD independent of PD
	Endogenous LGD	Exogenous LGD

Meaning of independence of PD and LGD
Two variables are said to be independent if having knowledge of the value of one of them tells us nothing about the value of the other. Each of the variables has an expected value, distribution and set of likely values. Consequently, this means that the LGD distribution does not have any effect on the distribution of PD or *vice versa*.

The CreditRisk+ model treats recovery as a constant. The CreditMetrics model allows LGDs to be stochastic but independent of default rate, and therefore an increase in default rates leaves LGD rates unchanged.

Recovery rating
Credit rating agencies such as S&P's and Fitch have started offering recovery rating services since the last few years. Recovery rating complements the traditional corporate credit ratings. A separate recovery rating from the issuer rating is as per the Basel II requirement of a two dimensional rating.

S&P's methodology for recovery rating
Recovery ratings are estimates of post-default recovery of nominal principal. Recovery rating is based on a fundamental analysis of key factors that drive post-default recovery.

An issuer credit rating is an opinion of the creditworthiness of an obligor with regard to a specific debt issue. Issue ratings take into account the ranking, payment terms and recovery prospects of the issue and may be rated in the same way as the corporate credit rating – lower for junior debt or higher for well-secured debt. In all cases, however, issue ratings remain anchored to the corporate

Table 5 Recovery rating scale

1+	Highest expectation full recovery	100% recovery of Principle	Issue rating = issuer rating + 3 notches
1	Full recovery expected	100%	Issue rating = issuer rating + 2 notches
2	Substantial recovery	80–100%	Issue rating = issuer rating
3	Meaningful recovery	50–80%	Issue rating = issuer rating
4	Marginal recovery	25–50%	Issue rating = issuer rating
5	Negligible	0–25%	Issue rating = issuer rating

Source: Adopted from S&P, Bank Loan & Recovery Rating 2006 Users guide and Fee Schedule

ratings. The assignment of RRs is a six-step process according to S&P's methodology:

❑ assess transaction structure;
❑ assess borrower's financial and other projections;
❑ simulate a default path;
❑ based on the simulation, forecast the borrower's free cashflow at default;
❑ identify priority debt claims and value;
❑ determine collateral value available to lenders.

The recovery rating scale (see Table 5) estimates the likely recovery of principal in the event of default and is de-linked from the corporate credit rating.

According to S&P's and Fitch, recovery rating is useful and important for the following reasons.

❑ Recovery rating is useful for borrowers to obtain more precise and better pricing for their debt. This facilitates investment by institutional investors in the collateralised debt, syndicated loans, and other instruments of lower grade borrowers.
❑ It also helps buyers that need recovery ratings and do not have skilled internal credit staff.
❑ Recovery rating helps investors to differentiate and take advantage of pricing anomalies between well-secured loans and those "secured" in name only.
❑ The recovery rating is applicable for lower speculative grade issues and distressed/defaulted structured finance securities.

The recovery rating is useful to differentiate between their repayment capabilities.

❑ For higher grades, the recovery rating is not relevant. Higher-grade firms have sufficient flexibility available in their capital structures and they are likely to borrow more before defaulting and entering into recovery mode. Therefore, no recovery modelling is possible.

Fitch methodology for recovery ratings

The following discussions are based on two papers published by Fitch: Recovery Ratings – Approach and Process for Corporate Finance, August 9, 2005 and the Role of Recovery Analysis in Ratings – Enhancing Informational Content and Transparency, February 14, 2005.

The recovery rating describes the recovery characteristics in case of default. While S&P's provides a cardinal scale of recovery, Fitch provides an ordinal scale of recovery from R1 (outstanding) to R6 (poor). Fitch explicitly publishes the recovery ratings for all issuers with rating B or below.

Fitch incorporates in its recovery ratings the trend observed in most jurisdictions about preference for the reorganisation of the bankrupt firm over the liquidation of the firm, especially if the firm has more than US$100 million debt or to protect employment opportunities. This is incorporated in the models through going concern assumptions and reorganisation expenses and concessions granted by senior claim holders. Fitch also addresses the correlation and impact of business, economic and industry cycles. Fitch has two different methodologies for recovery ratings. Each methodology differs substantially from the other.

For investment grade and high speculative grade the recovery rating is not based on the issuer-specific analysis. The rating is based on the instrument type and the security position, and reflects the average recovery expected for the instrument type. For North America, Fitch has published instrument-based recovery ratings for the following types of instruments:

❑ assets based;
❑ secured bank loans;
❑ secured bonds;

❏ unsecured debt (bank loans, bonds and other identified general claims);
❏ subordinated debt;
❏ preferred stock.

For the issuers with rating B and below, Fitch has developed the following three-step methodology.

❏ *Estimating the enterprise value*: This involves consideration of the enterprise as a going concern and application of the distress test depending upon the relative industry position. Cashflow is estimated in the light of what it will take for the company to apply for bankruptcy or for creditors to enforce their rights, which may be driven from covenants. Haircuts and discounts are applied to cashflows and a multiplier is applied for enterprise values (EVs). A sector may have a different valuation approach that may include traded assets valuation or balance sheet liquidation. Haircuts and multiplier differ according to sector.
❏ *Estimating the creditor mass*: This provides a good estimation for EAD. Some of the assumptions used include fully drawing the unused portion of committed lines of credit, a portion of lease obligations that are likely to be rejected in bankruptcy, assumptions on compulsory administrative expenses post-default and concession assumptions from senior secured creditors to junior creditors. Claims from the supplier and trade creditor are likely to be significant.
❏ *Distribution of enterprise value*: In the US, absolute priority rule is applied to the EV and liquidation value for the estimated creditor mass. Different types of facilities and securities obtain different recovery ratings.

Distressed recovery ratings to structured finance transactions
Fitch does not provide any issuer rating to the structured finance securities. Distressed Recovery Ratings (DRRs) estimate recoveries on a forward-looking basis considering time value of money. DRR provides an enhanced analytical approach to securities that are currently distressed or defaulted. Fitch DRR is based on the stress from a given default scenario. Generally the following components are considered:

❏ the time period for recoveries, which is considered to be maturity;

❑ both default and adverse movements in interest rates in a given stressed scenario;
❑ the evaluation of recovery cash flow components;
❑ discount factors;
❑ security balances.

FACTORS DETERMINING RECOVERY RATES

Loans and bonds have different information availability pre- and post-default. Since the bankruptcy process is very difficult to model, the market information availability post-default is a major determinant of the modelling approach. Information availability not only has a major impact on the model but also on the factors that are identified to be determinant of RRs.

Empirical studies have found that the following factors are broadly determinant of the RRs of loans:

❑ collateral value;
❑ seniority of the claim;
❑ firm characteristics;
❑ industry characteristics;
❑ macroeconomic conditions.

Security value

Empirical studies have found that the RR for secured and un-secured debt varies greatly. RRs for various assets classes have shown that the RR is either high and around 70–80% for secured exposures or low and around 20–30% for unsecured exposures. The recovery distribution is said to be "bimodal" or two-humped. The average recovery of the entire assets class or portfolio is therefore misleading. The RRs of secured and unsecured debt should be computed separately (see Table 6).

The special comments have made the following observations:

❑ the three-year average default rates for bonds are 1.25 times higher compared to loans;
❑ the loss severity of a bond compared to a loan is 2.25 times higher.

The value of collateral, as with the value of other assets, fluctuates with the economic conditions. With an economy going down, the collateral value also goes down. Conventionally, collateral value is

Table 6 RRs of secured and unsecured debt

	Bonds			Loans		
	Default rate (%)	LGD (%)	Credit loss (%)	Default rate (%)	LGD (%)	Credit loss (%)
Baa	1.6	62.6	1	1.6	50.3	0.8
Ba	5.3	84.6	4.5	10	31	3.1
B	21.1	76.7	16.2	24.3	30	7.2
Caa-C	51.7	77.7	40.1	59.3	34	20.2

Source: Adopted from "Credit Loss Rates on Similarly Rated Loans and Bonds", *Moody's Special Comments,* December 2004. The comments are based on North American based bonds and loans

assumed to be constant. The LGD at facility level can be constant only if collateral value is assumed to be constant. A constant LGD at facility level will mean that the average LGD is also constant.

Conventional credit models overlook the effect of economic conditions on collateral. They allow default to vary from year to year, but hold fixed the average value of collateral and the average level of recovery.

Seniority of the claim

Relative priority of debt is a very important factor for LGD. For example, preferred stock is the lowest seniority class in a typical capital structure, but it might hold the highest seniority rank within a particular firm that has no funding from loans or bonds in its capital structure. In addition, in cases where a firm issues debt sequentially in order of seniority, it may occur that senior debt matures earlier, leaving junior debt outstanding. LossCalc considers a debt's seniority in absolute terms, via historical averages, and in relative terms, when such data is available, within a particular firm. Both are predictive of LGD and are reasonably uncorrelated with one another. Each seniority class must compete with other classes for the available funds.

The order of priority varies in the detailed provisions of the insolvency law. Priority should not be assumed to be linked to security. In a number of jurisdictions having security is not sufficient, as the insolvency legislation may provide that other creditors are paid out at the same time or prior to the secured claim.

The importance of priority can be seen from Table 7, which shows a typical relative average RRs.

Table 7 Importance of seniority in the recovery

	Senior secured	Senior unsecured	Senior subordinate	Sub-ordinate	Junior subordinate	Preferred stock
Senior secured	100					
Senior unsecured	76	100				
Senior subordinated	49	58	100			
Subordinate	30	46	57	100		
Junior subordinate	27	35	54		100	
Preferred stock	13	11	24	20	37	100

Recovery rate difference between bonds and loans

Loans are typically senior to bonds. Therefore, a higher RR is typically expected for loans.

Firm characteristics

Every characteristic, mostly financial, that contributes towards reducing leverage and improving the quality of the assets helps in improving RR and *vice versa*.

Leverage or enforcement of collateral

Leverage indicates the extent of claimants for the assets in the event of default: lower leverage improves the enforcement of claim. Even for senior and secured debt holders, higher leverage definitely impacts the collateral enforcement and recovery since concessions are often extended to the junior and unsecured debt holders for obtaining their consent to various settlement schemes. Therefore, collateral enforcement is a challenge. Empirically, it is found that the leverage impacts recovery to the extent of 5–15%.

Safety, value and existence of assets

In a limited liability paradigm, the assets left behind after default are the only source of repayment. Therefore, safety, value and the existence of the assets are important factors in recovery. Safety, value and existence are ensured through selecting assets with the following features:

❑ assets whose quality is less likely to deteriorate over time (eg, land);

❑ assets that are less likely to disappear, such as physical assets;

❑ assets with a better market value compared to book value;

❑ assets having undergone better maintenance. This is possible only if the firm has been earning a good returns on the assets provided as collateral.

Michael Pykhtin (2003) argued that, apart from dependence on the systematic factor, the collateral value also depends on firm-specific characteristics. The general motivation for this dependence is that borrowers in financial distress often cut back on collateral maintenance and control. This behaviour tends to reduce the value of the collateral for borrowers with lower asset returns. Since asset returns have both systematic and idiosyncratic components, collateral should generally depend on both.

Industry characteristics

Empirical studies have established that RRs, such as default rates, are different for each respective industry. Every industry has different traits and can be in different stages of an economic cycle. Some of the industry traits relevant for the recovery are the growth rates, leverage, asset quality, etc. Firms in a growing industry with a better financial position or asset quality are likely to get sold as going concerns (restructured) rather than the distressed sale of assets.

North American empirical studies have found that the industry of the obligor matters. Tangible asset-intensive industries, especially utilities, have higher RRs than service sector firms, with some exceptions such as high-tech and telecom sectors (see Table 8).

Different recovery rates at the industry level

Various experts have found that RRs are different for different industries. In a recent study, Acharya *et al* have found that when an industry is under stress, the RR goes down by 10–20%. The type of industry has a major impact on the RR and Acharya *et al* found that RR varies from 37 to 74%.

Long-run recovery rates

Edward Altman of New York University, Moody's Investors Service and S&P's have published long-run average RRs on

Table 8 Recovery rates across the industries

Industry	Mean recovery	Standard deviation	Industry	Mean recovery	Standard deviation
Utilities	74	21	High technology	47	22
Insurance	37	27	Aero/Auto	52	21
Telecom	53	8	Building products	54	25
Transportation	39	27	Consumer	47	21
Financial institutions	59	25	Leisure/Media	52	26
Healthcare	56	22	Energy and natural resources	60	25

Source: Acharya *et al* (2003)

corporate bonds for many industries. These sources also calculate RRs by priority level. The observations cover a sufficiently long period, dating back as far as 1970 in the case of Moody's, so as to make the averages statistically valid.

Macroeconomic conditions

The empirical analyses indicate that systematic risk is a major factor influencing RRs of bonds and loans.

Economic downturn has a major impact on the LGD. During a recession, the LGD goes up by up to one-third. It is now established that macroeconomic conditions cause an increase in the default rate and a decrease in the RR. The simple explanation for this is that with economic downturn the value of the collateral deteriorates and the RR declines. There are various measures for macroeconomic conditions, such as growth domestic product (GDP) growth, stock market returns, etc. To identify the economic downturn, Basel has recognised GDP growth as one of the macroeconomic factors.

Owing to the impact of the systemic factors, RRs exhibit a high level of volatility. Some of the collaterals such as real estate, shares, bonds, etc, are usually affected more by the macroeconomic conditions than other collaterals such as stock and receivables.

Impact of the business cycle

Empirical studies have found strong evidence that recoveries in recessions are lower than during expansions. The impact of recession is higher on lower-grade debt than on investment grade debt. Speculative bonds are subjected to higher systemic risks.

Empirically, it is found that there is a negative correlation between recoveries and aggregate default rates. The business cycle has a twin impact on recovery: due to recession, the businesses of borrowers are impacted which results in an increase in the default rate. Under the increased default rate environment, the number of defaulted assets to be sold increase substantially. This results in a lower value for collaterals, depressing the recovery.

Insolvency legislation in the jurisdiction

Insolvency legislation has an impact on the all aspects of loan, capital structure of the firm, loan covenants and recoveries. Understanding the legislation is a must for proper issuer and recovery ratings.

Frye model for impact of systemic factors on recovery

In the first version of his model, Frye argued that the macroeconomic systematic factors driving defaults also drive recoveries. To describe this effect quantitatively, Frye proposed a Merton-type one-factor model. In this model, systemic factors linearly determine the collateral value and the collateral value determines the recovery. In the next version of his model, Frye modelled recovery directly to the systemic factors.

Case study of the LossCalc LGD model

LossCalc is a statistical model to measure immediate and one-year LGD for bonds, loans and preferred stock.

LossCalc is built on a global dataset of 3,000 recovery observations for loans, bonds and preferred stock from 1981 to 2004. This dataset includes over 1,400 defaults of both public and private firms – both rated and unrated instruments – in all industries.

The model has identified nine factors, grouped into the following five categories, to be helpful in predicting LGD.

❑ *Collateral*: There are four types of collateral – cash, all assets, property plant and equipment, and support from parent/subsidiaries.
❑ *Debt-type and seniority grade*: The former differentiates between loans, bonds and preferred stock, etc, and the latter between secured, senior unsecured and subordinate debt, etc.
❑ *Firm-specific capital structure*: Adjusted leverage and seniority standing, firm-specific Moody's KMV distance to default, etc.

❑ *Industry*: This includes historical average industry recoveries, industry or region aggregated distance to default, etc.
❑ *Macroeconomic factors*: Industry or region's aggregated distance to default using local data (for the industry and region).

This improves the calibration of historical RRs. The model addresses the following three problems in the area of recovery:

❑ there is an insufficient LGD dataset to build and validate the LGD model;
❑ the bankruptcy process is complex and not suitable for modelling;
❑ not much empirical research has been conducted to identify predictive factors.

Business requirements for a system to capture recovery information

Data quality, default designation and loan repayment are the primary issues in recovery management. There are two approaches to record default and aggregate repayment received towards recovery. The first approach is to build capabilities into each of the product processors separately. The second approach is to build a recovery system to manage the recoveries in all the systems.

Since banks have separate product processors, it is very difficult to build a single view of all the defaults and recoveries for a customer. A separate recovery system helps in consolidating default for various facilities (products) that are otherwise scattered across various product processors. A separate system helps in recording recoveries, losses and costs, and in analysing the recoveries on various dimensions. Recovered cashflow may be attributed to interest, principal and liquidation of collateral. Non-cash recoveries may need to be converted into cash recoveries.

A discount rate appropriate for the riskiness of the distressed instruments should be used to discount cashflows, as opposed to the original contract rate on the loan or a cost-of-funds rate.

In general, there are four stages in the life of assets:

❑ loan origination;
❑ loan approval;
❑ loan collection;
❑ loan recovery.

Recovery starts when an account goes into collectability status 3, ie, it has defaulted. The loan recovery system tracks the loan from default to write-off. A loan can be restructured (upgraded to non-default status) at any time. The aims of the recovery system are as follows:

❑ record claims/recovery;
❑ compute or simulate interest and penalties;
❑ generate two accounting views of the facility, one for a customer depicting what is due from the customer and a second internal view depicting the actual accounting position including loan loss provisioning or amounts written off.

Interface requirements
After a facility or obligor is marked as a default, the amount received from the customer should be recorded as recovery. The recovery system should have the following capabilities.

❑ The system should be able to consolidated the following amounts and details from the loan processing system, trade finance system and the treasury system from the date of past due:

    ❑ principal outstanding and past due;
    ❑ interest outstanding and past due;
    ❑ penalties outstanding and past due;
    ❑ collateral details including valuation;
    ❑ guarantor detail;
    ❑ foreclosure detail;
    ❑ (CIF) and group CIF of the customer;
    ❑ loan repayment schedule;
    ❑ details of the loan;
    ❑ branch, region, account officer, industry classification.

❑ The system should be able to set two views of the account.

    ❑ The customer view, which includes the following features:

        ❑ The account should not reflect the loan loss provision and the fact that the loan is written off;
        ❑ the repayment received should include assets settlement;
        ❑ the repayment received is first applied according the policy of the bank;

❑ a statement on overdue principal, interest, fees and penalty accruals, interest on interest, penalties, interest and repayment received;

❑ The internal view should include the following features:

  ❑ the amount of outstanding loan loss provision/written off, and repayment received and not yet applied should be shown separately;
  ❑ repayment should be appropriated towards principal first and then towards interest;
  ❑ the amount of original principal, overdue interest, penalties and interest, fees and penalty accrual.

❑ The system should contain the capability of simulating interest and penalties for:

  ❑ different rates for different periods;
  ❑ different methods of calculation of interest (simple, compound);
  ❑ different methods of calculation of penalties, fees and charges;
  ❑ different methods of appropriating repayments from a date or from a collectability status.

❑ The system should have the capabilities to record and manage:

  ❑ collateral valuation and collateral and guarantee tracking;
  ❑ repayment methods, repayment schedules, follow-up dates and actions;
  ❑ assets settlement, repayment received and appropriated;
  ❑ appropriation of repayment into principal, interest, penalties and fees, and reduction in outstanding due to settlement, forgiven or loss;
  ❑ status of the account, including collectability status;
  ❑ loan loss provided and applied;
  ❑ links to the history in the group companies;
  ❑ record expenses incurred on recovery.
  ❑ LGD and RRs computation.

LGD UNDER BASEL II

As with default rate measurement, the Basel II treatment for LGD depends on the risk asset class, IRB approach, and credit risk mitigation supplied.

For retail exposure, the bank provides its own estimates of LGD for both the foundation and advanced approach. A minimum of five years of data is required to estimate LGD for retail portfolio and LGD is estimated at a pool level. For corporate, sovereign and supervisory categories, LGD is 45% for senior claims and 75% for subordinated claims LGD under the Foundation IRB (FIRB) approach. Under the advanced IRB approach, the bank provides its own estimates of LGD for corporate and sovereign assets class.

Collateral type: There are two types of eligible collaterals – eligible IRB collaterals, which include receivable, specified commercial and residential real estate, and eligible financial collaterals such as cash, eligible financial securities and bonds.

The Accord recommends the following best practices for eligible collaterals.

Best practices for eligible collaterals other than financial collaterals:

❑ The claim on the collateral must be legally enforceable and all legal requirements for establishing a claim must have been fulfilled.
❑ Collateral should not be valued at more than the fair value on the date of valuation.
❑ The bank should monitor collateral on a frequent basis, at least once a year. For a volatile market, more frequent valuation is required. Statistical methods or the services of a qualified professional must be used.
❑ Junior liens should be recognised only if there is no doubt about the legal enforceability of such claims.
❑ Property taken as collateral must be adequately insured. Other claims such as taxes on the property also need to be monitored.
❑ Receivable means – claims with an original maturity of less than one year, repayment from commercial or financial flows from the underlying assets of the borrower. It is a self-liquidating debt arising out of the sale of goods or services. It does not include securitisation, sub-participation or credit derivatives.
❑ For receivable as collateral, the bank must have a sound process for determining the credit risk.

Treatment of collateral: The Accord prescribes the following two approaches.

❑ *Reduce the exposure*: The reduced exposure is computed as follows:

$$Reduced\ exposure = \max\{0, [Ex(1 + He)] - [C(1 - Hc - Hfx)]\}$$

where He is the haircut for exposure, C is the current value collateral, Hc is the haircut appropriate for collateral and Hfx is the haircut appropriate for foreign exchange.

❑ *Divide the senior exposure into collateralised and un-collateralised portions*: Different LGD rates are applicable for collateralised and un-collateralised portions. Both eligible and non-eligible collaterals reduce the losses from defaulted assets. Eligible collaterals also reduce the capital required.

Credit risk mitigation and LGD

There are various approaches to recognise the credit risk mitigation. If the credit risk mitigation is partial or in different currencies, the exposure can be split into covered and uncovered. The treatment differs with the type of protecting instrument, ie, collateral, guarantee, credit derivative, and between the supervisory supplied and bank determined LGD.

For both the supervisory supplied LGD and the advanced approach, the risk mitigation provided by guarantees and credit derivatives reduces the LGD in one of the following three ways:

❑ LGD for the secured credit is lower than the LGD for unsecured credit. Gurantees and credit derivatives secure the credit;
❑ by reducing the exposure by the estimated protection provided (reduction may not be 100% of the collateral value) and applying the applicable LGD estimates to the remaining exposure as unsecured;
❑ by replacing the LGD applicable to the underlying transaction with the LGD applicable to the guarantor or the protection provider.

For the advanced approach, the bank can choose to reflect the risk mitigation effect on either PD or LGD.

The Accord differentiates between a senior claim and an subordinated claim.

For banks using their own estimates of LGD, the risk mitigating effects of guarantees and credit derivatives can be reflected through either adjusting the PD or the LGD estimates.

Facility rating

For banks using the advanced approach, facility ratings must exclusively reflect the LGD. Facility ratings should reflect any and all factors that can influence LGD including, but not limited to, the type of collateral, product, industry and purpose.

There is no specific minimum number of facility grades. The only requirement is that a bank must have a sufficient number of facility grades to avoid grouping facilities with widely varying LGDs into a single grade. The criteria used to define facility grades must be grounded in empirical evidence.

Banks must also collect and store a complete history of data on the LGD and EAD estimates associated with each facility and the key data used to derive the estimate and the person responsible. Banks must also collect data on the estimated and realised LGDs and EADs associated with each defaulted facility. Banks that reflect the credit risk mitigating effects of guarantees/credit derivatives through LGD must retain data on the LGD of the facility before and after evaluation of the effects of the guarantee/credit derivative. Information about the components of loss or recovery for each defaulted exposure must be retained, such as amounts recovered, source of recovery (eg, collateral, liquidation proceeds and guarantees), the time period required for recovery and administrative costs.

LGD under economic downturn conditions

Paragraph 468 of the framework document requires that the LGD parameters used in Pillar 1 capital calculations must "reflect economic downturn conditions where necessary to capture the relevant risks". The purpose of this requirement is to ensure that parameters are embedded in the LGD forecasts if credit losses are expected to be substantially higher than average. However, if the increase in credit loss is only due to an increase in the default, and RR is independent of increase in default rate, the LGD parameters need not incorporate the impact of economic downturn. This means that there can be an economic downturn without a change in the LGD and such an economic downturn need not be recognised in a LGD computation.

Economic downturn conditions are to be tested with respect to asset class. The same economic conditions impact each asset class differently. Therefore, for an economic condition to be designated as a downturn, a higher than average LGD is a pre-requisite. For example, a higher unemployment rate or negative GDP growth can be designated as an economic downturn if accompanied with higher LGD or distressed LGD risk drivers for a given risk class. The downturn conditions may also vary geographically. The same risk drivers may also be determinant of PD; this is called adverse dependencies, ie, both PD and LGD are dependent on the same risk drivers and both worsen at the same time. The following steps are used to incorporate the impact of adverse conditions on the LGD estimations.

❑ Identify the downturn conditions.
❑ During a downturn condition, the absolute value of LGD typically increases for two reasons: an increase in the number of defaults and a reduction in recovery due to a reduction in collateral value. The impact of the economic downturn conditions on the LGD may be estimated or established as follows:

  ❑ compare the loss rates (LGD) during downturn periods with average long-term loss (LGD) rates (this is the absolute change in the loss);
  ❑ compare observed RRs for defaulted exposures to the LGD from given typical collateral values (due to adverse dependencies between RRs and defaults);
  ❑ compare RR forecasts derived from robust statistical models that use both "typical" assumptions about collateral value changes and appropriate "downturn" conditions (this is due to a reduction in collateral values).

❑ Identify and establish the adverse dependencies between PD and RRs. The adverse dependencies might be identified by some or all of the following steps.

  ❑ The process starts with the identification of risk drivers determining the RRs.
  ❑ The second step is to identify the relationship of the identified risk drivers with PD.

❑ The next step is to identify the impact of the identified risk drivers on PD and RRs under adverse economic conditions. This can be performed by using the following measures:

❑ a comparison of average RRs during normal economic conditions with RRs observed during the downturn periods;
❑ analysis of collateral values during economic downturn;
❑ a statistical analysis of the relationship between observed default rates and observed RRs over a complete economic cycle;
❑ the stressing of appropriate risk drivers to forecast losses or other risk components.

Computation of downturn LGD

Regulators do not identify a downturn period but require banks need to identify a downturn period and document it with evidence of larger losses. The economic downturn period is different for banks and products within the same bank. The economic downturn period may extend to several quarters before and after peak LGD. During the economic downturn period, some products may exhibit higher than average losses while others may exhibit lower than average losses.

PD and LGD losses do not peak during the same period. It takes many months before the economic downturn is observed across the credit cycle. An adverse dependency between PD and RR can only be established using statistical tools.

The following steps are used for identification of downturn period.

❑ First identify the downturn period.
❑ Estimate the LGD during the period at the portfolio level on a balance-weighted basis.
❑ Apportion the LGD to the product and segment level. The following approaches are employed for apportionment at the product level:

❑ the actual LGD for each product;
❑ the long-run default weighted average LGD;
❑ apportionment of the actual LGD on the basis of the proportion of long-run default weighted LGD of the product in the portfolio.

REFERENCES

Acharya, V., S. T. Bharath, and A. Srinivasan, 2003, "Understanding the Recovery Rates on Defaulted Securities", Working Paper, London Business School.

"The Role of Recovery Analysis in Ratings – Enhancing Informational Content and Transparency", *Credit Policy*, Criteria Report, *Fitch*, 2005.

"Recovery Ratings – Approach and Process for Corporate Finance", *Corporate Finance*, Criteria Report, *Fitch*, 2005.

"Structured Finance Distressed Recovery Ratings", *Structured Finance*, Criteria Report – Exposure Draft, *Fitch*, 2006.

Frye, J., 2000, "Collateral Damage – A Source of Systematic Credit Risk", *Risk Magazine.*

Pykhtin, M., 2003, "Unexpected Recovery Risk", *Risk*, August.

8

Credit Risk Fortification

INTRODUCTION

Banks have established credit policies to address target markets, portfolio mix, price and non-price terms, the structure of limits, approval authorities, exception processing/reporting, etc. The risk policies, as a part of existing credit policies, measure credit risk in a very preliminary and crude way and apply credit risk related controls only as preventive controls (exposure limits and ordinal ratings) and not as proactive risk management controls. In order to measure credit risk, credit policies and processes need to be upgraded and augmented with credit risk management focused policies, processes, standards and measuring tools. Discussions on credit risk policies are out of the scope of this book. Based on the various surveys published by the Basel Committee and the personal experiences of the author with various banks, this chapter focuses only on the *incremental* policies, processes, standards and measurement tools pertaining to credit risk.

Credit risk fortification

As identified in figure, in the business model for banks is changing. This changed business model has permanently impacted the organisation structure for risk. Therefore, in addition to what has been covered in the other chapters, banks are required to incorporate the impact of credit risk measurement on the accounting policies and suitably reorganise their credit management organisation, building the organisation units required for discharging credit risk management responsibilities.

Risk measurement capabilities and technologies have permanently changed the asset valuation and accounting processes, the implementation of limits across the banking and trading books and actively-managed credit risk portfolios.

In this chapter, we shall *cover only those areas which have not been covered by other chapters* but are a must in order to support and protect the credit risk management framework built into each of the other chapters and also to fully extract the benefits of the credit risk management. This is the reason why this chapter has been named "credit risk fortification".

CREDIT RISK POLICIES, PROCESSES AND STANDARDS

The credit risk management process can broadly be divided into a four-step process: identification, measurement, monitoring and the reporting of credit risk.

Credit risk policies should cover credit risk in all banking activities including lending, trading, investments, liquidity support/ funding and asset management. Credit risk policies and practices should be consistent with applicable accounting frameworks and appropriate supervisory guidance. There may be a difference between the accounting and regulatory capital frameworks. The senior management of the bank is responsible for the compliance with both the frameworks. Management often work with the regulators and accounting standard bodies to synchronise the two frameworks.

Policies to identify credit risk

The identification of credit risk broadly covers the criteria for risk segmentation, the definition of default, rules for concentration, implementation of limits and acceptable risk mitigation techniques. It also includes the unbundling of credit risk into risk components (exposure at default (EAD), PD, LGD and correlation) measurement.

Risk segmentation

Risk segmentation has a major impact on the risk quanta, risk measurement methods and processes and the risk components to be measured. The widely accepted criterion is to segment exposures on the basis of orientation, product, granularity, collaterals offered and loan characteristics. The Accord provides guidance on defining retail segments, segregating Small and Medium Enterprises exposures into retail and corporate, risk measurement for specialised lending such as project financing, high-volatility commercial real estate exposures and object finance exposures. The bank should have a policy to segment the exposures into risk segments and such policy should be consistently applied.

Risk segments are further divided into portfolios to bring homogeneity and better risk measurement. Although offering finer risk segmentation provides homogeneity for the risk measurement and finer risk segmentation reduces the data points needed for risk measurement. Risk characteristics, assets and product characteristics and the size of exposures are the general elements considered for segmenting exposures. The Basel Accord, the documents on validation techniques and IRB approaches (published by Basel and national supervisor), Basel documents supporting Quantitative Impact Studies (3, 4, 5) and the advanced notice on the proposed rulemaking (ANPR) published by US regulators are some of the useful reference documents for framing risk segmentation policies.

Definition of default
We have explained the definition of default in detail in the "Probability of Default" chapter. A strict definition of default increases the volatility of loss (unexpected loss), while a slack definition may not be acceptable to banking regulators. Therefore, a balanced and consistent approach is needed. A definition consistent with the local supervisor, accounting standards and legal requirements may be adopted. The definition of default may vary for the different risk segments (if permitted by national regulators), although banks generally employ stricter internal default definitions.

Risk concentration
The bank must have effective policies, procedures and measurement tools to monitor single-obligor and sector concentrations on an aggregate basis. This covers the following issues.

❑ *Defining credit risk concentrations.*
❑ *The treatment of individual entities in a connected group*: This includes the circumstances under which the same rating may or may not be assigned to some or all related entities.
❑ *The implementation of limits*:

315

❑ aggregate limit for the management and control of all of its large exposures at the obligor and group levels;

❑ country and sector level limits;

❑ compliance to large exposure limits specified by the national supervisor;

❑ credit category level limits;

❑ a granularity level for the size of one individual exposure relative to the total pool, the size of the pool of receivables as a percentage of regulatory capital or the maximum size of an individual exposure in the pool.

❑ *Computation of risk contribution*: Banks with advanced risk measurement tools measure average and marginal risk contribution.

Compliance to limits

There are three types of limits to comply with:

❑ controlling exposures through fund based limits, pre-settlement limits and settlement limits;

❑ economic capital based limits to control credit risk exposures on both individual counterparty and portfolio basis;

❑ rating grade based limits to control exposures within each credit grade, within speculative grades and default grades.

Policies for measuring credit risk

Risk measurement methodologies differ with the risk segment.

Underwriting policy

The underwriting policy includes credit ratings and loan approvals. It must reinforce and foster the independence of the rating process from the relationship managers and sales teams. It must also reinforce the segregation of the rating process from the loan approval process for high ticket exposures. The policy also deals with credit committee structures, authorisation powers and handling exceptions. The credit rating process has been explained in the "Credit Rating" chapter. The underwriting process is followed not only for the new credit but also for the existing credits and includes the process for credit under default to classify them under special mention categories.

Risk mitigation techniques

We have explained in detail the collateral, guarantees, securitisation, credit derivatives, credit insurance and netting in the "Credit Risk Mitigation" chapter. Policies, processes and standards are required for the following issues.

❑ Design contracts and documentation for credit derivatives, securitisation and loan trading has been an area of major challenge. Formulating clauses for events and restructuring has been difficult.
❑ Accepted role of the rating agencies in the credit risk mitigation and transfer techniques.
❑ Counterparty risk in the unfunded risk transfer.
❑ Accounting, valuation and measurement of risk mitigation instruments.
❑ Capital computation and regulatory treatment for risk transfer instruments.

Policy on credit derivatives

The policy on credit derivatives covers the following issues:

❑ strategy and limits for different types of credit derivatives (CDs) (we have explained the types of CDs in the "Credit Risk Mitigation" chapter);
❑ managing basis risk, residual risk, legal risk and counterparty risk;
❑ measuring, monitoring, reviewing, reporting and managing the associated market, liquidity and operational risks such as credit risk, market risk and liquidity risks;
❑ standardisation of contract and contractual terms;
❑ the selection of reference assets and obligations;
❑ valuation and accounting;
❑ internal controls, audit, record keeping and settlement.

Country risk

Banks with cross border assets have to develop a country risk management framework. The framework includes an internal country risk rating system, which taps into the expertise of the bank relating to the markets it operates and where the country assessments are made independent of business decisions. Using the country assessment, country limits are set. Day-to-day operational country limits, called working limits, are also imposed to manage the shape and growth of the cross-border exposures as they build up. A rigorous environment scanning process is built for proactive action to roll back country exposures, as and when warranted. We have dealt with sovereign risk in detail in the "Credit Rating" chapter.

Risk measurement systems
The banks frame policy, processes and standards in the following areas.

❑ Type, design, calibration and benchmarking of credit scoring models, market information based models, credit rating systems, portfolio management systems, capital measurement systems, reference data, mapping of portfolio data to reference data and the types of models to be used for each risk segment type.
❑ The meaningful distribution of exposures across grades with no excessive concentrations, on both borrower-rating and facility-rating scales.
❑ Quantitative and qualitative information and credit risk factors to be used in measurement models.
❑ Assumptions and parameters (eg, holding period, observation period, confidence interval, etc.), performance over time, model validation and stress testing.
❑ Continuously monitoring the changes in the PDs associated with internal and external ratings.

In a credit risk grading system, a bank should address the definitions of each credit risk grade and the delineate responsibilities for the design, implementation, operation and performance of the system to minimise and manage any conflict of duties.

Credit risk organisation
Managing conflict of duties and the risk management functions necessitates building credit risk organisation. We have dealt with this in detail in a separate section in this chapter.

Standards to measure expected loss and loan loss provisioning
The standards cover concept, measurement, and application of expected loss (EL). The expected loss needs to be synchronised with the accounting loan loss provisioning. If the EL is less than the loan loss provision (LLP), revisit mapping of credit grades calibration to the losses. If the LLP is less than the EL, revisit the credit

quality. National supervisors allow the excess LLP over EL to be considered as a part of tier II capital, only after such a revisit.

We have dealt with the accounting policies in the credit risk environment in detail in another section of this chapter "Accounting Policies under Risk Measurement Regimes".

Impaired assets

The security held plays an important role in impairment definition. According to paragraph 452 of the Accord, default occurs when the obligor is unlikely to pay in full without recourse to realising the security (if any) or the customer is past due for more than 90 days. Assets are marked impaired when security is insufficient for the repayment of both principal and interest and facility is overdue for 90 or more days.

In the United States, the definition of non-accrual/impaired assets is standardised. Assets are marked impaired if payment in full of principal and interest is not expected and where principal and interest have been in default for 90 days, unless the asset is both well-secured and in the process of collection.

Defining impaired asset

Impaired assets definition includes how and when a bank determines an asset to be impaired. It also includes propagation of

---

Sample policy on cross impairment/default
Where a customer has multiple facilities and one of the exposures is impaired, all exposures to the same customer will be classified as impaired except when there is sufficient security to ensure ultimate collectability of all principal and interest on both impaired and performing exposures.

Where the customer belongs to a group, all facilities to the related parties in the group shall be classified as impaired except in the following circumstances:

❑ there is no cross collateralisation of facilities or cross guarantee arrangements between the related parties;
❑ cross-collateralisation and guarantee arrangements exist but, in aggregate, there is sufficient security among the group of related parties to ensure ultimate collectability of all principal and interest on both the impaired and performing exposures.

These standards do not cover retail exposures.

default. If one of the assets of a customer is classified as an impaired asset, all exposure to the customer should be classified as impaired.

Measuring Impairment: The policy to measure the extent of the impairment on individual exposure is dependent on the following factors:

❑ present value of expected future cash flows;
❑ fair value of collateral less costs to sell; or
❑ the loan's observable market price.

Measuring impairment in the assets assessed on the portfolio basis: Typically, homogenous exposures within retail and SME segments are managed on the portfolio basis using statistical techniques (please refer the "Credit Scoring" chapter – payment and behavioural models are applied to assess the impairment). Portfolio managed exposures are often not subject to formal regular review other than in cases where payments are in arrears. Typically, assets with payment in arrears up to a certain number of days are not included in impaired assets. After the asset is assessed as impaired, the asset can be written off directly or can be assessed on the basis of security whether the asset is sufficiently secured.

Provision or reserves against losses on impaired assets: The reserve approach and method will differ from portfolio to portfolio. For consumer and residential mortgage loans, the reserve amount will typically be determined through a formula, either based on the aging of the portfolio or on the bank's experience with the particular type of loan.

Provisioning approaches are based on the following factors.

❑ Solely on the number of days past due.
❑ On regulatory or internal loan classification.
❑ For larger loans, based on the loan review process.
❑ Recovery ratings.
❑ The market value of the collaterals or defaulted bonds/securities.
❑ Capital provision on the past due loans. According to paragraph 75 of the Accord, a loan past due for more than 90 days attracts capital on the unsecured portion netted for specific provisions. The capital charges for the unsecured portion are related to the specific provision.

Restructuring

The following concessions are an indicator for the exposure being classified as restructured:

❏ a reduction in the principal amount of the facility, or the amount payable at maturity, as set down in the original loan agreement or reduction in the accrued interest rate;

❏ at least 200 basis points (bps) lower interest rate than that being charged to other similar customers and facilities;

❏ a reduction of accrued interest, including forgiveness of interest;

❏ a deferral or extension of interest and/or principal payments;

❏ an extension of the maturity date or dates at a stated interest rate.

The restructuring policy includes the following factors:

❏ approval authorities and reporting requirements;

❏ the minimum age of a facility before it is eligible for restructuring;

❏ the delinquency levels of facilities that are eligible for restructuring;

❏ the maximum number of restructuring per facility;

❏ a reassessment of the borrower's capacity to repay;

❏ status after restructuring and during the watch period.

Sample policy for status after restructuring: Regulatory rules mandates a watch period before upgrading delinquent customers to collectability status 1. Table 1 will determine the collectability status after restructuring.

Table 1 Collectibility status after restructuring

Collectability status before restructuring	Collectability status immediately after restructuring	Collectability status after 3 months or 3 installment periods, whichever is later after restructuring (after the watch period)	
		Performing No Days Past Due (DPD)	Non-performing – applicable according to DPD after the date of restructuring but not better than the following status
2	2	1	2
3	3	1	3
4	3	1	4
5	3	1	5

If there is sufficient evidence to demonstrate a relative improvement in the condition and debt service capacity of a customer (apart from performance to date) that would warrant a return to accrual status prior to the three instalments or three months, whichever is the later threshold, then the facility may be upgraded to performing status. This includes additional permanent sources of income such as the signing of a lease or rental contracts, or an equity injection.

Charge-off policy

Charge-off policies are determined by the same factors as an LLP policy. In addition, the legal bankruptcy framework also impacts the policy. Charge-off is faster in an efficient bankruptcy legal framework (such as the United States) as opposed to frameworks in other countries where bankruptcy processes are not very efficient. The following factors determine the charge-off policy:

❑ the bankruptcy process is completely over;
❑ the bank is nearly certain that it will not be able to recover anything;
❑ a time period prescribed by regulation has elapsed;
❑ the charge-off is accelerated where tax authorities allow faster charge-off.

Policies for monitoring credit risk

Internal risk ratings are an important tool in monitoring credit risk. Internal risk ratings should be adequate to support the identification and measurement of risks from all credit exposures, and should be integrated into an institution's overall analysis of credit risk and capital adequacy. The ratings system should provide detailed ratings for all assets, including those assets that require special attention or problem assets. The monitoring and review of loan loss reserves should be included in the credit risk monitoring.

Independent risk review teams should conduct regular reviews of the risk quantum and risk processes. These reviews provide senior management with objective and timely assessments of the effectiveness of credit risk practices and ensure that group-wide policies and guidelines are being adopted consistently across different business units, including any relevant subsidiaries.

Past due

The bank must have clearly articulated and documented policies with respect to the counting of days past due. Past due computation is useful for banks for their collection strategy, the classification of loans to non-performing categories, loan provisioning and determining default. The banks have developed automated systems to compute past due for exposures.

The loan review process

The loan review process is comprised of three sections: LLP, active portfolio management and recovery and legal action.

For the loan review process, the bank should have an effective loan-grading system to identify the differing risk characteristics and loan quality problems in an accurate and timely manner. In addition, the bank should have sufficient credit risk management processes to ensure that all relevant loan-review information is appropriately considered when estimating losses.

Validation policy

Within each risk segment, different risk measurement systems are established to measure credit risk. Each of the following risk segments needs a separate validation policy and methodology:

❑ corporate;
❑ SMEs;
❑ specialised lending;
❑ purchased corporate receivables;
❑ sovereign;
❑ bank;
❑ equities;
❑ residential mortgages;
❑ qualifying revolving retail;
❑ other retail.

Validation policy includes back testing, definitions, methods and data for estimation and the validation of the required risk component, ie, PD, LGD and/or EAD, including assumptions employed in the derivation of these risk components.

Review of risk assessment process

The bank should conduct periodic reviews of its risk management process to ensure its integrity, accuracy and reasonableness. The following areas should be included in the review:

❑ appropriateness of the bank's capital assessment process given the nature, scope and complexity of its activities;
❑ identification of large exposures and risk concentrations;
❑ accuracy and completeness of data inputs into the bank's assessment process;
❑ reasonableness and validity of scenarios used in the assessment process;
❑ stress testing and analysis of assumptions and inputs;
❑ model risks and model assumptions.

Reporting credit risks

We will now outline some of the techniques used to capture all material risks.

Assumptions on holding periods, haircuts, or volatility

Every model and measurement makes various assumptions and the various types of assumptions made by the publicly available models are explained in the "Credit Risk Portfolio Models" chapter. These assumptions need to be continuously reviewed to check if all the material credit risk components are being measured.

Stress tests

In the past few years, various "shocks" in the financial market have adversely impacted the creditworthiness of borrowers. As a result, the stress testing of credit risk has assumed increasing importance. A rigorous forward-looking stress testing that identifies possible events or changes in market conditions that could adversely impact the bank should be performed.

Credit risk concentrations are also tested using stress testing. Banks generally use a combination of "top-down" and "bottom-up" credit risk stress testing approaches to assess the vulnerability of the portfolio to "exceptional but plausible" adverse credit risk events.

Disclosures

Market disclosures, the third pillar of the Accord, are a primary driver in the credit risk measurement and management structure. Disclosures are aimed at providing assurance and insight to the various stakeholders and regulators of risk management policy, process, standards and measurement tools. The effectiveness of credit risk measurement and management is substantiated through the quantitative disclosures on various risk measures, risk mitigation and securitisation. For more details, readers may like to refer to the Accord.

ACCOUNTING POLICIES UNDER RISK MEASUREMENT REGIMES

Every financial intermediary manages credit risk – the difference is in the level of granularity in the risk measurement. In this section, we will discuss the changes in the accounting policies when credit risk is measured at the level of granularity recommended by Basel II.

The existing loan provisioning norms are based on the past due or the EL in the next one to three year period and are not based on the expected economic loss. Critics argue that the existing accounting policies and other processes cannot identify the borrower's financial problems at an early stage because none of them are based on identification of the economic value of the loan assets. The economic value of the assets is implemented through fair value accounting of financial instruments. The banking industry has not yet agreed to adopt the fair value accounting for assets in the banking book.

Accounting policies aim to reflect the correct financial assets and liability values in the financial statements. Risk management principles and practices play a key role in financial instrument valuation. The role played by risk management has implications for accounting policies dealing with valuation of defaultable assets and credit loss provisioning.

For reasons of prudence, the traditional accounting policies encourage the overestimation of credit loss. Prudent accounting policies aim at a reduction in the income volatility and build margins for errors in the assets valuation that might have crept in due to absence of good and verifiable quantitative valuation and risk measurement systems.

Risk management has enabled assets valuation in a more refined, granular and verifiable manner, reflecting economic valuation.

Through stress tests, sensitivity analysis, external and internal benchmarks and validation processes, the best risk management practices aim to reduce the valuation errors. An improved risk and value measurement is the key differentiator. Therefore, the risk management implementation has a major impact on the traditional prudential accounting policies.

Fair value accounting for defaultable assets

The primary criticism of risk management has been that it has not developed significantly over the previous century during which accounting prudence has been in practice. The impact of risk management on the accounting policy and standards is not a one-way flow. The provisions of IAS 39 encourage banks to adopt the sound risk management practices and sound practices for managing banking and trading books. The IAS 39 also reduces the differences between the banking and trading books.

Both accounting frameworks and Basel II recognise credit grading as a tool to assess credit risk. Both frameworks recognise movements in credit risk, and not just severe credit deterioration. In choosing to measure an asset at fair value under IAS 39, banks need to satisfy the following two conditions:

❑ assets are managed and evaluated at fair value according to the risk management or investment strategy policy;
❑ to minimise any attempt to reduce earnings volatility, reduce basis risk or reduce any un-hedged risk factors within the hedged risk position.

Fair value measurement of financial instruments under IAS 39 has raised several practical implementation issues. In the absence of active markets, there are practical difficulties in obtaining or computing reliable fair values for non-marketable financial instruments that are otherwise held at cost.

Leading accounting standard-setters are currently discussing the advantages and disadvantages of moving toward a greater use of fair value accounting for financial instruments. Without prudent and balanced standards for the estimation of fair values, particularly when active markets do not exist (such as is often the case for loans), the use of a fair value model could reduce the reliability of financial statement values and increase the volatility of earnings and equity measurements. The Basel Committee believes that fair value accounting is appropriate under situations such as, for example, financial instruments held for trading purposes. However, sufficient regulatory data to provide the appropriate guidance on the estimation of fair values and on the treatment of the value adjustments needs to be provided before this system of accounting can be extended to all banking book financial assets and liabilities.

Synchronisation of loss provisioning with loss rates

Traditionally, the link between LLP and actual losses has been weak due to a lack of information systems and models to measure assets quality. To safeguard the banks, and as a measure of prudence, banks have built loan loss reserves through general LLPs.

With the development of risk management practices, risk management information systems and models, banks are now in a position to establish a relationship between the LLPs and historical losses. The relationship between loss provisioning and historical loss has received recognition from the Basel Accord by way of allowing the excess LLP to be a part of tier II capital.

Loss rates are determined by measuring historical loss data over a period that may be determined by regulatory instructions or the risk policy. Loss is computed for a portfolio based on the product and geographical structure.

Charge-offs and recoveries rates are considered in determining the loss rates for each of the loan grades. The rates may be adjusted for changes in the economic factors since the losses were suffered. The determined loss rates are to be supported by documentary evidence. The Basel Accord recommended documentary evidence is as follows.

❑ Computation of historical loss rates. This includes loan volume, delinquencies, restructurings and concentrations, historical charge-off and recovery history.
❑ Computation of loss rates.
❑ Underlying assumptions for the computation of loss provisioning.
❑ Collateral valuation process and underlying assumptions.
❑ Compliance with supervisory requirements.
❑ Benchmarking against the bank's peers.

Under the IRB approach, as a part of pillar 3 on market disclosures, banks are required to disclose actual losses, which includes charge-off and specific provisions at each of the portfolio levels and are required to compare it over a period. Banks are also required to discuss the reasons for such variation (such as changes in the EAD, PD and LGD).

Accounting policies for the impaired assets

For impaired assets, banks need to formulate accounting policies on the following issues.

❑ Interest previously accrued but not received, is reversed or capitalised.
❑ The treatment of cash received on an impaired asset. This is primarily driven by the contractual terms with the obligor. In the absence of contractual terms, a bank should have a policy to apply the cash received. In general, the cash received is applied in the following order:

  ❑ statutory charges;
  ❑ penalty interest and fees;
  ❑ overdue interest and fees;
  ❑ current interest and fees;
  ❑ principal.

❑ If interest accrual continues, it is provisioned fully. In particular, this is required to provide the same view of the customer account internally and externally. If interest is not accrued and posted to the account during the default period, some of the courts do not permit the charging of interest.

Considering the impact of collateral on the LLP is very difficult and varies across the regulatory regimes. As we have seen in the "Credit Rating" chapter, the best practice is to have a two-dimensional rating – the obligor rating and facility/recovery rating. The collateral values get fully reflected in the recovery rating. While defining the default and interest rate accrual, the best practice is to follow the obligor rating and for the recovery and loss provisioning follow the recovery ratings. This will also be in tune with fair value accounting. The only problem is for non-financial collaterals – collateral valuation and recovery rating is difficult in such cases.

The accounting policies are supported by impaired assets policies.

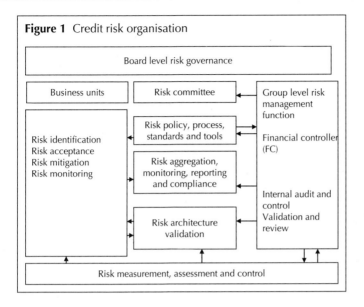

Figure 1 Credit risk organisation

CREDIT RISK ORGANISATION

The purpose of discussing credit risk organisation here is to bring home the importance of managing a conflict of duties in the credit risk management and in building and strengthening organisational units for credit risk functions.

Credit risk organisational units

These are organisational units (see Figure 1) that are required to be set up for discharging responsibilities in the areas of setting up credit risk policies, building and validating credit models and managing and measuring portfolios. In addition, some of the existing units are to be strengthened.

❑ *Risk governance*: This is separated into three levels – at board level, top management and business level. In the document on credit risk management, Basel has published standards for risk governance at the board and top management levels.
❑ *Risk measurement, assessment and control*: Risk management organisation at the business level includes independence and autonomy of risk measurement, veto power with the risk management, control over new products, passive versus active risk management and mitigation. Reorganisation and strengthening of risk assessment

units is required to manage and avoid a conflict of duties. For example, a conflict of duties in a credit rating and loan approval process is managed by involving an organisation unit that is independent of Relationship managers and sales managers.

❑ *Risk control process*: The segregation of risk management from the business management process, measuring the risk and adherence to risk limits. In addition, this also includes the measurement of risk limits and the setting up of granular limits. The existing limit management unit has to be strengthened to measure the economic-capital-based risks, to implement the centralised limit management system and to measure and manage limits at portfolio level.

❑ *Risk policies and standards*: Framing risk management policies is the key responsibility and distinguishing mark of risk management organisation. The depth of the risk management function is determined by the depth of the risk management policies and standards.

❑ *Risk architecture*: The management information systems (MISs) unit requires a major strengthening to support credit modelling. Tier I and II banks have already established credit risk architecture departments with the skills and experience of working with various credit products, software programming and credit risk quantification. This unit also includes risk quantification experts for measuring risk components, correlation and economic capital. Back testing, validation and stress testing also form part of the risk architecture.

❑ *Risk mitigation*: Collateral management and margining and management of work out and distressed positions.

❑ *Risk aggregation*: Management monitoring, reporting and compliance. This includes measuring risk at portfolio and limits management at country, sector and portfolio level.

Best practices have now emerged on managing "Material non-public" information by the credit market participants. This includes written policies and procedures for dealing with the private information acquired about the obligor, the working of independence compliance functions, the segregation of organisational units and duties, building information walls, record-keeping and training and education.

Maturity level in credit risk management organisation

The assets size, types of products, existing organisation structure of the bank, geographical spreads (spreads across the countries under different regulatory regimes) and availability of information systems are the five key determinants of credit risk organisation.

There can be various shades of credit risk organisation. No two banks have or can have the same credit risk organisational structure: every bank has to evolve its own. While evolving organisational structures, banks need to consider the following factors.

❏ Broad principles to avoid conflict of duties. This is reflected in the reporting structures, compensatory controls, human resources and compensation policies.
❏ Strengthening or building the skill base for the following reasons.

   ❏ Internal rating and risk assessment for new assets, existing assets or impaired assets or assets in a banking or trading book.
   ❏ Credit risk measurement. There can be specialised teams for each risk measurement model the bank is using. The team will require IT skills, model building skills and model validation skills.
   ❏ Active portfolio management. This includes portfolio measuring, monitoring, securitisation, sell-off, credit derivatives

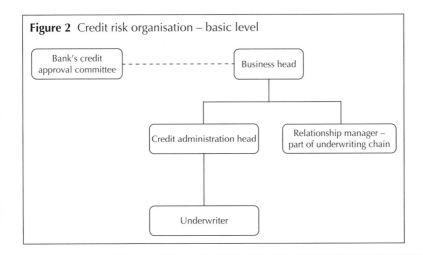

Figure 2 Credit risk organisation – basic level

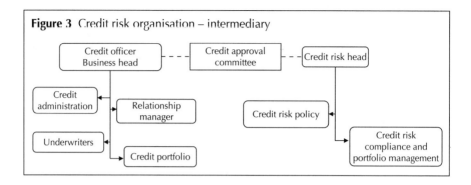

Figure 3 Credit risk organisation – intermediary

and credit insurance. This includes working closely with a structured products division if it is a separate division.

Larger banks with larger assets sizes that are spread across countries have developed federated risk management structures. Part of the risk is managed within the line of business or country with a centralised corporate level or group level policies and risk measurement. In general, the credit risk committees and risk officers, or the credit risk officers, work as a bridge and coupling to move the entire credit risk organisation. At a maturity level, there are three levels of maturity.

Basic level of maturity

At basic level (see Figure 2), there is not much of the credit risk management. Credit risk is generally managed as a part of the traditional credit administration through collaterals and limits. This type of organisation is and can still be seen in smaller banks.

Intermediary level of maturity

At an intermediary level, most of the credit risk functions exist but the intensity of function varies (See Figure 3). Conflicts of interest are also not fully managed. The credit risk organisation will often report to a business head, who is also referred to as the credit officer in some banks. However, not all risk management functions may report to a business head. In general, a risk management department exists, which is responsible for working on compliance with the Basel I norms, measuring limits for compliance to legal limits and issuing some credit risk policies such as impaired assets,

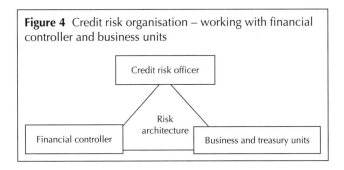

Figure 4 Credit risk organisation – working with financial controller and business units

pre-settlement limits, etc. This type of organisation exists in tier III banks and for some assets classes/risk segments in tier II banks. For example, retail businesses generally manage the risk management function within the business but the risk manager of the retail may have a dotted line relationship to the risk officer of the bank.

Matured risk management organisation
There are many characteristics for a matured credit risk management organisation. The first and foremost is the recognition of risk function and credit risk officer (CRO) separate from the financial controller (FC) and from the business heads (see Figure 4) and, in particular, the treasury head. The common areas between the FC and CRO are the capital computation and accounting policies for the recognition of credit risk in the banking and trading book. With the business heads a CRO shares risk measurement and management in the business.

To be an effective credit risk officer (see Figure 5), a manager should be supported by various capabilities described under the risk architecture. Another most important characteristic is the establishment of a risk committee at board and top management levels. The credit risk committee should have the following functions:

❑ establishing credit risk policy, credit risk measurement guidelines, credit risk management organisation structure, accounting loss/provisioning policy, risk mitigation/collateral management policy;
❑ the identification, measurement and monitoring of credit risk portfolio, including special loan and asset review situations, specific credit concentrations and credit trends affecting the portfolio;

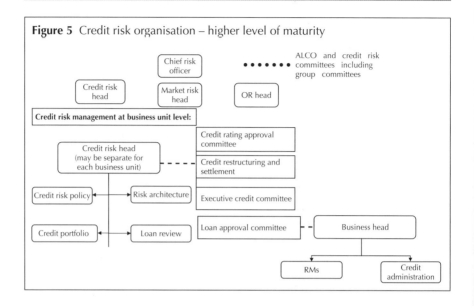

Figure 5 Credit risk organisation – higher level of maturity

☐ establishing credit limits and credit policy at the sector, business and country levels.

In case the bank is part of a financial group, a federated risk management organisation structure generally exists at the group level, as recommended by Basel. Credit risk is measured and managed in every asset. The bank has an active credit portfolio management functionality with separate divisions for structured finance and credit derivatives. Loan review structure is well established and the bank has developed information systems and the skills to manage the problem assets.

Principles for avoidance of conflict of duties

Independence of risk policy setting, risk measurement and risk monitoring is required from the risk takers. The front-end staff responsible for bringing business or selling the bank's products and services is the primary driver for credit organisation. One of the adopted best practices is to involve the relationship or sales managers in the initial assessment of the credit proposals. Proposals should be subsequently subjected to the strong credit risk assessment review process by independent underwriters. This makes RMs and sales staff responsible for the credit quality and helps to manage any conflict of duties through compensatory controls. Front-office staff should also be required to monitor the credit quality on a continuous basis. A 100% segregation of duties is neither feasible nor desirable. Instead, strong compensatory controls should be implemented. It is best practice to manage credit risk policy independent of the front office, although the front office may be influencing the credit risk policy within the prudence risk practices. It is also recommended practice to maintain and manage an independent risk-monitoring unit.

In addition, it is recommended practice to approve and review the risk management policies at the board level committee. The risk measurement and risk-monitoring processes should be completed at top management level or lower and not overridden by the board members or the committee.

Credit risk may be managed within a product, business or geographical level; for example, the Basel Accord is implemented at the group level. The best practice is to have an integrated organisation framework.

❏ Consistency in policy and procedures to identify, measure, monitor and control credit risk across products and activities, business, countries and the group. The Basel Accord allows different levels of credit risk measurement sophistication within a risk segment across different asset portfolios and across countries.
❏ Risk should be managed in all of the bank's activities and products.
❏ Risk should be managed at both the individual credit and portfolio level.
❏ Credit policies and credit risk policies should be communicated throughout the organisation.

Remuneration policies should be consistent with the credit policies and should not weaken the credit risk management process.

The credit review department, operating independently of the business units, audits asset quality, the accuracy of gradings, self-assessment and the state of credit risk management, and reports the results directly to the board of directors and the management committee.

MANAGING CREDIT LIMITS

The management of credit risk through the use of credit limits is as old as the granting of credit itself. However, the technology, risk management techniques and blurring boundaries between the trading and banking book and measurement of risk in terms of economic capital is changing the type, granularity and frequency of limit monitoring and management. Limit management differs across the institutions and assets. Limit management is broadly governed by the following factors.

❑ *Type of limits*: Different types of limits can be set by product, asset class, business lines, ratings, geography, industry, lending office, group, obligor or duration.

❑ *The size and nature of the institution*: Limit management in tier I (multinational bank with assets size of over US$300 billion), tier II banks (regional banks with US$100 billion assets) and tier III banks (national banks with asset size of over US$20 billion) is different. Limit management also differs according to the product and customers. No limit management is required for the retail segment of national banks. Tier I and II banks may require an aggregation of limits extended to directors and owners as a part of the retail offerings with their group limits. The aggregation of treasury and corporate limits may not be required for tier III banks that generally do not have corporate clients as their treasury clients. On the other hand, tier I may require a strong limit management in place for treasury and corporate clients since they deal in complex products.

❑ *Business strategy*: The portfolio management and limit system should be aligned to the business strategy of the bank. For example, a bank that is interested in collateralisation of its loan portfolio as a part of its business strategy should at a minimum measure and manage each of the concentrations like issuer concentrations, industry concentrations, and regional or country concentrations.

❑ *Credit risk appetite (economic capital/regulatory capital measures)*: Under the risk measurement regimes, limits are managed in terms of economic capital.

❑ *Competitive advantages*: In general, banks have expertise only in certain sectors or regions. This makes it difficult for the banks to avoid risk concentration. Although diversification by sector is beneficial, there is a diminishing marginal benefit to diversifying a portfolio by industry.

Issues in managing portfolio limits

No portfolio is homogenous: each portfolio has numerous products, exposures in various rating categories and other differentiating characteristics. Strictly speaking, the exposures to credit risks are not additive as they are for market risks. For example, within an industry, two exposures of US$10 million each in AAA and BBB are not same ie, they are not exposed to the same quantum of credit risk. Another example is that an unsecured facility and a secured facility are not exposed to the same credit risk. Economic capital is a common denominator to measure credit risk. Therefore, economic capital (EC), rather than the notional exposure, should work as a variable for managing the concentration or the risk. For rating grade, product types, tenors and collateral types, the measurement of economic capital can be a single measure.

A common approach to limits is to set them as a percentage of capital, with a variety of definitions available for capital – for example, tier I capital, tier II capital, regulatory capital, market capitalisation, or EC. Untill the time it becomes possible to compute economic capital for all types of instruments, portfolio limits can be managed with regulatory capital measurement. Typically, the legal lending limit is 15–25% of book capital. However, the problem is that the bank loses the notional amount and not the EC.

Banks have started adopting the Herfindahl–Hirschman index (HHI) to measure concentration risks and have started building the index. The HHI index can be built on the basis of exposure or the capital required.

It is necessary that the limits are actually enforced and the limit management team has the required authority, tools and skills to enforce limits.

Sector and geographic limits have the effect of limiting the loss from identified scenarios and are a powerful technique for managing "tail" risk and controlling catastrophic losses.

Centralised limit systems

Various facilities that are extended to the customers are managed by multiple product processors. Every product processor has some form of limit management. However, as an effective tool, obligor level limits should also be managed centrally. With the increased synchronisation of credit risk measurement techniques for banking and trading book products, a need for centralised limit management systems is increasingly felt by the banks, and banks have started central limit systems for monitoring aggregated exposures. Centralised limit systems (CLSs) capture the bank-wide exposures of obligors and their groups, control credit limits across the products and customers by managing sanctioned limits, drawing powers, exposures and collateral values from different product processors to enable bank-wide credit risk management, including monitoring limits, exposures and credit concentrations. CLS also provides facilities to implement omnibus limit management, pre-settlement and settlement limits.

The following impacts of a CLS system are felt by the risk management and credit administration processes.

❏ Designing and implementing risk and exposure assessment and control at the group level. Also defining the default propagation rules for a default within a group to other group companies/members.
❏ Risk management processes and controls to implement omnibus structure. The primary requirement is to rank the products on the basis of inherent credit risk.
❏ Standardisation of collateral information to capture and build master tables for information on collateral.
❏ Policies and processes for cross collateralisation.
❏ A mechanism to control and consolidate the total exposure.

ACTIVE PORTFOLIO MANAGEMENT

The major contributions of credit risk measurement have been the liquification of credit risk and also the convergence of the loans and bond markets. Credit risk measurement enables the segregation of the funding decision from the risk-taking decision. This segregation has helped broaden the range of players and attracts players with different choices. The segregation has attracted insurance companies, investment banks, hedge funds and other similar institutions. The segregation of funding decision from risk-taking decisions, the development of instruments for selling various quanta and shades of credit risk and the development of depth for these instruments has helped banks to develop active portfolio management in the credit risk. The syndication, securitisation and CDs are the three primary instruments for active portfolio management.

The risk measurement tools, instruments and markets enable banks to determine the relative values of different choices – owning credit risks, divesting credit risk through hedges, credit insurance or selling the credit risky assets. Loan portfolio is managed in the style of a bond portfolio – taking views on the credit spread, non-linear movements in the credit spreads and the optimisation of the portfolio for better returns. The entire science around the active portfolio management of credit risk is being developed. The issues are very complex since every type of instrument and every transaction is different and is exposed to different types of risks including different types and quanta of basis risks. Active portfolio management has also changed the entire gamut of wholesale banking and the relationship banking.

Policies, processes and methodologies for accounting, hedging, risk measurement, risk optimisation, capital measurement and disclosures for active portfolio management have to be developed. In the "Credit Mitigation" chapter, we have attempted to provide the first cut for each.

REFERENCE

Quantitative Impact Studies 3, 4, 5, 2002, 2004,2005, Basel Committee on Banking Supervision, published on BIS website.

Advanced Notice on the Proposed Rule Making – Risk Based Capital Guidelines for Corporate Banking (2003) **and Retail Banking** (2004), published by Federal Reserve System and other Agencies.

Credit Risk Portfolio Models

INTRODUCTION

Quantitative methods for portfolio analysis have developed since Markowitz's pioneering work in 1950. These methods have been applied successfully in a variety of areas of finance, notably to equity portfolios. However, the same development has not occurred for credit risk portfolios because of the analytical and empirical evidence in quantifying credit risk components and still more difficulties in modelling correlation among risk components. Primarily, there are two problems with credit risk.

❏ Credit risks are highly skewed and fat tailed. Therefore, a normal distribution cannot be applied and parametric models represented by the mean and standard deviation cannot be built for credit risk.

❏ Correlations are more important for credit risk than for market risk. However, owing to lack of data, it is almost impossible to estimate any type of credit correlation directly.

Both these problems can be solved by componentisation, approximation and substitution (CAS). In this chapter an attempt has been made to provide a general framework for a portfolio credit risk model. In a nutshell, to model portfolio credit risk, the risk components need to be modelled and measured. The next step is to apply a correlation to the components through the transformation equations and generate the component distribution. A portfolio loss distribution is generated from a component distribution. The loss distribution is used to measure expected loss (EL) and unexpected loss (UL) and economic capital. These measurements are then used for pricing and provisioning.

The first two sections provide an overview of the general techniques of CAS, a definition of the risk components (componentisation) and an identification of factors helping to measure credit risk components (substitution). In "Transforming observable risk factors into risk components" and "Modelling correlation", an attempt has been made to understand the approximation techniques used to model risk components and correlation. "Generating loss distribution" presents the portfolio credit loss distribution modelling techniques and "Application of portfolio loss distribution" describes the measurement of economic capital and risk-adjusted returns on capital (RAROC) and their uses in the decision-making process.

Each of the sections is substantiated and interspersed with case studies and examples from the CreditMetrics, CreditRisk+, CreditPortfolioView and Basel II credit models.

The framework divides the portfolio models into the following building blocks:

❑ Componentisation, approximation and substitution.
❑ Defining risk components and identifying risk factors.
❑ Transforming risk factors into risk components.
❑ Applying correlation to risk components – generating the component distribution. For each state of the world or scenario, risk indicators are mapped/modelled to default rates and are measured.
❑ Generating the loss distribution. The distribution of portfolio default is obtained for each state of the world. Then by multiplying by the weights of the each state of the world, aggregated portfolio losses are obtained.
❑ Use the loss measures – risk contribution, economic capital and RAROC.

The generalised framework explained above helps in comparing and benchmarking portfolio models and can be used as a baseline for building internal risk models for Basel.

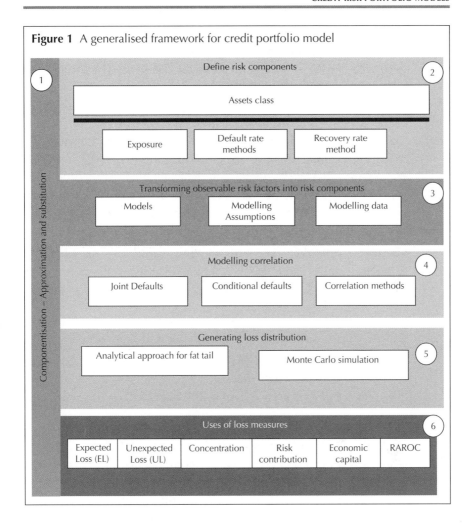

Figure 1 A generalised framework for credit portfolio model

It may be noted that this is a conceptual framework and may not necessarily mean separate steps in the portfolio modelling process. For example, in CreditRisk+ modelling correlation, generating the correlation distribution and generating the loss distribution is performed as a single step. The CreditRisk+ models the loss distribution of sub-portfolios as a mixed Poisson distribution.

Furthermore, this is a broad framework covering the entire gamut of activities to be performed for portfolio modelling. The commercial solutions available in the market cover only a portion of the activities and limited loan products. This does not mean that the activities listed here are not required, but it means that they need to be done outside the solution. A commercial solution generally covers loss distribution and uses of loss data and may not cover: "Defining risk components" and "Transforming observable risk factors into risk components".

Transforming risk factors into risk components is explained in detail in the section on probability of default (PD) and in the section on loss given default (LGD). Software and data sources are explained in a separate chapter on Software and Data.

COMPONENTISATION, APPROXIMATION AND SUBSTITUTION

Data availability and model complexity issues are addressed through CAS.

Componentisation

As explained in the introduction, credit risk is studied in terms of risk components. The Basel Accord recognises PD, LGD, exposure at default (EAD) and maturity as risk components. In addition, to compute portfolio losses, the correlation and UL distributions are required. All portfolio models estimate these six risk components separately. This makes it possible to work within the data constraints for any or all risk components.

Approximation

This approach is adopted to reduce the data requirements. This also reduces the estimation complexities. Risk components are measured at the approximate level. Techniques used to approximate risk components include the following.

❑ *Constant*: For example, the CreditRisk+ model and Basel II foundation approach assume LGD to be constant. The Basel II standardised approach assumes PD to be constant. Correlations are assumed to be constant across the portfolios. Even when default rates depend upon several stochastic background factors, the credit exposures are assumed to be constant and are not related to changes in these factors or changes in other risk components.

❑ *Average and constant*: Here, risk components are assumed to be constant with some value. Since measuring correlation is very difficult, Basel has adopted the asymptotic single risk factor (ASRF) model for correlation. ASRF is a single-factor model and therefore it is easier to model. The implementation of the ASRF model developed for Basel II makes use of average PDs that reflect expected default rates under normal business conditions. The Basel II model assumes the correlation to be average and constant for an asset class. LGD is assumed to be average (45%, 75%, etc) and constant in the foundation approach. Maturity is also assumed to be constant and have a holding value of 2.5 years.

❑ *Distribution*: The distribution of variables is approximated to the parametric distribution. Examples of some approximated distributions are explained in Table 1.

❑ *Banding*: Credit grades are nothing but the PD bands. So instead of separate PDs for each exposure, the credit grade PD is

Table 1 Approximation in distribution

Distribution	Component distribution	Portfolio loss distribution
Normal	Economic factors and unconditional default rates are assumed to be normally distributed	Loss is never distributed normally
Beta	Recovery rates	Loss distribution with thick tail
Gamma	Mixing variable is gamma distributed	The conditional default rates generated have negative binomial distribution
Poisson	Default rates and recovery rates are assumed to be poisson distributed	Unlike conditional default rate distribution, the loss distribution is not a poisson distribution
Binomial	Distribution of larger number of smaller exposures and independent defaults have thin tail and are equivalent to normal distribution. Smaller number with larger exposure and dependent defaults have thick tail Binomial is also used to model default and non-default	Thick tail is the most important for credit. Monte Carlo can generate a good binomial curve, analytical techniques are required for thick tail

applied. Exposures are banded by putting limits on the upper and lower bounds, and by making assumptions about the average exposure. As with PD grades, there are recovery grades. The Basel II model assumes a range of correlations. Mapping credit ratings to a set of default rates provides a convenient approach to setting the level of the default rate.

❏ *Modelling approaches*:

❏ Assumptions on variables: There are two approaches to modelling default rate and recovery rate.

❏ Discrete variables – Jump and hazard rate models assume the default to be discrete.

❏ Continuous variables – Diffusion models assume the default rate to be a continuous variable, such as in Brownian-motion-based models.

❏ It is easier to build closed form solutions with continuous variables. For example, the first-generation structural models started with continuous variables and in the subsequent generations discrete variables were introduced.

❏ Default *versus* credit quality migration.

❏ Assuming a model to be a default model reduces the data requirements and the model complexities to a great extent. For example, the CreditRisk+ focuses on default and therefore it requires relatively simpler equations and less data.

❏ Credit quality migration models are generally based on a transition matrix (TM). TMs are difficult to model and, in general, the Markov process is used. Modelling credit quality migration is very difficult. Only recently have reduced form models started to model credit quality migration based on state variables, TMs and share price changes.

❏ *Joint default* versus *conditional default*: Joint default is pair-wise correlation and represents all possible combinations. However, no data on the pair-wise default is available. Conditional default is the default under certain scenarios (and not all scenarios). This reduces the number of simulations and the data requirement to a large extent. The correlation is reflected by the extent to which the borrowers' conditional default rates vary together in different relevant scenarios.

❑ *Assumptions for correlation*: Measuring credit risk correlation is almost impossible – there is simply not enough available information. Today, correlation is measured by making various levels of assumptions and simplifications, and will continue to require some assumptions to be made. Let us look at the assumptions being made by the publicly available portfolio models.

❑ Basel II model: Basel has established a regulatory capital computation equation by making the following simplifications and assumptions.

   ❑ The Basel model is built on the ASRF model, which calculates the regulatory capital by considering the bottom-up approach. The ASRF model assumes that:

      ❑ there is a single systematic risk factor driving correlations across obligors;

      ❑ no exposure in a portfolio accounts for more than an arbitrarily small share of total exposure – higher granularity;

      ❑ model fails to detect the concentration and diversification and does not measure the risk contribution.

❑ Adjust the components: For example, in CreditRisk+ the exposures are netted for the recovery amount before modelling the loss distribution.

❑ Empirical studies: There are many variables in credit risk measurement and there is a paucity of data. Most credit risk measurement relies on empirical studies. Therefore, it is almost impossible to establish a cause–effect relationship between the risk and the risk drivers. The relationship is established through the statistical models and empirical studies. Basel prescribes an empirical-results-based risk measurement and calibration as the best practice; for example, "Each estimate of PD must represent a conservative view of a long-run average PD for the grade in question, and thus must be grounded in historical experience and empirical evidence".

❑ Scenario analysis: The approximation for components is built by building models for the scenarios instead of making the model applicable in all situations.

❑ Analytical approach: The analytic approach is used to approximate the default distribution, correlation and loss distribution. Depending on the availability of data and transformation formulae, approximation assumptions can be relaxed. Therefore, it is possible to model the loss distribution from a simplified model to a full-blown model.

To obtain good results, the approximation should not be far away from reality. Approximation definitely creates model risk and makes model validation difficult. Approximation also impairs the ability to cope satisfactorily with non-linear products such as options and foreign currency swaps.

Substitution

The problem of the non-availability of data is solved by identifying the next best available data (risk factor) that can represent the required credit component. The challenge is to discover the approximation and substitution parameters to solve the business problem. There are two types of substitutions:

(1) substitute the measurement of risk components with the risk factors and measure risk indirectly;
(2) substitute the risk component distribution with an assumed distribution of the risk factor.

The most important criticism is that the relationship established under the substitution may not be stable under the stress conditions. Following are some of the substitutions for risk factors.

❏ *PD*:
 ❏ Accounting ratios and other attributes are used as a substitute for the PD under credit scoring models.
 ❏ Credit spread is the substitute for the PD under the reduced form model.
 ❏ Equity prices and assets volatility are the substitutes under the option theoretic model.
 ❏ State variables, equity prices and TMs are used for measuring PDs and correlation in reduced form models.
 ❏ State variables are used in factor models.

❏ *Recovery rates*:
 ❏ There are three approaches to recovery rates – accounting losses, economic losses and market values losses.
 ❏ LossCalc v2 uses a linear regression model (explained in the Chapter on LGD) as a substitute for the recovery rate.
 ❏ Reduced form models use state variables as risk drivers.

❑ *Correlation*: A pair-wise correlation of 1,000 (*N*) firms will require 1 (N^2) million pair-wise correlations or rating migration data. This data is simply not available. Therefore, relationships between credit components or credit component changes with some observable data/parameters/proxy are to be established. Some examples of the substitutions used are as follows:

❑ CreditRisk+ uses the asset price correlation as a proxy for default correlation;
❑ CreditMetrics substitutes the rating or equity series correlation for the risk component correlation;
❑ the Basel model uses a single-factor model to build the correlation.

❑ *Substitution with an assumed distribution*: The credit component distribution and portfolio loss distribution are substituted with some known distribution.

This is not an exhaustive description of CAS, this section is only to whet the appetite of the readers. Every model uses these techniques. Readers need to extend and reuse the concepts explained here and continuously invent such techniques.

DEFINING RISK COMPONENTS
Basel broadly defines four risk components: EAD, PD, maturity and LGD. In addition, for a portfolio, risk correlation is also an important risk component. All the models and assumptions regarding credit risk measurement work only for an homogenous portfolio or for an asset class. Basel has defined a different asset class since the risk components' measurement method is different for each asset class.

We have explained how to build a homogenous asset class in the validation chapter.

Exposure
EAD is not the same as exposure. In general, a credit quality does not change to default suddenly; it changes over a period of time. There is a time lapse before credit quality changes are noted by the market, lender and others. Meanwhile, the obligor withdraws the lines of credits and commitments.

Empirical results show the correlation between PD and the magnitude of EAD. This results in an increase in EAD.

Off-balance-sheet items such as derivatives are exposed to pre-settlement risk. Derivative contracts keep fluctuating from a state of "in the money" to "out of the money". There are two elements in derivative contracts: current exposures and potential exposures. Therefore, EAD is generally driven by the following factors: accrual of interest; customer behaviour when default is approaching; and changes in the market value of the exposure and collateral.

The following paragraphs explain EAD for various products.

Loans and bonds
There are two types of loans/bonds.

❑ *Fixed*: Exposures on fixed-rate bonds or loans vary according to the maturity. Instruments with longer maturities have higher movements. Depending on the modelling assumptions, exposure can either be assumed to be fixed or at par with the face value or can modelled as market instruments.
❑ *Floating*: Exposures on floating-rate bonds or loans are always very close to par.

Receivables
These are generally non-interest-bearing trade credit subject to the risk of change in the credit quality of obligors. In general, receivables are treated along similar lines as loans and bonds. Since receivables are zero interest, they are also treated as zero coupon bonds payable on due dates. Therefore, receivables can be re-valued at the credit spread.

Loan commitments
This is essentially a loan and an option to increase the amount of the loan, up to the limit of the commitment. The borrower pays interest on the drawn amount and a fee on the undrawn amount in return for the option. In almost all markets negative convexity is observed. This means the amount drawn under the commitment is a function of obligor credit quality. With the deterioration of the credit quality, the obligor draws additional funds. Therefore, while modelling commitment, the following two factors are considered:

❑ the current undrawn amounts;

❑ expected changes in the amount drawn due to changes in the credit quality.

Financial letters of credit

A letter of credit (LC) is an option for the obligor to borrow. These instruments are typically priced as loans. The obligor almost certainly draws this amount in case of distress. A financial LC should be distinguished from a performance or trade LC (EAD/customer behaviour is different).

Market-driven instruments

With the development of swap markets, the measurement of credit exposure requires sophistication. Swaps typically have zero value at the outset. As time passes, the swap may be "in the money" or "out of money". Exposure can be divided into current exposures and potential exposures. Current exposure is the value of the swap at the current time t. The potential exposure represents the exposure on some future date, or sets of dates. Here the market risk is converted into credit risk.

❑ *Swaps or forwards*: This is the irrevocable commitment to purchase or sell some asset on prearranged terms.

❑ *Long options*: Options may also create credit exposure. The current and potential exposure will also depend on movements in the driving risk factors. There is no possibility of negative Vt values since options always have positive or, at worst, zero value.

❑ *Short option*: Assuming that option premium has been paid, the current and potential exposure can only be zero.

❑ *Expected and worst credit exposure*: Expected exposure is the expected value of credit replacement and worst credit exposure is the credit exposure at some level of confidence.

Default rate methods

Default rates have been explained in the chapters "Probability of Default" and "Credit Rating".

Recovery rate methods

LGD and recovery rate methods have been explained in the chapter "LGD and Market-information-based Models".

TRANSFORMING OBSERVABLE RISK FACTORS INTO RISK COMPONENTS

In transforming risk factors into risk components, the two most important variables to be considered are the meaning of the credit loss and the time horizon for the measurement.

❑ *Meaning of credit loss*: Loss is suffered if the borrower defaults or there is deterioration in the credit quality during the risk evaluation period. We have explained this in the first chapter.
❑ *Time horizon*: A key decision in the credit risk component and portfolio modelling is the time horizon. In general, the time horizon chosen should not be shorter than the time frame over which the risk-mitigating actions can be taken.

There are two approaches to the time horizon.

   ❑ Constant time horizon: In general, a constant time horizon of one year is considered. It should also be noted that one-year default rates have a significant variation:

      ❑ across the rating grade;
      ❑ across the industry;
      ❑ year on year.

   ❑ A hold-to-maturity or run-off time horizon.

Exposure models

The types of instruments have a major impact on the changes in the credit exposure. There are two types of instruments:

❑ loans or loans equivalent;
❑ market-related or derivative instruments.

Exposure model assumptions
❑ Empirical evidence shows that EAD estimation is volatile over the economic cycle. The empirical studies also show that there is a correlation between PD and the magnitude of EAD.
❑ Historical data: Estimates of EAD must be based on historical data for a time period that must ideally cover a complete economic cycle but must in any case be no shorter than a period of seven years. This can be five years for retail.

Modelling EAD
Loan or loan equivalent models: The loan equivalent (LEQ) model makes an attempt to predict or model EAD. LEQ can be defined as

the portion of a credit line that is an undrawn commitment, which is likely to be drawn down by the borrower in the event of default. In general, internal historical data is used to estimate LEQ. There is no agreement on the factors that can be used for an *a priori* estimate of LEQ. The general assumption is that as the credit quality of the borrower depreciates, their other sources of funding and credit may dry down. Therefore, they are likely to draw the unutilised lines with the bank to finance their operations and other commitments. The acceptable method of managing or minimising the additional withdrawal or increase in the exposure is to have rights through covenants. The remaining tenor of the commitment is one of the important dimensions used to measure LEQ, therefore LEQ should be tenor adjusted. LEQ is generally observed to decrease as the credit quality worsens and LEQ volatility also plays an important part in the LEQ model.

To prepare this model, the regression analysis equation is used. Table 2 shows the LEQ *versus* the time to default for revolving credit predicted by using a regression model.

Under the advanced internal ratings-based (IRB) approach, EAD may be determined by the bank via a model. The following studies are helpful in modelling EAD.

❑ For exposures like overdraft which have uncertain EAD and LGD, the loan equivalency factor (LEQ) represents a quantitative estimate of how much of the commitment will be drawn down by a defaulting borrower. LEQs are different across both credit quality and facility type.

❑ Empirical work on the LEQ is still sparse.

Table 2 Regression model predicted LEQ by facility risk grade and time to default of revolving credit

Facility risk grade	1 year	2 year	3 year	4 year	5 year
AAA	55	66	77	88	99
A	52	63	74	85	96
BBB	42	52	62	72	82
B	34	44	54	64	74
CCC	31	41	51	61	71

Source: Michel Araten and Michael Jacobs Jr., in RMA Journal May 2001

❏ As part of a broader study of loan performance, Asarnow and Marker (1995) have analysed the performance of large corporate commitments at Citibank from 1988 to 1993 and shown the importance of credit (debt) rating, particularly at the speculative end.

❏ Araten and Jacobs (2001) evaluated the behaviour of over 400 facilities from defaulted borrowers over a six-year period and found a highly significant increase in LEQs relative to time-to-default across all rating grades and a somewhat weaker relationship between LEQs and ratings grades. They have noted that, similar to LGDs, observed or realised LEQs are widely dispersed.

EAD – derivative instruments: The credit risk of derivative instruments has two features.

❏ It is a two-way credit risk that shifts from one counterparty to another depending on who owns the positive value.
❏ The credit risk keeps changing owing to market risk parameters.

EAD for derivative instruments consists of two components – current exposure and potential exposure. Current exposure is the present value of the discounted value of future flows indexed at market parameters. The potential exposure is the possible increase in the MTM value of the instruments owing to market movements.

The models to value current exposure and potential exposure differ from instrument to instrument.

Historical estimation of EAD

Historical estimation of EAD must be an estimate of the long-run default-weighted average EAD for similar facilities and borrowers over a sufficiently long period of time, but with a margin of conservatism appropriate to the likely range of errors in the estimate. The cyclic nature, economic downturn conditions and EAD volatility need to be considered, if necessary, instead of only the long run average. The impact of accounting and payment processing policies must also be considered.

❏ There are fewer empirical studies on EAD.
❏ For lines of credit where a borrower is theoretically able to draw down at will up to the committed line of the facility, the borrower

will typically draw down as much as possible on existing unutilised facilities in order to avoid default.

Basel recommendations
Under the IRB foundation approach, EAD is supplied by supervisory values. For instance, EAD is 75% for irrevocable undrawn commitments.

PD models
Several statistical methods are used to measure PD. We have explained these methods in detail in chapters on credit scoring, market information based models and credit ratings.

LGD models
Modelling LGD is explained in the chapter "LGD" and "Market-information-based Models".

MODELLING CORRELATION
"If obligor A's credit quality deteriorates, how well does the credit quality of obligor B correlate to obligor A?"

Correlation is a probability of joint default and conditional default. In credit risk measurement, correlation is very important for the following reasons:

❑ correlations of credit risk are positive and increase the risk of losses;
❑ correlation drives the shape of the loss distribution;
❑ correlation generates fat tails;
❑ correlation relates to external conditions and economic cycles.

Correlation is a measure of volatility, which is more important for credit risk than market risk. EL on equity price is to the tune of 15% of the portfolio value compared to 0.5% EL due to credit risk. Obviously, volatility on a larger base is always smaller while on a very small base it is exponential. Therefore, a volatility of 10% has a totally different meaning for credit risk compared to market risk. This will be clear from the following example.

Let the PD of an obligor A and B equal $P_A = P_B = 1\%$ and the correlation between A and B equal $\rho_{AB} = 10\%$.

The joint default probability is given by

$$P_{AB} = P_A P_B + \rho_{AB}\sqrt{P_A(1-P_B)P_B(1-P_B)}$$

The conditional PD is given by

$$P_{A/B} = P_A + \rho_{AB}\sqrt{P_A(1-P_A)(1-P_B)/P_B}$$

($P_{A/B}$ is the probability of default of A given B has already defaulted)

$$P_{AB} = 0.01 \times 0.01 + 0.1 \times 0.01 \times 0.99 = 0.00109 \approx \rho_{AB} \times P_A$$

$$P_{A/B} = 0.01 + 0.1 \times 0.99 = 0.109 \approx \rho_{AB}$$

The correlation coefficient dominates the joint default probabilities and the conditional default probabilities. Correlation is broadly equal to conditional PD. The joint default probability is a multiple of correlations.

Correlation and diversification

Default correlation measures the probability of both the obligors defaulting during the same time period. If there is no correlation, then the defaults are independent and the correlation is zero. In such a case, the probability of both borrowers being in default at the same time is the product of their individual probabilities of default. If two borrowers are correlated, this means that the probability of both defaulting at the same time is heightened, ie, it is larger than it would be if they were completely independent. Therefore, the volatility or the UL of a portfolio is not the weighted average stand-alone UL. As seen in the previous paragraphs, joint and conditional default is driven by a correlation coefficient.

Let us briefly consider what is meant by diversification.

❑ Diversification is a reduction in the UL and not the EL.
❑ Diversification can be achieved by changing or reducing the correlation. A reduction in correlation is achieved by selecting exposure in such a way that obligor credit losses have either lower correlation or negative correlation and/or the quantum of exposure to the correlated borrowers is reduced or controlled.

Case study

Let us consider two loans, each with a value of US$50, and a portfolio with a value of US$100. On average, one loan defaults every year (so the total loss is US$50). Correlation determines the UL. The portfolio manager aims to reduce the UL.

Case I – perfectly negative correlation (100% *negative correlations*): Every year only one loan can default. Negative correlation means that if one loan has defaulted the other loan cannot. Each loan has an equal chance of default. Both loans cannot default in the same year.

Case II – perfectly positive correlation (100% *correlation*): Either both loans default or neither defaults. On average, one loan defaults every year.

Case III – correlation (50% *positive correlation*): there is a 25% probability that both loans default and a 25% probability that both will survive. There is a 50% probability that one will default (or a 50% correlation that both will default or survive). Table 3 shows the computation of EL and UL.

With negative correlation, the UL for the portfolio is zero. With a perfectly positive correlation, the UL is 50%. With intermittent correlation (which is actually found in real life), the UL is 25%.

There is significant variation in the number of defaults from year to year. Within each year, different industry sectors show different default rates and default variation rates. This is due to the fact that each sector of the industry is affected to different degrees by the state of the economy and each firm within the industry is exposed to different idiosyncratic risks.

Table 3 Impact of correlation on unexpected loss

	Case 1	Case 2	Case 3	Explanation for Case 3
EL 1	50	25	25	$(25\% \times 50 + 0.5 \times 50\% \times 50)$
EL 2	0	25	25	$(25\% \times 50 + 0.5 \times 50\% \times 50)$
Portfolio EL	50	50	50	$(25 + 25)$
UL 1	25	25	25	$PD = (25\% + 0.5 \times 50\%) = 50\%$
UL 2	25	25	25	$UL = (50 \times \sqrt{0.5(1-0.5)})$
Portfolio UL	0	50	25	$\sqrt{25 \times 25 \times 50\% + 25 \times 25 \times 50\%}$

Modelling idiosyncratic factors

Every borrower is exposed to idiosyncratic and systematic factors. Normally, the influencing factors are assumed to be systematic factors and are modelled through statistical fit. Any statistical fit equation provides error terms, which actually represent the idiosyncratic factors. Therefore, the higher the idiosyncratic risk the lower the correlation effect and, conversely, the higher the systematic risk the higher the correlation effect. Hence, the quantum of idiosyncratic risks controls the correlation and therefore the loss distribution. Whenever it is difficult to model systematic risk factors generally, a ratio between idiosyncratic to systematic risks is considered.

As is common in finance, the asset value itself is not modelled. Rather a model is built for the changes, ie, the returns, on the logarithm of a firm's assets and this return is represented by a normally distributed random variable. This random variable has a mean of μ and a standard deviation of σ. Under the Basel II model, it is assumed that two types of factors have a continuous impact on asset returns:

❏ systematic factors – which impact equally on all assets (within the same risk grade and industry);
❏ idiosyncratic factors – effecting each return separately.

Idiosyncratic movements are assumed to be independent from the systematic factors and independent for different borrowers. All random variables are serially independent. Thus, within a risk segment, variance, the covariance and correlations between borrower are given by

$$Variance\ for\ each\ assets = \sigma^2$$

$$Covariance\ of\ the\ asset\ \sigma_{ij} = b^2$$

Correlation among the borrowers shows the proportion of the variance explained by the systematic factors. Basel assumes that around 20% is explained by the systematic factors, while 80% is explained by the idiosyncratic factors:

$$Correlation = Covariance\ \rho_{ij} = \frac{\sigma_{ij}}{\sigma_i \sigma_j} = \left(\frac{b}{\sigma}\right)^2$$

Correlation impacts the PD. Each of the default grades is impacted differently and, therefore, the default rate itself will have a probability distribution. The default rate without the correlation impact is called the unconditional PD and after the realisation of correlation it is called the conditional PD.

Some empirical observations regarding PD correlations

Empirical studies and intuition tell us that the higher the PD, the higher the idiosyncratic (individual) risk components of a borrower. The default risk depends less on the overall state of the economy and more on individual risk drivers.

Asset correlations increase with firm size. The larger a firm, the higher its dependency upon the overall state of the economy, and *vice versa*. Smaller firms are more likely to default for idiosyncratic reasons.

❏ Default correlations are generally low and are generally inversely related to rating class, ie, the higher the class the lower the rating.
❏ Default correlations generally increase initially with time and then decrease as the horizon is extended.
❏ Default among and between specific industries is inconclusive.

Conditional default rate

To understand this relationship, let us first understand the relationship between unconditional and conditional default rates.

While measuring and modelling PD, we measure only the unconditional default rate. This is the weighted average of the default rate under various economic conditions/scenarios. The economic scenarios actually determine the default rate. There are two ways to look at the difference in default rates:

❏ The first approach is to measure the default rate/loss as weighted average default rates under various scenarios and build a correlation equation to link this average to change under different economic factors.
❏ The second approach is to model default rate under different (bad) economic cycle.

Unconditional default rate

This is the first approach as explained in the above paragraphs. An obligor's unconditional PD is the PD before the horizon given all information currently observable. It is the one period PD. It is the probability of an asset's value falling below a threshold without information about the realisation of the common random factor. It is assumed that unconditional PD does not vary over time.

Conditional default rate

This is the second approach as given in the above paragraphs. The conditional default probability is the PD we would assign the obligor if we also knew what the realised value of the systematic risk factors during the time period under consideration would be. It is the PD given the realisation of the random factor. Conditional PDs change over time.

For example, the unconditional PD is the average value of the conditional default probability across all possible realisations of the systematic risk factors. Consider, for example, that there are three economic states, and look at their respective probability and PDs for a particular obligor (see Table 4).

In practice, however, the unconditional PD is given and conditional PD is to be estimated. Conditional PD is determined by the state of the economy, which is measured indirectly by considering economic factors. Therefore, conditional default rate is the rate driven by the state of the economy or factors.

Joint default rate

Joint default requires 2^N joint default data. The obvious source of information on default correlation is the historical incidence of joint defaults of similar firms in a similar time frame. This data is objective and directly addresses the modelling problem. Unfortunately, because joint defaults are rare events, historical data on joint defaults is very sparse. To gain a statistically useful number of observations, long time ranges (several decades) have to be considered and the data must be aggregated across industries and

Table 4 Conditional and unconditional PD

Economic state	Percentage chance of the state	Conditional PD (%)	Unconditional PD = weighted average of conditional PD
Bad	30	2.0	$=2\% \times 30\% + 1.5\% \times 40\%$
Neutral	40	1.5	$+0.4\% \times 30\% = 1.32\%$
Good	30	0.4	

countries. In most of the cases, direct correlation data is not available. The following equation defines a relationship between joint
and conditional default:

$$P_{AB} = P_{A/B} \times P_B$$

$$Correlation = \sigma_{AB} = \frac{P_{AB} - P_A P_B}{\sqrt{P_A(1-P_A)P_B(1-P_B)}}$$

Approach to correlation modelling

There is a dual problem with the credit risk correlations: the first
part involves linking the risk drivers to credit events; the second is
finding the correlation between the risk drivers.

In a nutshell, correlation measurement is to identify correlation
models and measure the correlation among the risk drivers:

$$Correlation = f(Risk\ drivers)$$

Default correlations cannot be successfully measured from default
experience. The historically observed joint frequency of default
between two companies is usually zero. Even grouping the firms
together and measuring the default correlation of the groups does
not help. The estimates using historical data are highly inaccurate.
No satisfactory procedure exists for directly estimating default correlations. Therefore, correlation is measured indirectly through the
factor model. There are two types of models to measure correlation
(see Figure 2):

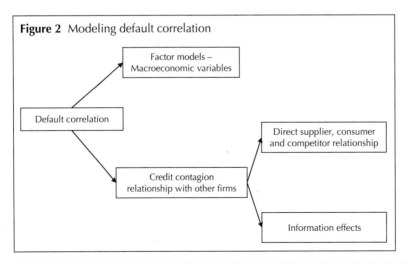

Figure 2 Modeling default correlation

❑ factor models;
❑ contagion models.

Default rates of firms are interdependent. The primary reasons for such an interdependence are the firms' sensitivities to the macroeconomic factors. These are common to all firms operating in an economy. Another reason for default dependence is the existence of business ties between firms. This is called credit contagion. The credit contagion works as a channel to propagate the economic distress across the firms. Contagion models are the emerging approach for modelling correlation. Separate contagion models have emerged from the hypothesis that macroeconomic models cannot quantify aggregated loss risk due to contagion. Allen and Gale (2000), Kiyotaki and Moore (1997) and Giesecke and Weber (2003) have proposed reduced form models for credit contagion modelling.

Contagion models
Under the assumptions of a large homogenous credit portfolio, the credit contagion effect is ignored and only the cyclical default dependence is modelled. If the contagion effect is ignored, loss volatility is modelled as volatility of macroeconomic factors. However, we know that, in practice, credit portfolios are rarely homogenous.

Modelling credit contagion is much more difficult than modelling asset correlation and, therefore, economic factors. The experts have modelled the credit contagion through the following two methods.

❑ *Linking the credit contagion to the asset value*: This is performed through an extension of the jump diffusion model. The jump in the value of a given firm triggers subsequent jumps in the firm values of other firms with a certain probability.
❑ *Liquidity state as the risk drivers for credit contagion*: This is performed by extending the default-barrier Bernoulli mixture model. The default barriers of a firm are made to be dependent on the liquidity state of the firm. This, in turn, is dependent on the default state of the counterparties of the firm. If the firm's liquidity position is stressed owing to higher default of the counterparties, it issues more debt to fund the required liquidity. This

increases the default barrier – everything else being equal, the firm is now more geared for, and therefore more likely to default.

Factor or macroeconomic models

Economic entities are related to each other since they are suppliers, customers or competitors to each other or they supply, consume or compete for similar resources. Owing to this relation, an economic or business cycle or an industry trend impacts entities in a similar way. This is called the Factor model.

The Factor model measures correlation between the PD of borrowers. In other words, economic factors impact the credit risk components such as default rate, recovery rate or EAD. Ideally, the impact of correlation on each of the risk components should be measured. In practice, owing to the paucity of data and supporting algorithms, the correlation or factor impact is considered only for the default rate.

Factors suitable for modelling

The macroeconomic factors and their impact on the risk components are both unobservable and very difficult to measure. Therefore, an attempt has been made by experts to identify observable risk factors and establish their relationship with the risk components. Risk drivers help in measuring correlation indirectly. The principle of measuring correlation indirectly is to assume credit events as dependent on the correlated drivers. Instead of credit events, the risk drivers are measured. In general, correlation among equity indices is considered to be a substitute for correlation for asset values.

Factors used to model correlation

All models assume the presence of factors creating correlations. However, they work on different correlation drivers to measure the correlations. The following factors are generally used:

❑ asset values/returns of the firm;
❑ indexes related to the country;
❑ indexes related to the industry;
❑ equity prices/returns and equity indices;

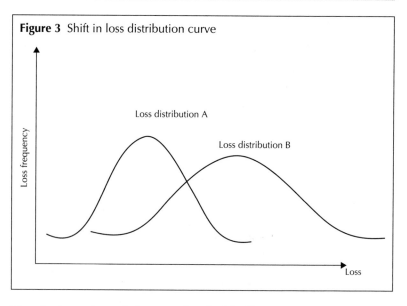

Figure 3 Shift in loss distribution curve

- ❏ a single systematic factor to be decided by users;
- ❏ multiple systematic factors to be decided by users.

Impact of economic cycles on the correlation
There are two approaches used for considering cyclical impact:

- ❏ *ex post*: the realised credit losses show a clear pro-cyclical pattern – increasing during recession and decreasing during expansion;
- ❏ *ex ante*: a shift in the estimated loss distribution curve – the entire loss distribution shifts (see Figure 3).

CreditPortfolioView considers the cyclic factors by converting the unconditional credit migration matrices into matrices that are conditional on macroeconomic factors. Each cell in the credit TM shows the probability that a particular obligor, rated at a given grade at the beginning of the period, will move to another rating by the end of the period.

The model uses a distributed lag model to forecast macroeconomic conditions based on both fundamental macroeconomic variables and idiosyncratic risk factors. Each transition probability is computed as a function of the macroeconomic forecast and diffused through the migration matrix. Different conditional TMs can be estimated for different credit horizons corresponding to fluctuations in the macroeconomic conditions.

Table 5 Conditional and unconditional TM

Initial credit grade	Transition grade	Unconditional TM	Conditional TM
C	A	0.01	0.0124
	B	0.04	0.0340
	C	0.80	0.7796
	D	0.15	0.1740

Source: Saunders and Allen, 2002

Table 6 Developments in non-traditional correlation

Correlation	Issues/studies
LGD and LGD	Weighted average recovery rates are lowest in the recession years Relationship between LGD correlations and bankruptcy rules (renegotiation) and macroeconomic shifts is not yet recognised
LGD and PD	PD and LGD are dependent on the same macroeconomic factor so worsen together
EAD with PD	There is significant increase in EAD when default rates increase

For example, Saunders and Allen (2002) show how the unconditional credit TM can be transformed into a conditional TM for a cyclical downturn (see Table 5).

Correlation beyond the PD correlation is not yet considered in the portfolio models. The Bank for International Settlements (BIS) in its working paper "A survey of cyclical effects in credit risk measurement models" has summarised the issues in correlation measurement (see Table 6).

Case study of systematic factors
Factors include economic, macroeconomic and industry factors. Under the KMV model (defined and explained in the chapter "Market Informartion based Models"), correlation between default events is modelled through correlations between their asset returns and thus implicitly through correlations between their equity returns. Instead of estimating equity return correlations directly for each pair of counterparties, the KMV model employs a factor model in which the common factors relate to geographical regions and industry sectors. By letting individual asset returns depend on the same common factors, correlation between default events is

induced. KMV distills the common factors by applying principal component analysis to historical equity returns of a large pool of individual companies.

CreditMetrics uses "joint migration matrices" providing the joint transition probability of migration to various credit states of the pair of obligors. CreditMetrics uses observable equity return correlations as proxies of unobservable asset value correlations. To model such correlations, it relies on a multifactor model. The model derives pair correlations from coefficients of the factor model. The joint migration matrices result from equity return correlations. This correlation matrix drives the default rate distribution.

CreditRisk+ considers the default intensities dependent upon the common factors. CreditPortfolioView derives default correlations from their common dependence on economic factors.

Correlation type

Generally speaking, there are two types of correlation

❑ *Joint default*: Default correlations between two borrowers are modelled directly by assuming a two-dimensional correlated Brownian motion for the returns on the values of the firm's assets that trigger the defaults.

❑ *Conditional default*: CreditPortfolioView or CreditRisk+ models correlations by borrowers' exposures to common risk factors. Given the realisations of the risk factors, asset returns and the defaults are uncorrelated. The Basel approach follows this idea. It assumes the existence of one non-observable risk factor that is responsible for correlation. The factor model for the Basel Accord starts with constant or given unconditional PDs, which in turn models the conditional PDs for a special realisation of the risk factor. The model does not prescribe how to model unconditional PDs.

Case study: Implied correlation

The collateralised debt organisation (CDO) market quotes provide a benchmark for correlation computation. This is analogous to the equity derivatives market, where quoting implied volatility is equivalent to stating the price, since all other variables are known. Increasingly, market participants are quoting the implied correlation rather than the spread or the price of a CDO tranche. The implied correlation of a tranche is the uniform asset correlation number that makes the fair or theoretical value of a tranche equal to its market quote. Currently, the most common models used to compute the correlation are variants of the one-factor Gaussian copula model. In this model, the correlation of default times is determined by the correlation of asset returns, so tranche values are directly related, albeit in a complex way, to the assumed asset correlations.

Case study: Copula – modelling correlation

The primary problem with the transformation model involves building a joint distribution function of two different types of distribution: one distribution may be a beta distribution and the other may be a lognormal distribution. A copula is one of the methods that may be used. The word copula is a Latin noun that means "a link, tie or bond". A copula helps to divide joint-distribution structures into two parts: the dependence structure and the marginal behaviour.

In general, there are two approaches to compute joint losses/correlations: direct estimation with data or indirectly through a structural model of firm valuation and default. The constraint against direct estimation typically is data availability.

One of the indirect methods that is gaining ground is called the marginal copula. A marginal copula is a function that links the marginals to the joint distribution.

Beyond allowing for aggregation of diverse marginal distributions which capture some of the essential features found in risk management, such as fat tails, copulas also allow for a richer dependence structure than allowed for by simple models such as the multivariate normal distribution models.

Mathematically, a copula is a function that allows one to combine univariate distributions to obtain a joint distribution with a particular dependence structure. There are various types of copula for different dependence structures. The most widely used copulas for joint default distribution are the Gumbel copula for extreme distributions, the Gaussian copula for linear correlation and the Archimedean copula and the t-copula for dependence in the tail.

An empirical distribution and a parametric test are the two approaches for marginal behaviour. Further discussions on copula are out of the scope of this book. Readers interested further may refer to articles on copula, given in the bibliography.

Case study: Credit Portfolio view

Credit Portfolio divides the portfolios into sub-portfolios. It models the default rates of sub-portfolios as a logit function of economic index E. The economic index is a linear function of economic factors and works in a similar way to a multifactor model. The step to estimate Ei helps to build a predictive model and based on a random value influencing credit risk components (such as default rate). Some of the key points of the technique are as follows.

❑ It is a framework of model fitting. It does not identify the economic index to be used. It provides a freedom to choose the economic index depending upon the sub-portfolio and the data availability.
❑ Default rates are the aggregate default rates at the sub-portfolio level and not the individual default rates.
❑ The economic index captures the economic and business cycle.
❑ It uses the Monte Carlo simulation to generate default rates and the migration rate distribution.
❑ Default rates are estimated from the economic index and not from the asset values/individual firm risk drivers.

Case study: CreditRisk+

CreditRisk+ is a framework to directly model the loss distribution for sub-portfolios as a mixed Poisson distribution whose mixed Poisson parameters are $q \times d$, q being the random "mixing variable" and d being the default intensity analogous to a default probability. The mixing variable q changes the default intensity though time.

To model correlations, q is linked to the economic factors through the linear relationship

$$q = \beta_1 X_1 + \beta_2 X_2 + \cdots$$

The sensitivity of q with each of the factors may be different for different sub-portfolios. Thus, the correlated loss distribution for each sub-portfolio is built.

Case study: Basel model for correlation

The one-period PD is called the unconditional PD in the notion of Basel II. It is the probability of the asset value falling below the

threshold given the parameters of the process but without information about the realisation of the common random factor. In the context of Basel II, in addition to this unconditional PD, a conditional
PD, given the realisation of the random factor, is important.

Basel has two types of correlation equations: fixed and linked to
PD. For example, fixed correlations are assumed for residential
mortgages to be 0.15 and for qualifying revolving retail exposures
to be 0.04.

Basel defines the correlation through the following equation:

$$Other\ retail\ exposure = 0.03\left(\frac{1-e^{-50PD}}{1-e^{-50}}\right) + 0.16\left(1-\left(\frac{1-e^{-50PD}}{1-e^{-50}}\right)\right)$$

$$Corporate\ correlation\ (R) = 0.12\left(\frac{1-e^{-50PD}}{1-e^{-50}}\right) + 0.24\left(1-\left(\frac{1-e^{-50PD}}{1-e^{-50}}\right)\right)$$

The other retail correlation function and the corporate correlation
function are similar. However, their lowest and highest correlations
are different (3% and 16% instead of 12% and 24%). Moreover, the
correlations decrease at a slower pace, because the "k-factor" is set
at 35 instead of 50. The residential mortgages have higher correlations not only owing to the strong correlation of mortgage collateral value and the effects of the overall economy on that collateral,
but also owing to mortgages usually having long maturities that
drive the correlations upwards.

The aim of the component distribution is to model the distribution of each of the default components. While the factors are normally distributed, conditional default rates and credit losses are
not. The transformation from normal distribution inputs to skewed
output takes place through the component transformation models.

Modelling correlation in portfolio models

There are broadly three steps in building correlation into a
portfolio:

❑ define the unconditional default probabilities – this may be
according to the credit grades or according to an individual
obligor;
❑ define a transformation model to transform unconditional
default probabilities into conditional probabilities;

Table 7 Modelling correlation

	Correlation factors	Correlation drivers	Correlation distribution
CreditMetrics market factor model	Multifactor model – various industry and country indexes	Equity value correlation as a proxy for asset value correlation	Obligors' risk rating or obligors risk rating class. Joint migration matrices
KMV market factor model	Multifactor model – various industry and country indexes	Asset value or assets returns. Correlations are assets value correlations	Expected default frequency distribution
CreditRisk+ probability generating model	User defined factors	User defined linear relationship between default intensity of the portfolio segments and factors	Portfolio loss distribution Analytical distribution of default
CreditPortfolioView default rate model	User defined economic indexes	Logit function to predict the index value	Segment default and migration rates
Basel II	Risk factor	Single factor in the economy	Conditional PDs distribution

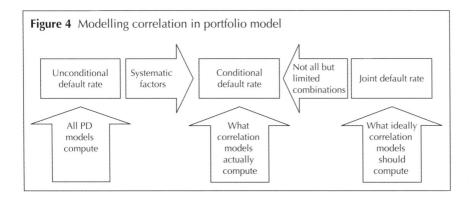

Figure 4 Modelling correlation in portfolio model

❑ generate the conditional default distribution.

Unconditional default probabilities are available from internal sources or external credit ratings. Within credit grades, they have the same unconditional default rate, but they will have different idiosyncratic or uncorrelated components. The unconditional default rates are distributed normally within a risk grade.

The transformation model defines a relationship between the correlation factors and the unconditional default rate. There are two types of transformation models.

❑ *Latent variable models*: A latent variable refers to a factor that is not directly observable. However, this may not always be the case. Examples are assets value as modelled in the KMV or CreditMetrics. Under these models, default occurs if a random latent variable Q falls below some threshold. Correlation between defaults is the correlation between the latent variables.

❑ *Bernoulli mixture models*: An example is CreditRisk+ where default events have a conditional independence structure conditional on common economic factors. Under these models, the conditional probability depends upon a set of economic factors while factors are independent of each other. Bernoulli mixture models are easy to simulate in Monte Carlo risk analyses. Mixture models are more convenient for statistical fitting purposes.

It should be noted that the correlation transformation model is at the heart of the portfolio model and is subject to considerable model risk. A small change to the structure of the model, the assumptions or the parameters will usually have a very large impact on the default distribution or loss distribution.

The transformation model is either a multifactor or single factor model. Basel uses a single-factor model. There can be two approaches to generate the component distribution:

❑ *The bottom-up approach*: Build the transformation equation and fine tune the equation to get the desired distribution;
❑ *The top-down approach*: Assume an input distribution and build a transformation equation that takes an input and gives the assumed distribution as output.

The second option is probably easier. The systemic risk factors are normal shaped, while the default rate is beta distributed. Empirical studies have also shown that the binomial distribution is a fit for

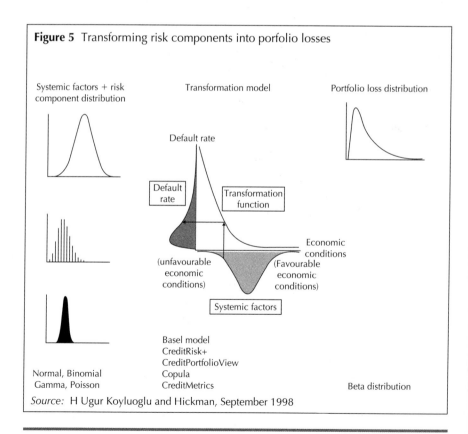

Figure 5 Transforming risk components into porfolio losses

Systemic factors + risk component distribution

Transformation model

Portfolio loss distribution

Default rate

Default rate

Transformation function

Economic conditions

(unfavourable economic conditions)

(Favourable economic conditions)

Systemic factors

Basel model
CreditRisk+
CreditPortfolioView
Copula
CreditMetrics

Normal, Binomial
Gamma, Poisson

Beta distribution

Source: H Ugur Koyluoglu and Hickman, September 1998

the exposure distribution while the beta distribution is a fit for the default and recovery rates.

Beta distribution for default rate and recovery rate

Default rates are a random variable or are driven by systemic factors that are normally distributed. Default rates themselves are not normally distributed. Conditional default is equivalent to increasing the default volatility. When this occurs, the default distribution is significantly skewed to the right.

The systemic factors are random and are usually assumed to be normally distributed. Since the conditional default rate is a function of these random systemic factors, the default rate will also be random. The default rate distribution is an explicit assumption in the actuarial model and an implicit assumption in the Merton-based and econometric models. For purposes of comparison, the default rate distributions implied by the latter two models can be derived easily.

The beta distribution is part of the parametric probability distribution with two degrees of freedom. It takes two parameters, $\alpha > 0$ and $\beta > 0$, where α and β represent the shape parameters – the steepness of the hump and the fatness of the tail. The beta distribution is characterised by a thick, skewed "tail", which corresponds to the general nature of credit loss or probability distribution frequencies.

The mean μ and variance σ^2 of the beta distribution are given by

$$\mu = \frac{\alpha}{\alpha + \beta}, \quad \sigma^2 = \frac{\alpha \cdot \beta}{(\alpha + \beta)^2 + (\alpha + \beta + 1)}$$

Tail fitting

With a beta distribution or any distribution, it is difficult to match exactly the statistics and tail of the distribution. A beta distribution with only two degrees of freedom is perhaps insufficient to give an adequate description of tail events in the loss distribution.

GENERATING LOSS DISTRIBUTION

The most important difference between equity price risk, foreign exchange risk, market risk and credit risk is the difference in the loss distribution. Equity returns are relatively symmetrical and can be well approximated by normal distributions while credit returns are highly skewed and fat tailed. Therefore, summary statistics such as mean and standard deviation are insufficient to define a credit loss distribution.

A credit loss distribution is typically a fat and long tailed asymmetric distribution owing to it having no upside opportunity and possibilities of severe losses where these risks are not easily diversified away. The credit portfolio loss distribution is highly skewed and leptokurtic.

The primary purpose in generating credit loss distribution is to measure the loss volatility and UL. The unexpected credit loss is a function of correlation, risk components (like PD and LGD) and confidence level and loss volatility. For a same EL, due to the different correlation, loss volatility can be different. Table 8 shows the losses at different confidence levels for a range of correlations and PDs (assuming LGD to be constant – which is the general and most accepted assumption).

At 99.99% confidence the portfolio for a 0% correlation and PD < 0.001% is about eight times less risky than for a distribution with 40% correlation.

Correlation has strong impact on the unexpected loss distribution shape. For a same expected loss, the unexpected loss can be different as shown in Figure 6. The loss rates of the dashed curve have higher loss volatility caused by a strong correlation among the individual exposures within the portfolio and with the systematic risk factor.

If the events of default on the loans in the portfolio were independent of each other, the portfolio loss distribution would converge, by the central limit theorem, to a normal distribution as the portfolio size increases. As the defaults are not independent and

Table 8 Credit losses – function of loss volatility and correlation

PD	Correlation (R) (%)	Confidence = 99%	99.9%	99.99%
0.01	40	4.5σ	11.0σ	18.2σ
0.001	40	3.2σ	13.2σ	31.8σ
0.01	10	3.8σ	7.1σ	10.7σ
0.001	10	4.1σ	8.8σ	15.4σ
<0.001	0	2.3σ	3.1σ	3.7σ

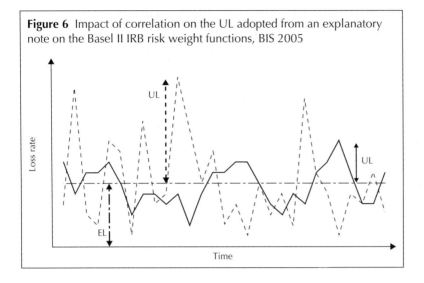

Figure 6 Impact of correlation on the UL adopted from an explanatory note on the Basel II IRB risk weight functions, BIS 2005

are beta distributed, the conditions of the central limit theorem are not satisfied and the loss distribution is binomial distribution.

Portfolio Loss Distribution – Distribution of loss frequency from a portfolio over a period of say one year. A portfolio has exposures which are exposed to risk components like PD, LGD and correlation. To understand loss distribution let us first understand the distribution of each component and the changes in the distribution during the time horizon.

Shape of Exposures Distribution – At best exposures are uniform with a very large number of obligors. The next best is exposures normally distributed with low obligor concentration. The next shape is beta distributed exposures with fat tails – some very large exposures.

Shape of Default Distribution – Shape of the default distribution is determined by the correlation and the number of obligors. When default events are independent, the default is binomially distributed. The actual shape of the binomial distribution depends upon the number of obligors. Law of large numbers also has an impact on the shape of the binomial distribution. Poisson distribution and Monte Carlo simulation techniques are widely used to model the default rates.

Shape of LGD Distribution – Most of the models assume LGD to be deterministic and are used to reduce the exposure to the extent of recovery. In general, the default rate estimate dominates the LGD estimate when estimating the expected loss of an exposure. CreditRisk+ uses such a fixed loss methodology: all exposures are input into the model net of any losses in the event of default.

A beta distribution is often used to model the uncertain recovery value. This distribution is useful because it can be bound between two points and can assume a wide range of shapes. All the popular commercially available portfolio management applications use a beta distribution to model the recovery value in the event of default.

Correlation – The systemic factors are assumed to be normally distributed. Normally distributed systemic factors are used as an input for generating distribution of exposures, default, and LGD and credit losses.

Loss Distribution – The loss distribution is binomially distributed. The exact shape of portfolio loss distribution and the default distribution (eg, the conditional default distribution explained in the previous section) may not be the same. Two distributions are the same only when the number of borrowers is very large and exposures and LGD are either uniform or normally distributed. These two distributions are different since the exposure size of the portfolio may also differ. There may be a smaller number of larger exposures or a larger number of smaller exposures. The loss distribution will not be the same in both cases. Therefore, differing exposure distribution and different LGD distribution will result in different loss distributions.

In general, existing models use the following two types of methodologies to model portfolio loss distribution:

❑ Analytical Methodology;
❑ Monte Carlo simulation.

Monte Carlo simulation
The Monte Carlo simulation is used to generate the loss distribution. Using a Monte Carlo simulation to generate a loss distribution is a complex and involved exercise for different types of model. Broadly there are two approaches to using the Monte Carlo simulation for credit risk portfolio models: option theoretic approach using assets value and econometric modelling of CreditPortfolio View.

Table 9 Monte Carlo complexities

Factors	Exposure	Correlation Structures	Default type	LGD
Increasing complexities				
Single	Uniform (fixed and equal)	Uniform, fixed and equal	Default mode	Constant and exposure reduced
Two factors	Fixed	Conditional	Default migration	Normal distribution
Multiple factors	Variable	Copula and joint correlation	Credit spread changes	Beta distribution

The Monte Carlo simulation generates random sets of future values of the economic factors and of the risk components from which correlation and asset values are generated. The variables that are usually generated in the Monte Carlo simulation are summarised in Table 9.

In practice, it should be noted that:

❑ only single-factor models are used – both the CreditPortfolioView and CreditMetrics models are single-factor models;
❑ models use uniform exposures and uniform correlation, with a random specific error, and are under default mode only;
❑ asset values are taken as a linear function of the factor;
❑ the error term represents the idiosyncratic risk.

In the case of CreditPortfolioView:

❑ a single-factor model drives an economic index directly related to the default rates of two portfolio segments through a logit model;
❑ the economic index is a linear function of an economic factor and idiosyncratic risk;
❑ a Monte Carlo equation is used to generate a random set of future values of economic factors and idiosyncratic risk;
❑ a logit function converts the economic factors and idiosyncratic risk into correlated default distributions (all inputs to the logit function are assumed to be normal);
❑ a larger number of trials provides the full distributions of the default rates and their sum over the entire portfolio.

The Monte Carlo methodology enables generation of the risk factors, according to the assumed correlation structure. Risk factors generated along with the assumptions about the exposure size is used as an input to generate the exposure distribution. Risk factors are used to generate the default probabilities for each set of exposures.

Complexities of Monte Carlo will depend upon how many variables are generated. Base line is to generate default distribution. The next level is to generate exposure distribution.

In this chapter, since we aim only to demonstrate the use of technique, we shall consider only two variables to be generated by MC technique.

The method is accurate (if used with a complete pricing algorithm) for all instruments and provides a full distribution of potential portfolio values, not just a specific percentile. The Monte Carlo simulation permits the use of various distributional assumptions (normal, t-distribution, normal mixture, etc). Thus, it can address the issue of fat tails, or leptokurtosis, but only if market scenarios are generated using appropriate distribution assumptions.

A disadvantage of this approach is that it is computationally intensive and time consuming, entailing revaluation of the portfolio under each scenario. The computing load becomes enormous for financial institutions with tens or hundreds of thousands of exposures. Even with the most powerful computers, a very long calculation time (in some cases, several days) is required before loss and risk distribution results are obtained. This becomes a bottleneck for financial institutions when they attempt to use simulation results for capturing day-to-day changes in the amount of credit risk.

Using Monte Carlo simulation

To use Monte Carlo simulation (see Figure 7) the following three types of transformation equation are required:

❏ a transformation of risk factors to generate various states;
❏ a transformation equation to generate either the conditional or joint distribution of loss for each state (conditional/joint default rates and conditional/joint LGD. Both may be modelled or only default rate may be modelled. Still there is no model in the public domain modelling both default rates and LGD in a portfolio model);

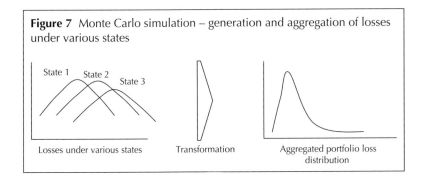

Figure 7 Monte Carlo simulation – generation and aggregation of losses under various states

Losses under various states   Transformation   Aggregated portfolio loss distribution

❏ aggregation of loss in each state to generate the portfolio loss distribution.

CreditMetrics and CreditPortfolioView use Monte Carlo techniques to simulate the loss distribution. In both cases, generating the portfolio value distribution requires generating risk drivers complying with a given correlation structure.

Analytical techniques for loss distribution or fat tail

Analytical techniques are widely used approximation techniques for analysing and modelling fat tails (another technique for modelling fat tails is the Extreme Value Theory).

We know that for a credit portfolio all the useful properties such as UL and diversification are found in the tail. The thick tail is the most important part of portfolio credit risk portfolio modelling. Owing to data paucity (conditional and joint default data) analytical techniques can be used to analyse the thick tail with approximation and restrictive conditions.

Analytical models enable us to simulate and visualise the effect of correlation usually under the restrictive assumptions. Analytical models are not full valuation models; they model simplified versions of the risk measures – generally diversification is measured through restrictive assumptions. The most widely used analytical model is CreditRisk+.

There are various levels of approximations – from a preliminary model to a full-blown analytical model.

In general, the following assumptions are made in an analytical model:

❏ exposures are assumed to be equal and no size discrepancies are allowed;
❏ fixed and uniform correlation across the firms;
❏ a single economic factor;
❏ defaults are assumed to be binomially distributed;
❏ LGD is assumed to be constant and exposure is reduced to the extent of recovery.

A starting point for this technique is to assume a uniform (equal) binomial exposure distribution. This also serves as a benchmark for measuring the effect of correlations on the portfolio risk. With the increase in number of obligors, the increase in loss volatility tends rapidly towards zero.

Analytical models generally assume a binomial distribution of the default and independent default events. The binomial distribution has the fat tail characteristics of credit risk; however, it does not have all the characteristics realistically correlated to the loss distribution. There are two approaches to model loss volatility in a binomial distribution.

❏ Portfolio loss is binomially distributed

$$Loss\ Variance = n \times E^2 \times d\ (1 - d)$$

$$The\ loss\ volatility = \sqrt{Loss\ Variance}$$

❏ Portfolio loss as a sum of random individual losses. Each random loss is $E \times Bernoulli\ variable$. The Bernoulli variable takes a value of 1 in the case of default and 0 in the case of no default. Losses are assumed to be independent. Therefore, the summation variance is the sum of the variance.

Diversification – Exposures are assumed to be constant. Therefore each additional exposure adds the same constant loss variance to the portfolio. However, loss volatility does not increase in the same proportion as it square root. Portfolio loss volatility is $\sqrt{p(1-p)/n}$.

Case Study: Moody's CDO rating methodology
Since 1996, along with other techniques, Moody's has been using the binomial expansion technique for CDO analysis. The entire portfolio is divided into a hypothetical portfolio of homogenous

and uncorrelated assets. Each of the portfolios is assumed to be an asset. Diversity score is the number of the assets in a hypothetical portfolio which has the same par value (total collateral par value divided by D). Each of the assets is also assumed to have the same probability of default as that of the pool it is representing.

Given the homogeneous nature of the hypothetical portfolio, the behaviour of the asset pool can then be described by $DS+1$ default scenarios (ie, with default occurring for 0 assets, 1 asset, DS assets), where the probability of each scenario is calculated using the binomial formula.

The long way out is to consider every possible combination of credit states across every obligor in the portfolio. However, it is computationally very complex and data is not available for every combination. A comparatively shorter route is to estimate the portfolio distribution by a process of simulation. Simulation reduces the computational burden by sampling outcomes randomly across all possibilities. Once the portfolio is approximated, it is possible to use summary statistics that describe the shape of the distribution. The shortest route involves the use of analytical techniques.

The crux of any portfolio model is the way in which credit risk components and correlation are modelled or their relationship simplified.

Moody's diversification method for asset-based securities is called the diversity index for the collateral, where the individual assets are distributed as companies across many industries. It assumes that firms in the same industry tend to be correlated and that firms in different industries are less correlated. Industry dispersion is used as a measure for their independence. The diversity score is a simple way to understand the degree of diversification by obligors and industry in any portfolio. The diversity index makes the assumptions that the correlation among the industries is zero and the correlation within an industry is high. There are empirical studies available to show that geographic diversification does not necessarily reduce the portfolio risk. The primary risk contribution is measured in the following two ways.

❑ The risk contribution of an asset is the change in the portfolio standard deviation due to the addition of the asset in aggregate to the portfolio. CreditMetrics uses this approach. This is the discrete approach to risk contribution, and is helpful for buy/

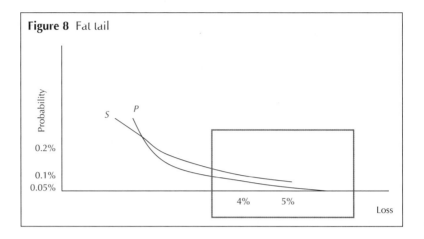

Figure 8 Fat tail

sell decisions and to quantify the effects of buy/sell decisions on the portfolio standard deviation.

❑ The risk contribution is the change in the portfolio standard deviation due to a small percentage change in the size of the asset in the portfolio multiplied by the size of the asset. CreditRisk+ and Portfolio Manager use this approach. This is a continuous approach to the risk contribution.

Diversification drives the tail frequencies

Owing to the tail dependencies of tail frequencies on the diversification, diversification has a direct impact on economic capital required. For example, the portfolio S has a 0.1% probability that its losses will exceed 5% of the portfolio while P has 0.05% probability that its losses will exceed 5%. The equivalent credit rating for Portfolio S is A and Portfolio S is the equivalent credit rating of B. If portfolio P and S are the only portfolios, the P's bank will be rated A and S's bank will be rated B. Therefore, diversification has a direct impact on the economic capital of the bank through the shape of the loan loss distribution tail (see Figure 8).

CreditRisk+ – a case study of analytical methodology
for fat tail
Correlation (or portfolio loss or the default volatility) is measured at different levels of granularity. In general, the existing models

measure the correlation at the approximation level rather than the finer granular level. For example, the actuarial models capture joint default behaviours through the use of default rate volatilities and sector weights. Volatility is measured indirectly through the ratio of default rate volatility to the unconditional default rates. At a more granular level, the default rate volatility can be differentiated by credit grades. Additional granularity can be obtained by slicing the default data by country and/or industry or assigning differentiated volatilities at the obligor level.

There are two important elements in the diversification or concentration.

❑ *Exposure to the same obligor*.
❑ *Exposure at the sector level*: Sector level analysis is an important ingredient of diversification. This is an efficient and practical way of managing a portfolio. Given this information, we can calculate the risk contributions of the sector to the overall portfolio risk. The ability to take sector correlations into account when calculating risk contributions is a prerequisite for active portfolio management. The risk contribution of a complete sector is obtained by adding up the risk contribution of all obligors in the sector. For a sector/industry approach, there are four approaches to estimate correlation:

❑ using an industry-specific time series of historically observed default events;
❑ using asset values of firms to derive pair-wise asset correlations, which are then linked to the relevant default correlations using the Merton model;
❑ using a factor model that relates the relative number of default events to macroeconomic drivers;
❑ using the binomial distribution to also estimate the correlation from the default volatility.

The binomial distribution is a well-known distribution for the number of defaults. This is also a loan distribution for uniform size exposures. The binomial distribution does not incorporate the exposure size changes. It works as a benchmark to measure the effect of correlations on the portfolio risk. It also illustrates how the increase in loss volatility as a percentage of the total portfolio exposure tends rapidly towards zero when increasing the number of obligors.

CreditRisk+ is the best example of an analytical method used to model the default and loss distribution. This makes it easy to manipulate and use numerical calculations algorithms to avoid full-blown Monte Carlo simulations.

The mixed Poisson distribution plays a pivotal role in generating independent default events. The Poisson distribution tabulates the frequency of observing k defaults when the default intensity or the number of defaults per unit time has a fixed value n. The default intensity is the Poisson parameter.

The model necessitates dividing the portfolio into segments by risk class and exposures net of recoveries. This allows us to bypass the limitations of the Poisson distribution that generates the distribution of the number of defaults, without considering size discrepancies. This makes it necessary to control both the size of exposure and exposures net of recoveries. The portfolio is divided into risk class and loss under default bands.

The mixed Poisson distribution uses a "mixing variable" q that allows the default intensity to vary according to the relevant factor and provides a function to model default intensity as a function of the economic factor.

In a special case, the mixing variable follows a gamma distribution and the resulting distribution is the negative binomial distribution. The gamma distribution allows variation in default intensity to fit the variation to the actual default rate volatility.

The time intensity parameters make defaults dependent on time, which is a useful feature for modelling the time to default. When aggregating distributions over all segments, it is possible to identify the time profile of default losses.

To model correlation into the model, the mixing variable is a linear combination of the economic and industry factors.

Note that risks measured through various methods are not equal. In the preceding paragraphs, we have described various approaches for measuring portfolio risk. However, it may be noted that risk measurements through different approaches do not produce a similar measure of the risks. The differences arise due to dissimilar default rates distribution differences in recovery rates distribution and assumptions, difference in the exposure and the effect of correlation parameter variation across commercially available and internally developed credit risk portfolio models.

In addition to parameter inconsistencies, the analytical assumptions are another source of differences, especially in the tail of the portfolio loss distribution.

APPLICATION OF THE PORTFOLIO LOSS DISTRIBUTION
Measurement of credit portfolio risk enables a bank to measure:

❑ EL;
❑ UL or loss volatility;
❑ the risk contribution of individual assets or groups of assets;
❑ economic capital;
❑ concentration;
❑ RAROC.

Traditionally, banks have managed the credit risk without measuring it. Credit risk has an impact on the credit pricing, loan provisioning and exposure limits. Traditionally, loan provisioning has been driven by regulatory requirements, credit pricing has been driven by the competition and crude exposure limits have been driven by the regulatory instructions and occasionally by linking limits to the capital of the bank. Moreover, the credit risk is managed not through the measure of the risk quantum but through stringent underwriting standards, limit enforcement and counterparty monitoring.

To understand how the measured credit portfolio loss can help in credit pricing, setting exposure limits and estimating loan provisioning, we need to first analyse the concept of portfolio loss.

Expected loss or average loss
Every credit portfolio suffers credit loss. Some of the borrowers will not pay back their obligations or will pay less than the obligations. This is equivalent to the "wastage" of input material occurring in a manufacturing firm. The amount of loss is measured in terms of a time unit, usually a year. So the entire loss during a year may be called "experienced" loss. However, the problem with the credit loss (the experienced loss) is that while the "raw material wastage" in a manufacturing unit is constant, the normal loss varies widely. Even for the same or consistent portfolio quality, the losses that are actually experienced in a particular year vary from year to year and depend on the number and severity of default

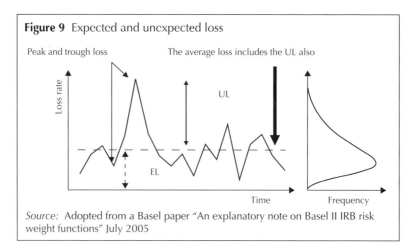

Figure 9 Expected and unexpected loss

Peak and trough loss   The average loss includes the UL also

Loss rate

UL

EL

Time   Frequency

Source: Adopted from a Basel paper "An explanatory note on Basel II IRB risk weight functions" July 2005

events. This wide fluctuation is managed in the following two ways:

❑ experienced loss is averaged out – especially over a business or economic cycle; this is called average loss or expected loss;
❑ variation or volatility in the average loss – this is measured through unexpected loss.

The division of losses into expected loss and loss variation helps in managing the losses better (see Figure 9).

Expected loss
EL is a forecasted average level of credit losses a bank assumes to experience. The EL includes peak (higher than expected loss) and through (lower than expected) loss. The expected loss is recovered as a cost of doing business and is managed through pricing and provisioning; only the recovery timings differ. Loss reserves may be built during the through loss period. Peak loss exposes the bank to the insolvency.

EL is the sum of credit loss. Ideally, EL is measured at the portfolio level. It is an average of the sum of the losses (over a time horizon of, say, one year)

$$EL_p = \sum Credit\ loss$$

We know that each exposure in the portfolio contributes towards the EL. The EL at the obligor level (stand-alone) is measured by

$$EL_i = EAD \times PD \times LGD$$

Therefore,

$$EL_p = \sum_{i=1}^{n} EL_i$$

Unconditional default rate and expected loss

The unconditional default rate is the weighted average default rate over an economic cycle. The EL is the average loss over an economic cycle. Therefore, EL rate represents the unconditional default rate.

Ex ante and *ex post* expected losses rarely converge. The average loss measures need to be continuously updated to reflect the business reality. This requires a continuous change in the provisioning and pricing. In practice, this may not happen or may happen with a time lag. To a great extent, provisioning is driven by the regulatory requirements and accounting standards. It may not be possible to reflect a reduction in average loss into the provisioning. However, with the adoption of an IRB methodology this is possible. Changes in the pricing upward may be a problem due to competition pressure. An increase in average losses is often provided from the profit. Banks generally attempt to implement stringent underwriting standards to control increase in the ELs.

Unexpected loss

The losses above the average level are referred to as ULs. This is also called loss variation. UL is generally measured through loss volatility. There can be two approaches to address the loss variation or loss volatility.

❏ The first approach is to increase the average loss to such a mark that there is either no peak loss or the difference between averages and peak is minimal. This can be done in two ways.

  ❏ By increasing the time horizon to measure and provide for the loss experience from one year to until the time-to-maturity. However, this is not allowed by regulators and accounting standards.
  ❏ By keeping the average and EL level high and recovering the higher average losses as a part of the pricing and provisioning.

The market does not allow ULs to be recovered from the pricing. This means that, owing to competition, recovery for the average loss level cannot be so high (therefore, high pricing) that there is no UL (no spikes). This means that a bank cannot over recover the loss such increasing the profits for a longer period. It should be noted that in the long run the entire loss (including losses in access of average loss) has to be recovered from pricing and profit. However, due to completion banks can only recover average losses.

❑ The second and more accepted approach, as prescribed by the banking regulators and acceptable to accounting standards and market/competition, is to cushion the ULs with capital. The protection provided by the credit capital goes far beyond the liquidity and it provides protection against insolvency. It is meant to absorb the ULs temporarily (until they are recovered through provisioning in the coming years).

Therefore, UL can also be defined as a likelihood that a bank will not be able to meet its credit obligations from its profits or loan provisioning and will require capital to pass over temporary variation in the EL.

Like EL, UL is measured on a stand-alone and a portfolio basis.

At an obligor level, UL(stand-alone) is measured by

$$UL = \sigma_j = EAD \times LGD \times \sqrt{PD(1-PD)}$$

The UL at portfolio level is given by

$$UL_p = \sqrt{\sum_{i=1}^{n} \sum_{j=1}^{n} UL_i UL_j \rho_{ij}}$$

Risk and return contribution – the four-way balancing principle

The EL and UL analysis leads us to a four-way balance: the acceptable market loan price determines the EL that can be recovered; EL determines the loan pricing and the worst or spike loss and therefore the UL; the UL determines the capital required; returns on the capital are an important element in loan pricing (see Figure 10).

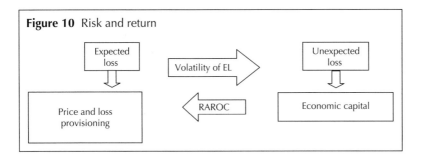

Figure 10 Risk and return

Risk contribution

Every exposure contributes a loss to the portfolio credit. The contribution to the portfolio credit loss is called the risk contribution. Credit loss is measured in two ways:

❑ loss volatility;
❑ economic capital.

There are two methods to measure the risk contribution:

❑ absolute risk contribution (ARC);
❑ marginal risk contribution (MRC).

While the ARC is the basis for the estimation of economic capital, the MRC is the basis for risk-based pricing.

Marginal risk contribution

The marginal contribution of an exposure or a sub-portfolio to the portfolio loss volatility or to capital is the difference of the portfolio loss volatility, or the portfolio capital, "with and without" the exposure or sub-portfolio.

Marginal risk contribution to loss volatility

The MRC to the loss volatility is the loss volatility minus the initial loss volatility. The marginal contribution to the loss volatility is lower than the absolute loss volatility contributions and stand-alone contribution to the loss volatilities. Therefore, they do not sum up to make the portfolio loss volatility.

Marginal risk contribution to capital

Capital is estimated from a loss distribution and the loss percentiles at various confidence levels. The MRC to the capital is the portfolio capital minus the initial capital used in the portfolio. Additional exposure brings in additional risks as well as bringing the diversification effect – ie, reduction in the capital required.

The diversification effect is often more powerful than the additional risk. Therefore, the MRC may be negative. While the contribution to the risk capital may be negative, the contribution to the loss volatility is not negative unless the correlation is negative. We know that, in practice, negative credit correlation does not exist unless it is with insurance or credit derivatives.

Uses and implications

The difference between absolute and MRCs and between the risk contribution for loss volatility and capital raises an important question of which risk contribution should be used when.

MRCs are relevant for adding or removing sub-portfolios or pricing a new transaction according to the incremental risk. There is no direct formula to convert absolute to marginal risks and *vice versa*.

The different vendor models estimate different types of risk contributions – the KMV model estimates ARC, CreditMetrics estimates the MRC.

Estimation of MRC is a tedious process of with and without estimations:

$$MRC_n = \sigma_{p+n} - \sigma_p$$

$$\sigma_{p+n} = ARC^p_{p+n} + ARC^n_{p+n}$$

$$\therefore MRC_n = ARC^n_{p+n} + (ARC^p_{p+n} - \sigma_p)$$

We know that the absolute loss volatility contribution of the portfolio in addition to one more exposure is always less than or equal to the portfolio loss volatility. We also know that stand-alone loss volatility is always less than ARC:

$$\therefore ARC^p_{p+n} \leq \sigma_p$$

$$\therefore MRC_n \leq ARC^n_{p+n} \quad and \quad ARC^n_{p+n} < \sigma_n$$

$$\therefore MRC_n \leq ARC^n_{p+n} < \sigma_n$$

Absolute risk contribution

ARCs are allocations of the portfolio risk to the existing individual facilities or sub-portfolio, embedding the correlation structure of the portfolio. Only ARCs add up arithmetically to obtain the overall portfolio risk, while the MRCs do not. ARCs and MRCs serve a different purpose. The MRC applies to a new transaction, and hence is useful for pricing. The ARC or capital allocation is a measure for *ex post* performance.

The ARC of an existing facility j to a portfolio P is the covariance of the random loss of this single facility j with the random loss aggregated over the entire portfolio (including j), divided by the loss volatility of this aggregated loss. The risk contribution sums up to make the loss volatility. The capital allocated on an individual asset sums to make the portfolio capital. It should be noted that the contribution to loss volatility does not depend solely on the obligor. It depends on both the existing portfolio and the obligor.

Marginal risk contribution *versus* absolute risk contribution

MRC is used for pricing while ARC is used for economic capital estimations. With the additional exposure in the portfolio, the absolute risk of the existing portfolio becomes reduced owing to diversification impact. This diversification benefit becomes reflected in the form of lower marginal risk. The incremental exposure is entitled to receive the benefit of the diversification.

The new facility diversifies the risk of the existing facility. Therefore, adding a new facility reduces the ARCs of the existing facilities. However, the *ex ante* based on the marginal contribution, and the *ex post* based on the absolute contribution differ for the same facility. Therefore, pricing is based on the MRC. Pricing is for the future portfolio while capital is for the existing portfolio. Consider the information presented in Table 10. In general, MRC means that the risk contribution to loss volatility is considered.

Table 10 Computation of ARC and MRC

Risk	Measure
Risk in the existing portfolio σ_p	300
ARC of the additional exposure ARC_{p+n}^n	27
Total portfolio risk (without diversification)	327
Less diversification impact (100 is reduced to 98)	2
Total risk of the enlarged portfolio σ_{p+n}	325
MRC σ_n	25

This is because many transactions are small compared with the reference portfolio and it is acceptable to use a constant ratio of capital to portfolio loss volatility. Pricing based on MRC charge is a mark up equal to the risk contribution times the target return on capital.

Concentration and diversification

Consider the following situation. An obligor has defaulted on US$1 million to a bank. The bank holds say a security worth US$400,000. So the recovery rate is 40%. Imagine the bank had an option to lend one US$ at a time. If the bank had lent US$1 for the same security, the bank would have got back US$1 in full. This is true until lending US$400,000, and – the recovery rate would have been 100% and EL and UL were zero. For the additional US$1,000, the marginal recovery rate is 0% and the average recovery would fall to 400/401%.

With every additional US$1,000 lent, the average recovery will continue to fall, until it reaches 40% at US$1 million exposure. This is true for each obligor in the same grade and across the grades.

The lesson from this situation is that every additional US$ lent is exposed to a higher marginal credit risk (EL as well as UL). This is different from market risk. In market risk, each US$ within the same exposure class is exposed to the same risk (VAR is the same for each US$). Readers are requested to develop this concept further in their thinking.

Concentration

In the light of the above example, concentration risk can be analysed from two perspectives–concentration with the same obligor or concentration within the group of obligors. Concentration with the same obligor (against the assets available to the creditor) has the impact of reducing the recovery rate (since the same asset is available for recovery) while the concentration within the group increases the default probabilities (and may also reduce the recovery rate) owing to contagion impact.

Concentration risk is the risk that too many defaults could occur at the same time. Diversification is loosely achieved through concentration limits.

Credit risk models do not distinguish between the concentration risk and the portfolio risk. There are two types of concentrations. The obligor and sector concentration are the main drivers for the credit risk. The concentration risk is generally measured and managed through stress testing.

❑ *Obligor concentration*: This is a higher exposure to the same obligor. This also represents the undiversified idiosyncratic risk. This form of credit concentration risk has been addressed via a granularity adjustment to portfolio capital. The obligor concentration is easier to identify (and to measure) than the sector concentration. As a portfolio becomes more and more fine-grained, idiosyncratic risk is diversified away at the portfolio level.

❑ *Sector concentration*: Borrowers have different degrees of sensitivity to systematic risk. As a consequence, defaults of different borrowers are usually not independent. Therefore, the systematic component of portfolio risk is unavoidable and only partially diversifiable.

❑ A large exposure to a single sector or to several highly correlated sectors can give rise to so-called sector concentrations. It is a difficult practical issue to identify those factors that are particularly important for a given portfolio.

The IRB approach assumes that (a) bank portfolios are perfectly fine-grained and (b) there is only a single source of systematic risk. To the extent that either assumption is violated, IRB capital requirements may understate the true economic capital requirement.

Diversification

Credit risk cannot be hedged, or "structured" away fully. Default risk can be reduced and managed through diversification. Default risk, and the rewards for bearing it, will ultimately be owned by those who can diversify it best. Since, the credit risk cannot be completely diversified, well developed tools are required to measure the credit risk. However, diversification measurement tools for credit risk have been underdeveloped. Thus, the financial industry has experienced many unexpected defaults in the past.

Diversification means aggregate risk in the portfolio is less than the average of each asset's stand-alone risk. Some part of each asset's stand-alone risk is diversified away in the portfolio.

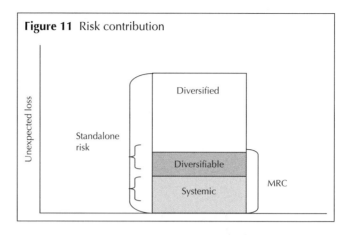

Figure 11 Risk contribution

Therefore, the stand-alone risk can be divided into the following two parts (see Figure 11):

❑ the risk that can be diversified away – and we know from portfolio theory that this is called idiosyncratic risk;

❑ the risk that cannot be diversified away – also called the absolute risk contribution – this is systemic risk. However, a part of the risk may be diversifiable. We know that the portfolio is not always the optimum portfolio. So some risk may always remain to be diversified.

As the holdings change, the risk contributions change. For instance, if the proportionate holding of this asset were increased in the portfolio, less of its risk would be diversified away and the risk contribution would go up.

Systemic risk is measured relative to the whole market of risky assets. The risk contribution is specific to a particular portfolio and within the portfolio to the particular asset. In any portfolio there are assets whose returns are small relative to the amount of risk they contribute; these are called mis-priced assets.

For a facility, the ARC is proportional to the stand-alone risk. The ratio of MRC to the stand-alone risk is called the "retained risk". This ratio is always less than one. A measure of the diversification is 1 – *Retained risk*. Diversification is a measure of the correlation of the asset with the portfolio (ρ_{jp}):

$$Retained\ risk = \rho_{np} = \frac{MRC(\sigma_{p+n} - \sigma_p)}{Stand\text{-}alone\ risk\ \sigma_n}$$

$$Diversification = 1 - (\rho_{np})$$

The change in the risk each asset contributes towards the portfolio risk has an impact on the measure of the risk reward relationship of other assets. Application of this process leads to higher diversification.

A large positive correlation between defaults results in a large probability of relatively small losses and a small probability of rather large losses. This also leads to a very high percentage of the time (around 80%), the actual losses will be less than the average loss. This may lead to belief that the portfolio is better diversified.

Economic capital estimation

Banks and other financial institutions are increasingly adopting the measurement of economic capital as a standard measure to determine the amount of capital needed to protect against financial distress in the event of unexpectedly large losses. Economic capital is a measure of UL defined as a difference between the estimated mean of the loss distribution and the estimated loss level corresponding to the chosen critical tail percentile. When calculating portfolio economic capital requirements, most models estimate critical values corresponding to extreme tail percentiles of a portfolio or whole-bank loss distribution. By their design, economic capital models are complex and are an extension of the credit portfolio risk models. They take portfolio credit risk distribution as the inputs.

Economic capital measures the economic risk on a portfolio basis and takes account of diversification and concentration; economic capital reflects the changing risk of a portfolio and it can be used for portfolio management.

The economic capital cushion for the UL cannot be absolute since the absolute cushion is economically inefficient. The two opposing forces of insolvency *versus* economic inefficiency have to be balanced. This is achieved through striking a balance at an acceptable level of insolvency frequency/probability. The UL will exceed the level of capital with only the acceptable level of probability.

In the case of regulatory capital, the regulators decide on the "acceptable level of insolvency probability or frequency". For

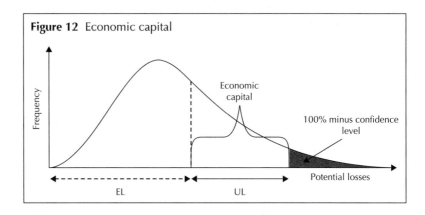

Figure 12 Economic capital

Table 11 Parameters for regulatory capital and economic capital

Modelling of risk components (see Table 13 for details)	The models in the public domain compute economic capital based on PD, LGD and EAD and correlation as risk components. Basel includes maturity also. Basel has three approaches to capital computation. Assumes correlation to be constant under all approaches.
Policy	Risk horizon, confidence level.
Default definition	Regulatory capital computation is computed only for the defaults. Economic capital also considers changes in the credit quality.
Correlation	Correlation for economic capital is measured as systemic and idiosyncratic risk contribution. While idiosyncratic risk should be diversified away, a part of systemic risk cannot be diversified away. For regulatory capital it is assumed to be constant for a risk segment and dependent on one factor.
Risk segmentation	Regulatory capital is computed for a given risk segment. Segmentation rules are defined in the Accord. Economic capital is computed on homogeneity of risk. (Both the approaches eventually converge).
Measurement	Regulatory capital follows Basel formula and can vary based on PD, LGD, EAD and maturity measurements and not only on correlation. Correlation is assumed to be constant for an asset class. Economic capital is to be computed as an unexpected loss at a certain % of confidence.

economic capital, the bank/financial institution decides on the acceptable level of insolvency probability or frequency. For economic capital, the insolvency frequency or probability is generally driven by the targeted debt rating for the bank.

Managing credit portfolios is the name of the game for estimating and managing the UL, economic capital and the required returns on the capital (see Figure 12).

Reconciliation between economic and regulatory capital

Ignoring regulatory capital is akin to ignoring taxation impacts in investment decisions: the analysis is, in practice, just irrelevant. The following statement distinguishes between the regulatory and economic capital regimes.

In the regulatory capital regime, "there is zero surplus regulatory capital and positive surplus economic capital", ie, the bank does not have additional capital but can take further risk (the bank does not mind taking risk, as long as regulatory capital is not increased). In the economic capital regime, the assumption is "there is zero surplus economic capital but positive surplus regulatory capital", ie, the entire risk is correctly measured and provided for through the economic capital. However, the bank has (a capacity to raise) additional capital to support the additional risk.

Economic capital can be synchronised with the eligible regulatory capital (tier I, II and III) the bank is holding in the following way (assuming economic capital required is larger than the capital bank is holding):

(1) If there is large mismatch in the regulatory and economic capital, the bank should rearrange its investment decisions (reduce the risk and economic capital);
(2) Increase the capital the bank is holding (tier I, II and III) to match the economic capital.

Addition of economic capital: If the correlation among the sub-portfolio is assumed to be zero, then the capital becomes sub-additive.

We know that by variance addition we obtain

$$\sigma_p^2 = \sigma_{sp1}^2 + \sigma_{sp2}^2 + 2\sigma_{sp1}\sigma_{sp2}\rho$$

If $\rho = 0$

$$\sigma_p^2 = \sigma_{sp1}^2 + \sigma_{sp2}^2$$

The primary uses of economic capital are as follows:

❏ for solvency, bank rating, internal and regulatory purposes;
❏ to measure performance at individual, divisional and business levels:

  ❏ RAROC;

❑ to manage and implement strategic decision support systems:

   ❑ strategic capital allocation;

❑ for pricing, profitability, limits:

   ❑ RAROC and hurdle rates;

❑ for active portfolio management:

   ❑ risk diversification, risk contribution and risk concentration measurements.

Risk adjusted returns on capital

RAROC is a profitability calculated on risk-adjusted capital.

Economic capital is the portfolio capital for the entire portfolio and is the capital allocation or risk contribution for sub-portfolio or an exposure. Which risk contribution to use (ARC or MRC) depends upon the purpose of the measure:

$$RAROC = \frac{(Revenue\ expenses - EL + return\ on\ ecomomic\ capital + transfer\ price)}{Economic\ capital}$$

$$RAROC = \frac{(income - EL)}{Risk\ contribution}$$

The EL has a statistical meaning for the portfolio EL while the EL for a single exposure does not provide much of a meaning to the RAROC:

$$Income = (Interest + fee\ income)\% - cost\ of\ debt\ (fund\ transfer\ rate)\% - operating\ costs\%$$

Computation of regulatory capital – a case study of the Basel model

This model considers a portfolio of a relatively large number of exposures. With this assumption, the idiosyncratic risks associated with individual exposures tend to offset one another and only systematic risks that affect many exposures have a material effect on portfolio losses. The systemic risks are modelled with a single factor.

The conditional EL distribution is built under ASRF. Conditional loss is estimated for conservative estimates of a single systemic factor. The average PDs estimated by the bank are converted into a conditional EL distribution by using a supervisory mapping function. This function is a result of the adoption of Merton's single asset model to credit portfolios as modified by Vasicek in 2002 for the assumptions that idiosyncratic and systemic factors are normally distributed.

Conditional PDs
The default threshold of the Merton model is related to PD and the systemic factors are related to PDs. Therefore, the default threshold can be inferred from PD and PD can be inferred from systemic risk factors. The correlation-weighted sum of the default threshold and the conservative value of the systematic factor then yield a conditional default distribution. In addition, a stress test is applied to identify the possible events that could have unfavourable effects on the bank's credit exposure.

Economic downturn LGDs
There are two approaches to estimate downturn LGDs:

❑ to extrapolate the downturn LGDs from the average LGDs;
❑ by internal assessment of LGDs during adverse conditions.

LGD estimation is an evolving area, so Basel has left a lot of discretion to the banks.

Correlation
The Basel model recognises that the quantum of correlation is different in different portfolios. While correlation may be higher for corporate exposures, it is lower for retail exposures. The Basel model is a one-factor correlation model and the factor is the overall economy condition.

Maturity impact
From the duration studies of fixed-income instruments, we know that longer maturity/duration instruments are more sensitive to price changes. Longer duration credit assets are more exposed to the downgrade risks. Loans with higher PDs have a lower market value than loans with lower PDs. Extending this further, longer duration/maturity instruments are exposed to higher credit risk. Consistent with this thinking, the Basel makes adjustment for

Table 12 Maturity Adjustment in the capital computation equation for a given PD

Default rate (%)	3 years	3.5 years	4 years	4.5 years	5 years
0.01	1.6934	1.8667	2.0401	2.2134	2.3868
0.02	1.6199	1.7749	1.9299	2.0849	2.2399
0.05	1.5306	1.6633	1.7960	1.9286	2.0613
0.1	1.4683	1.5854	1.7025	1.8196	1.9367
0.3	1.3777	1.4721	1.5666	1.6610	1.7554
0.5	1.3386	1.4232	1.5079	1.5925	1.6772
1.0	1.2883	1.3604	1.4324	1.5045	1.5766

maturity and PDs. The maturity adjustments are computed using the KMV Portfolio Manager model. The KMV portfolio model uses the parameters of the Basel model (99.9% solvency) and uses the capital market data for the time structure of the PD.

Table 12 gives the maturity adjustment for the default rate in percent and for maturity between three and five years.

The Basel model can be further understood by understanding the capital estimation equation:

Estimation of risk = UL capital requirement

$$= \left[LGD \times N \left\{ \sqrt{\frac{1}{1-R}} \times G(PD) + \sqrt{\frac{R}{1-R}} \times G(0.999) \right\} - PD \times LGD \right]$$

$$\times \left(\frac{1}{1-1.5b(PD)} \right) \times (1+(m-2.5) \times b(PD))$$

$$= LGD \left[N \left\{ \frac{G(PD) + \sqrt{R}G(0.999)}{\sqrt{1-R}} \right\} - PD \right] \times \left[\frac{1+(M-2.5)b(PD)}{1-1.5b(PD)} \right]$$

where
❏ $b(PD) = (0.11852 - 0.05478 \times \log(PD))^2$;
❏ LGD estimates are used to estimate UL and EL;
❏ the confidence level is 99.9%;
❏ R is the measure of correlation, which differs according to the asset class;
❏ the function $b(PD)$ is used to smooth the maturity adjustment over the PD;

Table 13 Modeling risk components under IRB approaches

Quantitative input	Foundation IRB	Advanced IRB
PD	To be modelled by bank	To be modelled by bank
LGD	Supervisory values set by Basel Committee (constant value)	To be modelled by bank
EAD	Supervisory values set by Basel Committee (constant value)	To be modelled by bank
Maturity (M)	Supervisory values set by Basel Committee, or at national discretion, provided by bank based on internal (own) estimates (with an allowance to exclude certain exposures) (constant)	Provided by bank based on own (internal) estimates (with an allowance to exclude certain exposures)
Best estimate of EL	To be modelled by bank	To be modelled by bank
Potential LGD	To be modelled by bank	To be modelled by bank

❑ $PD \times LGD$ estimates the EL; the capital is required only for the UL and the EL is reduced from the overall loss;

❑ the function $G()$ represents the inverse normal function;

❑ the function $N()$ represents the normal distribution.

The Basel approach does not prescribe conversion of average LGDs into conditional LGDs for loss or capital estimations. Instead, banks are asked to estimate LGD for higher severities when the adverse factor is realised.

The conditional EL distribution is built by multiplying conditional PDs with downturn LGD. The total conditional loss ($EL + UL$) must be supported by the sum of capital, provisions and write-offs.

For the wholesale market
The IRB approach is based on measures of UL and EL. The risk-weight functions produce capital requirements for the UL portion.

For defaulted exposures, two additional quantitative inputs are required:

❑ the best estimate of economic loss (BEEL), which is the estimate of the EL that the banking organisation expects to incur, given current economic conditions and the unique characteristics of the defaulted exposure;

❏ the potential LGD (PLGD), which represents a margin above the BEEL, to account for any uncertainty in the ultimate recovery on the defaulted exposure during the recovery period. The PLGD replaces the LGD once the exposure goes into a default status.

REFERENCES

Allen, F. and D. Gale, 2000, "Financial Contagion", *Journal of Political Economy*.

Araten, M. and M. Jacobs, 2001, "Loan Equivalents for Revolving Credits and Advised Lines", *Published RMA Journal*.

Asarnow, E. and J. Marker, 1995, "Historical Performance of the U.S. Corporate Loan Market: 1988–1993", *The Journal of Commercial Lending*.

Basel Committee on Banking Supervision (BCBS), 2005, "An Explanatory Note on Basel II IRB Risk Weight Functions", http://www.bis.org/bcbs/irbriskweight.pdf

Giesecke, K. and S. Weber, 2003, "Cyclical Correlations, Credit Contagion and Portfolio Losses", published on the website of Stanford University.

Kiyotaki, N. and J. Moore, 1997, Credit Cycles. Working Paper, London School of Economics.

Markowitz, H. M., 1952, "Portfolio Selection", *The Journal of Finance*.

Saunders, A. and L. Allen, 2002, "A Survey of Cyclical Effects in Credit Risk Measurement Models."

Ugur, H. and A. Hickman, 1988, "A Generalised Framework for Credit Risk Portfolio Models", September.

Vasicek, O., 2002, "Loan Portfolio Value", *Risk*, December.

10

Validating the Risk Measurement Process

INTRODUCTION

In a study released in April 1999, the Basel Committee concluded that it was premature to consider the use of credit risk models for computation of economic capital, which can be taken as regulatory capital, *primarily because of difficulties in calibrating and validating credit risk models.*

The objective of credit risk measurement is to measure credit risk for each exposure. The problem with credit risk is that it cannot be measured. The difficulties and problems of credit risk measurement are addressed by building various intermediary steps and making assumptions at each step.

RISK COMPONENT MEASUREMENT PROCESS

Figure 1 shows the intermediary steps: as a first step towards credit risk measurement, banks are empowered by the Basel II Accord to build models for risk components such as EAD, PD and LGD. These risk components are key input parameters for the computation of regulatory capital. As not many portfolio models are in use, the literature currently available prescribes validation of risk components only.

The aim of the validation process is to validate end-to-end risk component measurement, which includes the validation of relationships between each possible pair: risk segment, risk component, risk factors and models.

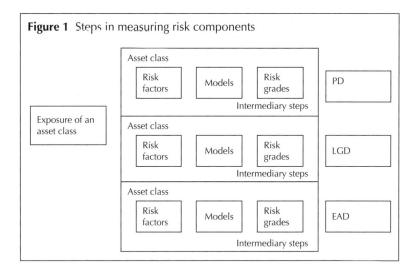

Figure 1 Steps in measuring risk components

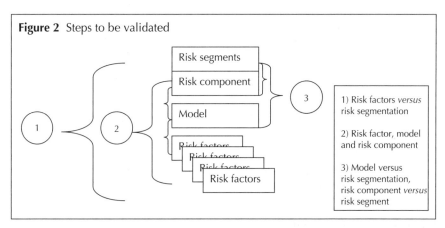

Figure 2 Steps to be validated

The relationships between each pair (see Figure 2) have to be tested for benchmarking, back testing, scenario analysis and stress testing. We will now discuss the relationships between the risk measurement pairs.

Risk factors *versus* risk segment

Risk segmentation is a process to classify exposure into an asset class. Asset class segmentation processes separate exposures into segments with homogeneous risk characteristics that generate reliable estimates of risk components. This is a crucial step in the risk measurement process as the applicable risk factors, risk measurement models and risk component distributions are determined by the risk segment and sub-portfolio within a segment. Each risk segment has exposure classification standards. The standards support consistency in assumptions, data and parameters used in risk measurement models.

Risk factor versus model versus risk component

This is the most import relationship to be validated. This also includes:

❑ risk factor *versus* model;
❑ model *versus* risk component.

Models measure a relationship between risk factors and risk components. The relationship is established or assumed based on the reference or historical data, and economic and statistical relationships between the data. However, the relationship keeps changing due to changes in the business environment, economic cycle, and correlation and changes in the data. Models also approximate and substitute the relationship due to data constraints or to avoid the complexities of computation. The accuracy and performance of the model is measured to validate the relationship.

We have previously discussed PIT and TTC measurement of PD rating. The desired stability of risk measure determines the choice of risk factors and risk models. Under the validation process, the choice, stability and suitability of risk factors are assessed and reviewed for the desired stability, model accuracy, performance, assumptions and substitutions.

Model *versus* risk segment

There are two approaches: PD computed at the obligors' level and PD computed at the pool level.

The calibration accuracy of a model is determined by the quantum and quality of data. As explained in the "Probability of Default" chapter, to address the data related challenge, internal data is supplemented by external data, models are calibrated using reference data, portfolio data is mapped to reference data and models are recalibrated.

It is required that the expected performance of each risk grade and sub-portfolio (risk segment) is consistent with the expectation underlying assignment of exposures to the segment. The most accurate approach to quantifying pooled PDs depends to a large extent on whether and how pooled PDs are expected to vary over a business cycle.

❑ *Model* versus *risk components*: Models have the required discriminatory power and are calibrated accurately. The bank needs to ensure that the risk measurement is valid over time. Therefore, relationships are tested periodically.

❑ *Risk components* versus *asset classes*: It is necessary to establish internal tolerance limits for differences between expected and realised outcomes. Validation will measure the differences against the tolerance limit. If no changes in the tolerance limit are required, then other pairs may be reviewed (such as model *versus* asset class, risk factors versus models and risk factors *versus* asset class).

If a relationship between more than one pair is developing, then the bank may take a holistic view, ie, review the relationship between the pair that is being developed. Each pair will also be benchmarked with the external systems (to the extent available) but all the pairs may not be available.

VALIDATION TECHNIQUES

Validation is a process to assess the performance of risk component measurement systems consistently and meaningfully (see Figure 3). Validation aims at assessing the systems to ensure that the realised risk measures are within an expected range. Validation also aims to use appropriate validation tools and to establish validation standards within the bank. Validation also aims to establish consistency and accuracy of the systems. Validation standards are called qualitative techniques and statistical methods are called quantitative techniques.

Quantitative techniques are insufficient for credit risk models

A major obstacle to back testing of credit risk components is the scarcity of data, caused by the infrequency of default events and the impact of default correlation. Due to correlation between risks components in a portfolio, the observed risk components can systematically exceed the critical risk component values if these are determined under the assumption of independence from the credit risk components.

All tests based on the independence assumption are the lowest estimation of risk components. On the other hand, tests that consider correlation between risk components will only allow the detection of relatively obvious cases of rating system mis-calibration.

The same exposure does not measure the same risk by two different banks. The difference is not only due to the different statistical techniques used but also due to the following differences:

❑ assumptions in the model about the risk factors, risk components, relationship between risk factors and components, risk component distribution and correlation;
❑ parameters used in the models;
❑ data used to calibrate the models.

Consistency in risk measurement is the primary determinant for quantitative accuracy and is assessed through qualitative techniques.

Qualitative techniques
Risk measurement standards provide consistency across time and asset classes. Documentation of assumptions about risk factors and risk components, the data used and model parameters are helpful

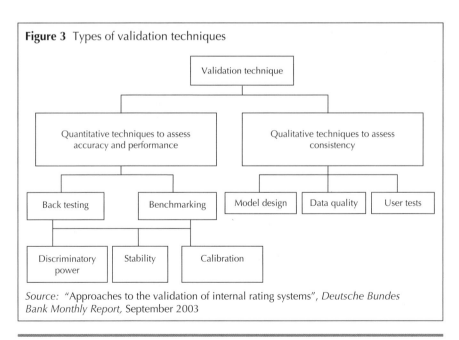

Figure 3 Types of validation techniques

Source: "Approaches to the validation of internal rating systems", *Deutsche Bundes Bank Monthly Report*, September 2003

in providing measurement consistency. Qualitative validation techniques aim to assess measurement consistency.

All supervisory instructions have identified assessment of model design, data quality and user tests as qualitative techniques.

Importance of documentation

The validation exercise can be defined in three words: document, document and document! Validation is a process to validate relationship between various intermediary parts built to measure risk. The ultimate purpose is to validate the risk measure for an exposure. Therefore, documentation plays a critical role in the validation process for the following individuals:

❑ bank supervisors;
❑ internal reviewers of the quantification model and risk segmentation system;
❑ model validation teams.

Some of the processes recommended for documentation are as follows:

❑ all major decisions in the design and implementation of the quantification process;
❑ all data sources, characteristics, assumptions on distribution and overall processes governing data collection and maintenance;
❑ assessment or review of the quantification process.

Validation of each step in the risk component measurement process

The following three validation techniques are recommended by bank supervisors and the Accord:

❑ review of model design, data quality and uses of the models;
❑ benchmarking;
❑ back-testing.

In addition, stress testing and scenario analysis are recommended. Table 1 shows a non-exhaustive list of risk measurement standards.

Table 1 Risk measurement standards.

Validations	Risk segmentation	Converting risk factors into risk components	Distribution of risk components	Data maintenance
Development evidence and logic Model design Data quality User tests	Revalidation of componentisation, approximation and substitutions	Identify risk components for each risk segment Identify risk factors and borrower characteristics to build homogenous risk segment to measure risk components Risk factors should be able to be able to measure risk consistently over time. Establish standards for each statistical technique. For example, univariate techniques should reduce the number of risk factors. Multivariate techniques should capture the combined effect of multiple factors	The assumption, approximation and applicable range for the relationship of risk factors with risk components keeps changing. Each should be revalidated Measurement should work well outside the developmental sample Document evidence to map reference data to the portfolio data	Every model assumes risk component distribution. The assumption may not be valid for a changed economic cycle or changed data (borrower/ obligor/exposure) set Validate distribution and other assumptions against empirically and external benchmarks	Assumptions and data about economic cycles, correlation and measurement horizon has a major impact on the loss distribution Historical data for the minimum prescribed period are used Reference data must closely resemble the bank's existing portfolio Reference data includes defaulted and non-defaulted data
Reasonableness of the system	Homogeneity is a primary driver for risk segmentation Some of the tests for	The aim is to determine the accuracy of the calibration	Power and accuracy of a model are interrelated No one test is sufficient	Data used by models is accurate and complete Data should include the period of stress

Table 1 (continued)

Validations	Risk segmentation	Converting risk factors into risk components	Distribution of risk components	Data maintenance
Back testing	homogeneity include the following: □ measures and values of the risk factors □ granularity/exposure □ borrower and facility related risk characteristics □ characteristics predictive of losses and risk components	Changes in the bankruptcy process impacts the accuracy of the model Mapping between the reference data and portfolio data should be revalidated periodically		The changes in the bankruptcy process impact the usability of historical data Establish sampling standards. The sample should represent the target population
Comparing realised outcomes with the predictions.	Bank should establish that the chosen criteria for segmentation are able to differentiate credit risk across the portfolio and captures changes in the direction and level of credit risk	Migration between the risks grades to be benchmarked against the industry migration tables	Check the model for over-fitting Use outcome-based methods to validate models	The relationship between internal portfolio and reference data has to be re-established continuously
Benchmarking	Choice of risk factors, assumptions underlying the model, risk components measures, all are to be benchmarked			

MODEL RISKS

The near failure of LTCM (Long Term Capital Management) has demonstrated that model development and use is an error-prone process. Model risk is the risk of occurrence of a significant difference between the risk measured by the model and the actual risk. It is a general term for the financial loss resulting from the use of a mathematical model for risk measurement.

Model risk is an addition to the complexities involved in credit risk measurement. The possible sources of model risk are as follows:

❏ inaccurate inputs, inaccurate choice of risk segment and risk factor;
❏ invalid model assumptions, invalid extrapolation or interpolation, break down of patterns and fat tails;
❏ invalid calibration of the model;
❏ low discriminatory power of the model;
❏ software bug.

A model consists of three components: an information input component, which delivers assumptions and data to the model; a processing component, which contains the theoretical model and transforms inputs into estimates via the computer instructions (code); and a reporting component, which translates the mathematical estimates into useful business information.

Model validation not only increases the reliability of a model, but also promotes improvements and a clearer understanding of a model's strengths and weaknesses among management and user groups. Table 2 shows a validation of model components.

Since models are software programmes, the information systems (IS) audit techniques can be used for model validation. The depth and extent of the IS audit approach is driven by the materiality concept. In the following paragraphs, an attempt has been made to

Table 2 Validation of model components

Inputs	Data – internal and external sources
	Assumption – about risk factors
Processing	Model purpose
	Logic
	Code and mathematics
	Benchmarking and back testing results
Output	Reasonableness of model result

apply the IS audit standards (as prescribed by IS audit control association standards on model review) and techniques to the credit risk model which are in sync with the Basel Accord.

Information systems audit techniques for model review
Independent review

Risk models are complex. Complexity of the model does not mean that the model builder should be involved with the model validation. One of the basic ingredient to achieve independence is to have good documentation supporting the working, algorithms and testing of the model. As far as possible, personnel validating the model should be independent of the personnel who constructed the model. If comprehensive independence is not possible, users and senior management should be informed.

Responsibility

Responsibility for the model validation should be defined. This should provide independence to the model auditors/reviewers. The charter should also ensure that no model can enter into production unless validated/audited.

Applicable regulations

Model validation is a specialised job. The team should have the requisite skills in credit risk, statistical techniques and software and IS audit techniques. Model development or acquisition is similar to the development or acquisition of any other software. Therefore, the relevant provisions of the CoBIT (Control Objectives for Information Technology) framework synchronised with the Basel and National supervisory instruction on model validation should be followed.

What to review

❑ Key assumptions.
❑ Development evidence, including the software quality control process, cost–benefit analysis for various assumptions and approximation and an independent review.
❑ Model documentation, which should be sufficient to enable independent review, training of new staff and clear thinking by the model developer (this means the documentation should be more thorough than the user manual).

❑ Documents created during the model construction:

  ❑ since models are generally complex, the document should be in sufficient detail to allow the precise replication of the model;
  ❑ summary overview of the procedures used and the reasons for using those procedures, and a description of model applications;
  ❑ limitations of the model application;
  ❑ validation/testing procedures – those performed and their results.

Review assumptions underlying the model

Computer models are built with an array of assumptions and the model is therefore correct within those assumptions. Frequently, those assumptions are the output of another model. If need be, the model producing the assumptions may also be validated using the principles elucidated here. The other practical way to validate the assumptions is against publicly available inexpensive data and this requires expertise in the area of risk management. The bank also may be holding the information/data on the assumptions internally. However, these assumptions need to be verified. If material, these should be highlighted in the validation/audit report. The auditor ought to verify that the senior management is aware of the assumptions and their implications.

❑ *Review model input data*: If there is a data problem and management still decides to use the model, there should be clear policies from the senior management for audit of the decisions taken based on the model. The builders of the model should document such lacunae before hand. The job of the reviewer is to validate the lacunae mentioned by the builders and ensure that there are no further lacunae in the system.

❑ General ledger entries are to be verified with other management information systems.

❑ General audit software such as ACL can be used for the following tasks:

  ❑ to test the quality of input data;
  ❑ to ensure that the internal data agrees with the transaction processing systems data;
  ❑ to match the external data with other sources (to test the reliability of the external data).

❑ *Review processing component*: This consists of a code or program and theoretical logic. The choice of theory is an art form. The theory is a great simplification of the real world and is represented in the form of mathematical equations.

❑ Code validation techniques include the following.

  ❑ White box test – code verification. However, this technique is also not foolproof.
  ❑ Parallel simulations. Model computations are compared with results from a second, well-validated "benchmark" model.
  ❑ Availability of code documentation is a crucial requirement.

❑ The technical documentation should cover interrelationships between modules, flow charts and "pseudo code". Documentation also covers the change control documents.
❑ The litmus test for the availability of documentation is whether an entirely new team can take over the development and maintenance of the model without any help from existing team members.

❑ *Review processing component in the model purchased off the shelf*: One common misconception is that validation of vendor models is not necessary since the model is being used by other banks for similar purposes. The recommended practice for the off-the-shelf systems is as follows.

❑ Vendors need to demonstrate that they have followed the model validation practices as required by the Basel Accord / National supervisors.
❑ Banks should obtain from vendors descriptions of key model parameters and the sensitivity of model results to changes in these parameters and their statistical weights.
❑ Bank should test the model performance (both accuracy and power) against the bank's own portfolio. The bank should have a clear strategy to test the performance of vendor model to compensate for various non-transparencies. This also includes documenting a clear demonstration of linkages between the model and the bank's risk measurement methodologies.
❑ Regulators are unlikely to insist on disclosure proprietary data or information. However, descriptions of the general

nature, model characteristics and sources of development data need to be disclosed.

General audit tools are useful for validation of processing logic and off-the-shelf models.

❑ *Review theory behind the models*: This requires several questions regarding statistical theory behind risk measurement. Recommended methods are as follows:

   ❑ compare the model with other models being used in the industry;
   ❑ compare the model with how the model will be used by the bank (with all its pros and cons) and how it fits into the business strategy of the bank;
   ❑ benchmarking of model is also helpful.

❑ *Review output*: Auditors have to verify that the model has been tested as required by the regulators. Furthermore, those models that are in use may be tested to verify whether the estimation made by the models actually materialised over the period of time. Models should also be tested for "what if" or sensitivity analysis. The "what if" analysis also helps in assessing the robustness of the model. In addition, model output may be benchmarked against the output of other models. The model is also validated by ongoing forecast versus actual comparisons.

❑ *Corporate governance for model risks*: Regulators require the following validations to be performed:

   ❑ decision makers understand the meaning and limitations of the model's results;
   ❑ when the model is in use for some time, the estimates generated should be tested against the actual outcomes;
   ❑ the input to the model should be audited by independent units;
   ❑ the person responsible for the model should have seniority commensurate with the materiality of the risk.

Best practices for model evaluation as per the actuarial standard of practice 38 "Using models outside the practice of actuary"

❑ Understanding the model:

   ❑ model components;

- inputs;
- model outputs.
- Appropriateness of the model for the intended application:
  - applicability of historical data;
  - development of relevant fields.
- Model validation:
  - user inputs – data quality and relevance;
  - model outputs:
    - results from alternate models;
    - comparing results produced with the historical actual results;
    - consistency and reasonableness relationship of various outputs;
    - sensitivity of the model outputs to the input and model assumptions.
- Documentation of the model.

Challenges in validation of credit risk models

The typical challenges facing credit risk modelling are as follows.

Model calibration qualities

Discriminatory power: Ideally, a model with a good discriminatory power means that there is no difference between *ex ante* and *ex post* measurement of risk components. However, in practice, such systems do not exist. For example, a rating system is said to have a high discriminatory power if "good" grades subsequently turn out to contain only a small percentage of defaulters and a large percentage of non-defaulters, with the converse applying to the "poor" grades.

Over fitting and stability of models: Due to data paucity, credit risk models typically have a lower dataset. Models should be tested not only on the development dataset but also on an independent dataset ("out of time" and "out of sample" validations). Otherwise there is a danger of over-fitting of the model to the development dataset. Over-fitted models perform poorly on independent datasets. Credit risk models are prone to instability due to a large impact of correlation on the risk measurement. Stability means the model adequately represents a cause–effect relationship, ie, between risk factors and risk components. The model therefore avoids spurious dependencies based on empirical correlations. In contrast to stable systems, unstable systems frequently show a sharply declining level of forecasting quality over time.

Accuracy: The accuracy of mapping between risk factors to risk components. Risk calibration problems are very complex due to the correlation angle. In practice, *ex ante* and *ex post* risk measurements always differ. The key question to be decided by the validation process is whether the deviations are purely random or whether they occur systematically. Ultimately this is also the purpose of validation.

Back testing is an insufficient measure for model validation: The back testing method is based on the binomial test. One of the assumptions of the binomial test is that the defaults of the borrowers constitute independent events. However, we know that defaults are correlated across the rating grades owing to cyclical influences. The problem would have been solved easier if we had correct measures of correlation. Therefore, back testing may not be sufficient since the correlation might have changed during the period due to systematic deviations.

Initially, during the early use of the market risk management model, back testing was a prudential validation practice. With the LTCM near failure, stress testing was included in the prudential practice. However, both these methods are insufficient for credit risk model validations.

The need for benchmarking
This is the examination of the performance of risk rating systems relative to the comparable risk rating systems and models.

This helps in checking systematic deviations of the bank's internal estimates from the estimates in the benchmark portfolio. Benchmarking serves as a useful complement to the validation process. However, the usefulness of this approach depends on the choice of a suitable benchmarking portfolio. The choice of a benchmark rating is likewise generally not an easy task.

Benchmarking and back testing are interrelated: back testing is the benchmarking of historical performance. Back testing is an assessment based on historical data and compares the historical performance with realised performance.

Models are machines: taking inputs and generating output. If there are a sufficient number of data points, and the relationship between the input and output is known, benchmarking may not be needed.

For credit risk, the relationship between the input (risk factors) and output (risk components) is not known, is not linear and is not stable (due to correlation). While market risk is symmetric risk, credit risk is not. The aim of the validation framework is to validate a credit risk model's forecasted loss distributions with actual credit losses observed over multiple credit cycles.

Therefore, models are not only to be calibrated accurately for the relationships between the inputs and outputs but also to be assessed for the risk measurement against the other models:

❑ benchmarking is a complementary tool for the validation and calibration of PD estimates;
❑ benchmarking involves the comparison of a bank's PD estimates to results from alternative sources;
❑ benchmarking internal PD estimates with external and independent PD estimates provides credibility to the PD quantification process.

Among the tools that could potentially be used are the following:

❑ benchmarking against the external/third-party PD/LGD estimates;
❑ benchmarking against the market-based models or market-based proxies for credit quality, such as equity prices, bond spreads or premiums for credit derivatives;
❑ internal migration matrices with external or third-party migration matrices;
❑ predicted PD rate for each rating against internal or external expert judgments;
❑ analysing the rating of similar firms;
❑ combining the adjustant risk grades and benchmarking the default rate of combined grades with the PD rate of combined grades.

Case study: Benchmarking the internal PD of listed companies to an external structural model
This is based on the Research Memorandum 06/2005 published by the Hong Kong Monetary Authority. The paper proposes a structural model to generate the PD of the obligors that are listed companies. The PD computed by the existing internet ratings-based (IRB) model

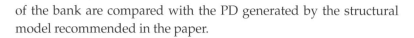

of the bank are compared with the PD generated by the structural model recommended in the paper.

❏ The paper proposes a structural model where the underlying stochastic variable is the leverage ratio of a firm, which is mean-reverting to a time-dependent target leverage ratio. The model generates term structures of PDs consistent with the term structures of actual default rates of credit ratings, as reported by S&P's.

❏ The term structure generated by the model is mapped to the default rate term structure of different ratings. Regarding the mapping process of the benchmarking model, the idea is to associate a company with an external credit rating by mapping the term structure of PDs of the company generated by the structural model to the term structures of default rates reported by a rating agency.

❏ Based on the mapping, a rating is assigned to each of the obligor companies. The implied PD from the model is the actual one-year average default rate of the benchmark rating.

❏ By comparison of the benchmark PD estimated from the structural model with the bank's one-year PD of the company based on its IRB system for a number of companies, the result will help to validate the model.

QUANTITATIVE MODEL VALIDATION TECHNIQUES

Model validation is an approach for validating the relationships of a model with risk factors, risk measures, grades or risk components and with asset class.

Aims of model validation

The validation process is primarily driven by three aims: consistency, accuracy and power.

Validating consistency

Risk management and measurement processes and various assumptions used to measure credit risk need to be applied consistently within an asset class, across the asset class and over time. Consistency is achieved through the documentation of the risk measurement and management process and assumptions, the qualitative validation techniques and the benchmarking. Every model approximates and substitutes risk factors and their relationships with risk components. Such approximations and substitutions need to be consistently applied and validated. Consistency is required to build an homogenous portfolio. A consistency is also required in implementation of rating process. The validation of the rating system can be further broken down into three components.

❑ The evaluation of the rating system design or model design.
❑ An assessment of the estimates of the risk components. In both cases, qualitative and quantitative methods can be applied.
❑ Validation of the rating system and the estimates of the risk components (PD, LGD, and EAD).

This involves important issues such as data quality, the internal reporting, how problems are handled and how the rating system is used by the credit officers. It also entails the training of credit officers and a uniform application of the rating system across different branches.

Validating accuracy

Accuracy is a very important element in validation, since credit risk models are used to generate the credit decisions. Model accuracy is measured in two ways.

❑ *Type I errors*: The model indicates low risk, when in fact risk is high. This results in credit loss.
❑ *Type II errors*: The model indicates high risk, when in fact risk is low. This results in potential loss of business or some financial loss due to selling of investments at a distress price.

These two types of errors are interrelated. As with Heisenberg's principles of uncertainty, minimising one type of error increases the other type of error.

A perfect model will have zero type I and type II errors. Some of the ways to assess the models based on Table 3 are as follows:

❑ true positive (TP) as a percentage of total default $= TP/D \times 100\%$;
❑ false negative (FN) as a percentage of total default $= FN/D \times 100\%$;
❑ false positive (FP) as a percentage of non-default $= FN/N \times 100\%$;
❑ true negative (TN) as a percentage of non-default $= TN/N \times 100\%$.

Table 3 Type I and type II errors

	Actual default	**Actual non default**
Bad	Actual default (TP)	Type II error (FP)
Good	Type I error (FN)	Actual non-default (TN)

Table 4 Validating accuracy of PD

Calibration errors	Are the central default probabilities (or expected loss levels) corresponding to each grade adequately established?	Comparing expected and actual default/loss rates.
Rating errors	Are assets graded consistently with their inherent loss characteristics?	Comparing the ratings awarded to the same company or set of companies by a variety of institutions, including rating agencies.
Granularity errors	Does the number of gradings allow sufficient differentiation of the exposures in a portfolio?	Breaking down the exposures by bucket on the rating scale.
Rating stability	Is the proportional relationship between the average Expected Default Frequencies that define rating categories consistent throughout the cycle, both overall and within market segments?	Review of the stability of ratings throughout the economic cycle or during economic shocks.

Compared with the evaluation of the discriminatory power, methods for validating accuracy are at a much earlier stage of development. Table 4 indicates the validating accuracy of PD.

Validating power
Power describes how well a model discriminates between defaulting ("bad") and non-defaulting ("good") obligors. Power is the ability of the model to identify and measure the risk factors and risk components as well as being the ability of the model to distinguish between two exposures to a greater detail.

Power indicates how easy it is for users of a model to tell the difference between defaulting and non-defaulting firms. It is also the ability to reject the bad quality loan request and produce more exact estimates of PD, and a measure of *ex post* default being near to, or lower than, *ex ante* default estimation.

A primary measure of credit default model performance is the degree of power the model exhibits at ordering firms from "worst" (most likely to default) to "best" (least likely to default).

The result of the lower power model is that the poorer credits end up going to the bank with the weaker model and the better credits go to the bank with the more powerful model. The bank using the weaker model in effect creates adverse selection for itself.

Numerous methods exist for the assessment of the discriminatory power. The most common techniques are the cumulative accuracy profile (CAP) and the accuracy ratio, which condenses the information of the CAP into a single number.

The main difficulty posed by this method is the need for a large time series of internal ratings and default or loss events.

Quantitative validation techniques
The validation techniques differ according to the type of model. Quantitative validation techniques are applied for a statistical model.

Validation techniques for different types of model
Primarily there are three types of model with the following validation techniques.

Statistical models:
❑ comparison of expected defaults with realised defaults;
❑ calibration/accuracy tests;
❑ power tests;
❑ likelihood tests.

Validation of expert judgment models:
❑ feedback from credit officers and relationship managers;
❑ an intuitive relationship between the risk factors and risk components;
❑ abnormal defaults;
❑ benchmarking migration matrices.

Vendor models:
❑ the model-provided default rate or risk rating are aligned to the external rating;

❏ risk factors provided by the model are relevant for the risk component estimation;
❏ a sound theory supporting the model.

Data sampling framework

The statistics techniques used for credit risk models can be highly sensitive to the data sample used for validation. Some ways of reducing data sensitivity are as follows.

Data is required for two purposes: model building or estimation and model validation. For both purposes the data should be different but statistically similar. Due to the paucity of data, it is a challenge to build two different sets of data. The sampling techniques (see Figure 4) used to address the challenge are as follows.

❏ *Out of sample*: This is the basic resampling technique, where validation data is chosen completely at random from the full dataset. This approach makes the following assumptions:

  ❏ that the properties of the data stay stable over time (stationary process);
  ❏ the distribution of data is reproduced in the sample.

❏ *Out of sample and out of time*: Here the data sample is segmented into two time zones.
❏ *Out of sample and out of universe*: Here the dataset is segmented into a two-sets one set for model estimation and one set for model validation. Both of these sets are exclusive sets containing no firms in common. Therefore, model validation data are out of

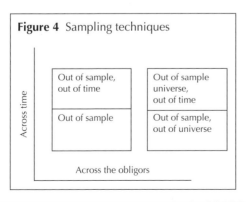

Figure 4 Sampling techniques

423

universe. An example of out of universe validation would be a model trained on manufacturing firms but tested on other industry sectors. This approach validates the model homogeneity.

❑ *Out of sample, out of time and out of universe*: As a first step, the sample is divided in time. For example, all of the data is divided into two groups: 1995–2000 and 2000–2005. Data for each time period is then segmented on the basis of industry, for example. Some of the segments are used for model estimation and some are used for model validation (but not for both).

These techniques are used for re-sampling to leverage the available data and avoid over-fitting of models. Model performance is measured with each sample. The variability measure indicates whether the model performance differences are statistically significant.

"But how do we balance the data division between model estimation data and model validation data?" Ideally, data should be in such volume that the question of dividing data between the two segments should not be important.

Every institution faces the following dilemma:

❑ more data for data estimation and less for data validation – this means it is extremely difficult to evaluate the true model performance due to a severe reduction in statistical power due to less data available for validation;

❑ more data for model validation and less for model estimation – in this situation the model is over-fitted and an estimation of model parameters is a serious concern.

Calibration

Models measure risk factors and provide a measure for risk components. While calibrating the model, risk factors are converted into a single risk number or indicator called the risk grade. Each risk grade has a historical or externally mapped risk measure, default rate or PD rate. Calibration describes how well a model's predicted probabilities agree with the actual outcomes; it is therefore different from testing for power. Calibration is tested on the basis of likelihood measures and is a two-step process:

❑ first, the model score is mapped to the empirical or historical default rates;

Table 5 Ordinal ranking

Rating grade	Assigned PD (external/ predicted) (%)	Default rate (historical /actual) (%)
1	0.03	0
2	0.05	0.0
3	0.10	0.02
4	0.13	0.15
5	0.25	0.35
6	1.17	0.23
7	1.42	0.96
8	2	2.31
9	13.10	27.5

❑ second, the difference between the historical default rate and the benchmarked external data (for example, rating agency default rates) is adjusted.

Validating calibration is also a two-step process:

❑ first, confirm the ordinal ranking of PD grades;
❑ second, check the consistency of the actual observed PD and the PD assigned.

The default rate measured by the model may be systematically lower or higher than the corresponding external benchmark. However, the model should preserve the ordinal ranking of obligor riskiness. For example, in Table 5 the ordinal ranking for the rating grade six (highlighted) is not preserved.

The assigned PDs may be very different from the historical PDs. For example, for risk grade five onwards there is a substantial difference, while for all grades the PD rate is different from the default rate. Under calibration, we need to check statistically (eg, likelihood test or hypothesis test via Monte Carlo) if the assigned PDs are plausible.

Accuracy of rating migration – a measure of PD model accuracy
Different models or models with different accuracy levels will generate different migration matrices. Migration matrices can help to differentiate between the different accuracies of various models. Since each matrix has more than 100 cells, it is very difficult to distinguish migration matrices generated by different models. Jafry

and Schuermann have proposed a method to distinguish rating migration matrices by providing a methodology for validating the accuracy of different models that are used to generate the migration matrices.

They attempt to find answers to the following three questions.

❑ How would one measure the scalar difference between matrices (matrices from different models)?
❑ How can one assess whether those differences are statistically significant?
❑ Even if the differences are statistically significant, are they economically significant?

For comparison of migration matrices, they propose three methods:

❑ cohort;
❑ duration – parametric – time homogeneity;
❑ duration – non-parametric – no time homogeneity.

Statistical significance is tested using matrix norms, eigenvalue and eigenvector analysis tests. Economic significance is tested by measuring the credit portfolio risk.

Impact of correlation on validation
As we have seen in the "Credit Rating" chapter, there are two approaches to measuring PD: PIT and TTC. PIT, being the conditional measurement, obviously has lower correlation measures than the TTC, which measures unconditional correlations.

Correlation is embedded so much into the PD measures that it is very difficult to segregate correlation measures from PD measures. Since the correlation estimates are different, the two risk ratings measure PDs at different levels and the two philosophies should be consistently applied. Furthermore, the Basel models estimate the capital assuming the TTC approach. Therefore, the assumptions and measures have a major impact on the risk component validation.

Monte Carlo simulation for incorporating default correlation to validate calibration of PDs
The Basel II Accord has prescribed the minimum number of rating grades, thus increasing the granularity of the system. However,

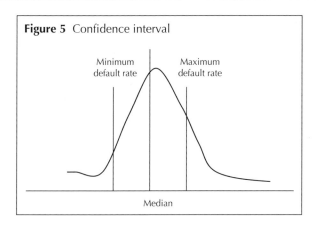

Figure 5 Confidence interval

Minimum default rate

Maximum default rate

Median

Table 6 Likelihood testing

PD rate	Number of defaults = 1	Number of defaults = 2
$\mu = 0.01$	$(0.01)^1 \times (0.99)^{99} = 0.0037$	$(0.01)^2 \times (0.99)^{98} = 0.000037$
$\mu = 0.03$	$(0.03)^1 \times (0.97)^{99} = 0.0015$	$(0.03)^2 \times (0.97)^{98} = 0.000045$
$\mu = 0.05$	$(0.05)^1 \times (0.95)^{99} = 0.0003$	$(0.05)^2 \times (0.95)^{98} = 0.000016$

	Number of defaults = 3	Number of defaults = 5
	$(0.01)^3 \times (0.99)^{97} = 0.0000004$	$(0.01)^5 \times (0.99)^{95} = 0.00000000004$
	$(0.03)^3 \times (0.97)^{97} = 0.0000014$	$(0.03)^5 \times (0.97)^{95} = 0.00000000135$
	$(0.05)^3 \times (0.95)^{97} = 0.0000009$	$(0.05)^5 \times (0.95)^{95} = 0.00000000239$

there may not be enough data in each grade to validate with statistical certainty. Therefore, while testing a null hypothesis through Monte Carlo, confidence intervals (CIs) are set (see Figure 5). The setting of CIs involves a trade-off between type I and type II errors. The assigned PD rates are accepted to be correct only if the historical rates are observed to be within a given CI; ie, the historical rates are neither less nor more than rates around an assigned PD with a given confidence level.

Likelihood testing
The likelihood that the mean PD rate μ is a valid PD rate is given by $\mu^{number\ of\ default} \times (1 - \mu)^{number\ of\ non\ default}$. For example, the situation for a portfolio with a total of 100 exposures is given in Table 6 (likely models are highlighted)

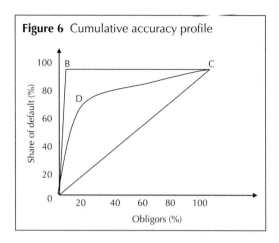

Figure 6 Cumulative accuracy profile

❑ For an historical default rate of one, the model with mean rate of one is the likely model;

❑ for an historical default rate of two, the model with mean rate of three is the likely model;

❑ for an historical default rate of three, the model with mean rate of three is the likely model;

❑ for an historical default rate of five, the model with mean rate of five is the likely model.

Cumulative accuracy profiles
CAPs are used to make visual assessments of model performance (see Figure 6). Similar types of tools are used under various names. CAPs are also called power curves.

The steps for constructing CAPs are as follows.

❑ Obligors are ordered by model score, from riskiest to safest, and defaults are mapped to the scores.

❑ A curve is constructed by plotting percentage defaults against the percentage of companies representing such model scores or defaults.

❑ In a good model, defaulters concentrate at the riskiest scores and smaller percentages of the population cover most of the defaults. The percentage of all defaulters identified (the *y*-axis in Figure 6) increases quickly as one moves up the sorted sample (along the *x*-axis).

❑ If the model assigned risk scores randomly, we would expect to capture a proportional fraction of the defaulters generating a straight line.

❑ The ideal power curve is also a straight line but captures 100% of the defaults within a fraction of the population equal to the fraction of defaulters in the sample. For an ideal model, the power curve would be OBC covering most of the defaults in a smaller percentage of the population. The line OC represents the random curve. Statistically, default companies should be fully distributed across the exposure in a random curve.

❑ The steeper the CAP curve at the beginning, the more accurate the rating process.

In reality, rating classifications are neither perfect nor random. The corresponding CAP curve therefore runs between these two extremes. Using the CAP curve, the discriminatory power of a rating process can be aggregated into a single figure. The accuracy ratio of the model or model power is equal to the area under ODC divided by the area under OBC. This is also called the Gini coefficient (GC).

For the validation of PDs, there are in general two stages: validation of the discriminatory power of a rating system and validation of the accuracy of the PD quantification.

Compared with the evaluation of the discriminatory power, methods for validating the accuracy of the PD quantification are at a much earlier stage of development.

While one of the methods is back testing, a major obstacle to the back testing of PDs is the scarcity of data, caused by the infrequency of default events and the impact of default correlation. Even if the five year requirement of Basel II for the length of time series for PDs is met, the explanatory power of statistical tests will still be limited. Statistical tests alone will be insufficient to establish supervisory acceptance of an internal rating system.

Other statistical tests for validating power
The following tests are some of statistical methods used to measure the power of the model.

❑ The Kolmogorov–Smirnov (KS) statistic is used as a relative indicator of curve fit.

❑ χ^2 is a non-parametric test of statistical significance for bivariate tabular analysis. A non-parametric test, such as χ^2, is a rough estimate of confidence; it accepts weaker, less accurate data than parametric tests as input (such as t tests and analysis of variance) and therefore has a lower status in the pantheon of statistical tests.

❑ CAPs, also called Lorenz curves, are a graphical representation of the cumulative distribution function of a probability distribution, which is a graph showing the proportion of the distribution assumed by the percentage of the values. This idea was developed by Max O. Lorenz in 1905 for representing income distribution. The GC is the area between the line of perfect equality and the observed Lorenz curve, as a percentage of the area between the line of perfect equality and the line of perfect inequality.

❑ The Mann–Whitney U test is a non-parametric statistical significance test for assessing whether the differences in medians between two observed distributions are statistically significant. If the difference in median is significant, the two samples are different. Using the information on the risk factors, credit scores are generated by the model. Over a given period either the population changes or economic factors change. Both of these outcomes result in a change in the credit scores. The Mann–Whitney U test is used to establish whether the development data and current data are two independent datasets or whether the sample is drawn from the same dataset.

Accuracy *versus* power

The requirements of higher model power are greater for a portfolio with a lesser number of exposures than with a higher number of exposures. With an increase in the number of exposures, the model power goes down. Therefore, if

Number of retail exposures > Number of SME customers > Number of corporate customers

is true for a bank, the model should have a higher power for corporates (see Figure 7).

Power and accuracy are related: power limits the accuracy (see Figure 8). A high accuracy without high power is difficult to be achieved. A perfect model will be able to accurately predict the

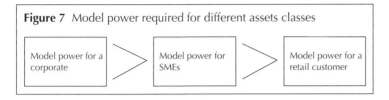

Figure 7 Model power required for different assets classes

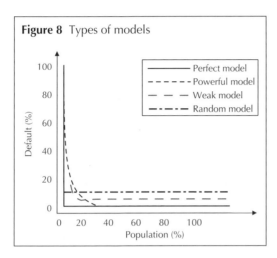

Figure 8 Types of models

default rate and therefore will be able to identify all obligors who are likely to default. For the obligors selected for extending credit, the default rate will be zero. A powerful model will classify "to be" defaulted exposures in the low rating classes. For the better category rating class, the default rate will be zero. A weak model will be able to identify only a few exposures in the lower rating classes. Therefore, each rating class will have a positive default rate.

A random model will not be able to distinguish between the bad and good exposures resulting in an average default rate for all categories.

STRESS TESTING

Under the Market Risk Amendment (1996), banks were first permitted to use internal models to calculate regulatory capital for market risks in the trading book, based on back testing as a model validation technique. It is only after the LTCM near failure that supervisors have started prescribing and emphasising stress testing. Some of the national supervisors have published consultative papers on stress testing credit risk.

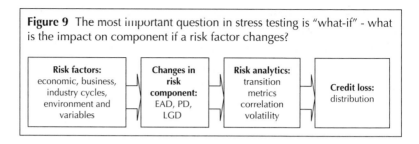

Figure 9 The most important question in stress testing is "what-if" - what is the impact on component if a risk factor changes?

Stress testing is "a generic term describing various techniques used by financial firms to gauge their potential vulnerability to exceptional but plausible events". Stress tests summarise the bank's exposure to extreme but possible circumstances. Stress testing is a useful tool for measuring and managing tail risk. We know that credit risk lies in the tail; therefore, stress tests are important for credit risk. Stress testing supplements the other measurement techniques for credit risks.

Typically, the results of a small number of stress scenarios are computed on a regular basis and monitored over time. The time series of stress test results are one of the key components of the risk management.

The stress testing process can be summarised as follows (see Figure 9):

❑ identify stress factors relevant for credit risk;
❑ identify risk factors;
❑ measure changes on the risk components and risk analytics;
❑ measure credit risk loss.

Stress testing is a combination of simulation and analytical techniques are used.

For changes in the risk factors, risk components are analysed using analytical techniques and credit loss or risk is estimated.

In stress testing, a bank alters assumptions about one or more financial, structural or economic variables to determine the potential effect on the performance of an asset, concentration or portfolio segment. This can be accomplished with "back of the envelope" analysis or by using sophisticated financial models.

The method employed is not the issue; rather the issue is answering the critical "what if?" question. Regardless of the sophistication of the method, all banks derive benefit from stress testing. Some form of stress testing is commonly employed when assets are subjected to hypothetical "rate shock" scenarios to determine their exposure to changes in interest rates. Similarly, the securitised consumer portfolios also use stress-testing techniques to better gauge their risk and to determine the level of credit enhancements.

One of the tests likely to be insisted on by supervisors may be a 30–60% change in the PD. This change is equivalent to a 1–2 notch downgrade of borrowers. It was found that this increases the mean increase in the risk, and therefore capital requirement, by about 8% to 15% (0.64–1.2% of additional capital of 8%).

Until recently, stress tests considered only a single risk factor change. This, implicitly, means a null correlation between the risk factors. In reality, however, simultaneous changes in the risk factors are observed. Univariate stress tests should therefore be supplemented by multivariate stress tests, in which more than one risk factor is changed at a time. When, one stressed factor is applied, the other relevant risk factor should not be left unstressed. For example, if interest rates are stressed, foreign exchange rates should also be stressed for the impact of interest rate changes. Multivariate stress tests are becoming a reality.

Types of stress testing

One-factor stress tests measure the effect of drastic changes in individual risk factors on risk component and portfolio credit risk.

Multi-factor stress tests attempt to simulate reality more closely an examine the effects of simultaneous changes in multiple risk factors.

Under each method, there are two approaches to the building of scenarios: historical scenarios and hypothetical scenarios. The main challenge in building a scenario is to identify the relevant risk factors and modelling those factors.

Stressing concentration risk in the credit portfolio: In general, there are two types of concentration: concentration in the single name and concentration in highly correlated industry sectors. The Basel Accord does not provide a quantitative framework to measure concentration risk and makes the following assumptions:

❑ bank portfolios are perfectly fine grained   no name concentration;
❑ there is only a single source of systematic risk – no sector concentration.

Under paragraph 773 of the Accord, banks are required to stress test the credit concentration for capital adequacy. Sector concentration is assessed through multi-factor stress tests.

Limitations of stress testing

❑ A stress test estimates the exposure to a specified event and does not estimate the probability of such an event occurring.
❑ The test relies on the judgment and experience of the risk manager.
❑ It requires high computational resources to revalue complex options-based positions.
❑ For the time being, market and credit risks cannot be integrated in a systematic way in their stress tests.

BEST PRACTICES
Data paucity issue

Default events are infrequent and default model outputs for consecutive years are highly correlated. It is an ideal and recommended practice to create a model using one dataset and then test it on a separate "hold-out" dataset composed of completely independent cross-sectional data. While such out of sample and out of time tests would unquestionably be the best way to compare models' performance, default data is rarely available. As a result, most institutions face the following dilemma: if too many defaulters are left out of the in-sample dataset, the remaining dataset will be smaller. This will seriously impact the model estimation and may result in over-fitting.

If too many defaulters are left out of the hold-out sample, it becomes exceedingly difficult to evaluate the true model performance due to severe reductions in the model's statistical power. In light of these problems, an effective approach is to rationalise the default experience of the sample at hand by a combination of risk grades.

❑ The number of occurrence events in the sample determines the stability of the model. A sample with 20 defaults, and a sample of one tenth size and two defaults are totally different.

❑ The sample is drawn from two universes that are not the same. A different asset class, industry and time create a different universe.

❑ The primary differences between the two samples are the variability in risk factors and the variability in the number of events in the sample. The variability is due to the sample size and sample universe.

❑ The performance is measured by checking the model results against the benchmarks. With this approach, the model's power can be differentiated through benchmarking.

❑ The results should be interpreted considering the impact of sample size and variability in the sample.

❑ The sample size should support the distribution assumptions.

❑ The use of closed-form approaches without the supporting sample size is not a recommended approach. It is likely to under or overestimate the variable.

❑ Using appropriate resampling techniques reduces the variability to the greatest extent possible. The ultimate objective is to synchronise the variability in the sample to the variability of the universe.

❑ Parametric and non-parametric techniques should be applied to reduce the sampling bias.

❑ Data should be pooled with other banks or market participants.

❑ Other external data sources can be used (example sources are identified in the "Software and Data" chapter).

❑ Market-based information measures of risk can be used.

❑ Sub-portfolios can be combined to build a larger sub-portfolio.

❑ The multi-year PD can be computed and then the multi-year PD annualised.

❑ The upper level of PD estimates of the rating agencies for higher grades can be used.

Validating credit scoring

The two most important properties of model validation are consistency in the risk measurement process and the validating statistical methods.

Validating consistency

Validating consistency means:

❑ defining the purpose of the credit scoring model – discrimination versus prediction;

❑ selecting a sample that reflects or represents the targeted population – a reference dataset;

❑ ensuring that the modelling technique is consistent with the purpose;

❑ ensuring that the risk factors reflect the lender's knowledge and historical experience;

❑ fitting the model and checking for model misspecification or over-fitting of the data;

❑ developing outcome based methods of verifying model.

Data selection
Different datasets for training the model and validating the model are required. However, both datasets should be comparable statistically.

Validating statistical methods
Validating statistical methods has the following procedure.

❑ Identify missing data.

❑ Identify omitted variables. These are risk factors that are relevant for the risk measurement but are either assumed to be constant or omitted. Some of the omitted variables can be repayment options in the product or business cycle and correlation assumptions.

❑ Verify the correlation assumption to check if correlation was constant across the time for which risk component was measured.

❑ Considering the purpose or use of the model, assess if the statistical method used is the best practice method for the purpose.

❑ Assess and re-perform the performance measurement results, compare the results of the KS test, CAP, likelihood tests, goodness-of-fit test and other parametric and non-parametric tests.

❑ Assess the out of sample analysis results.

Retail segment
Retail exposure, being more granular and possessing homogeneity, will have a lower loss volatility. Risk in retail exposures is measured through credit scores, which are more static in nature than the credit ratings as scores are not mapped to the default rates. Not all characteristics captured for credit scoring are sensitive to economic cycles. As we have seen in the "Credit Scoring" chapter, there are two types of models: models with factors that have

statistical significance and models with risk factors that have an economic relationship with the risk components. Models using statistically significant risk factors keep losing the calibration with a changing dataset (changed portfolio). Models using economically relevant risk factors may lose calibration due to changes in the economic cycle. Therefore, two types of validation are required.

❑ *Shifts in the credit scores due to changes in the data population*: This is done by comparing the score distribution for the developmental sample and current data population. This can be tested through the Mann–Whitney U test.
❑ *Change in the rank ordering*: This is tested through either the area under the ROC curve using a KS statistical Lorenz curve or GC, or using a χ^2 test.

Validating credit rating
In addition to the methods explained in the previous paragraphs, the following steps also need to be considered.

❑ *Data cleaning*: No data supplier provides data directly relevant for credit risk measurement. Data is collected and compiled for various other purposes. Therefore, irrelevant data must be deleted, for example, default data due to fraud and void bankruptcy, etc.
❑ *Choice of variables, financial ratio*: We have dealt with this in the "Credit Rating" chapter.
❑ *Segmentation on the basis of industry*: We know that default rates and correlation varies based on industry sector.
❑ *Assumption on distribution of risk rating*.
❑ *Weights and relevance of qualitative factors considered in credit rating*: We have dealt with this in detail in the "Credit Rating" chapter.

REFERENCES

Jafry, Y. and T. Schuermann, 2003, "Measurement, Estimation and Comparison of Credit Migration Matrices", Wharton Financial Institution Center.

Hong Kong Monetary Authority, 2005, "Benchmarking Model of Default Probabilities of Listed Companies" Research Memorandum 06/2005.

Deutsche BundesBank, 2003, "Approaches to the Validation of Internal Rating Systems", Monthly Report, September.

11

Software and Data

OVERVIEW OF SOFTWARE AND DATA FOR CREDIT RISK MANAGEMENT

Vendors have been developing software tools to assist institutions in their risk measurement and management functions. Broadly such products are divided into two types:

❑ vendor models;
❑ vendor data.

However, in practice this distinction does not hold good. Models come bundled with vendor-supplied data, which is also used for development and calibration of the model. Similarly, datasets supplied by vendors often incorporate some type or degree of modelling.

Figure 1 depicts software systems required for managing credit risk in a bank.

Data collection systems

Data collection systems were initially developed to analyse the financial data of borrowers. These tools have now been extended to capture non-financial information such as firm specific data and management, industry and macroeconomic data. These systems are integrated part of credit scoring or rating systems. For example, the Financial Analyst system from Moody's, which has been used for more than a decade; its latest version is Risk Analyst, which computes credit scores in addition to storing financial and non-financial data.

Figure 1 Software systems. An overview of credit risk management systems

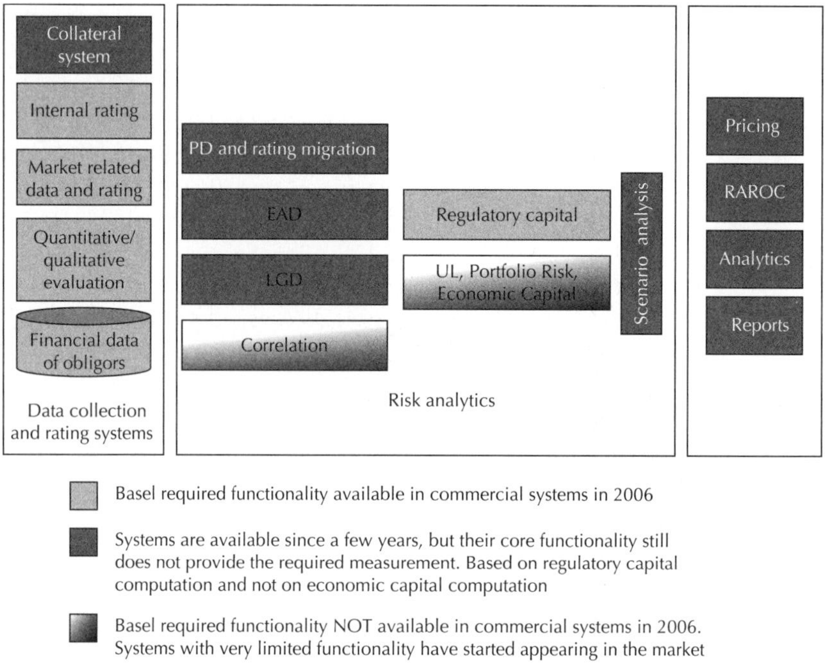

	Basel required functionality available in commercial systems in 2006
	Systems are available since a few years, but their core functionality still does not provide the required measurement. Based on regulatory capital computation and not on economic capital computation
	Basel required functionality NOT available in commercial systems in 2006. Systems with very limited functionality have started appearing in the market

Credit rating

Credit rating is the first step towards risk measurement. There are two aspects in credit rating estimation: the first is to rate the borrower; the second is to calibrate or map the credit rating model to the default or credit loss risk. For calibration or mapping, a default curve needs to be built. This default curve requires various inputs including historical data, peer data, market data on equity and bonds, correlation, spreads and rating migration.

For the past decade or so commercial credit scoring software has been available for retail and commercial borrowers. The offerings in the credit scoring or rating area have been strengthened to support the requirements of the Basel Accord. This has been done in three ways:

❑ by collating and supplying the borrowers' financial data and market data for their bonds and equity, and default data for both

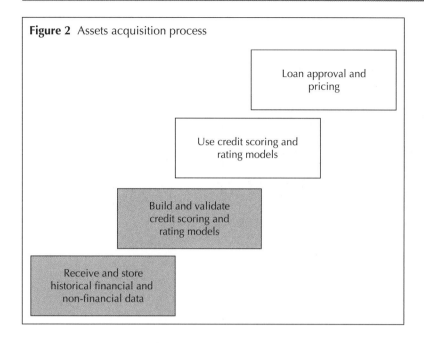

Figure 2 Assets acquisition process

private and public firms; these data can then be used for calibration and validation;

❑ by providing web-based access to a credit rating model specific to a region and an industry;

❑ by building a bank-specific model to comply with the Basel requirements, including supplying the stress test, rating migration, correlation and benchmarking data.

Collateral management system

Collaterals are recognised as risk mitigation instruments. Collateral management systems are required to capture the collateral details required for managing risk and computation of regulatory capital. There are two types of collaterals. Financial and physical. For the time being the Basel Accord recognises only financial collaterals in computation of regulatory capital. The systems required for financial and physical collateral are different. Financial collaterals are generally provided for trading assets while physical collaterals are provided for non-trading assets like loans. Financial collateral management systems are available as an integrated part of treasury systems. Product processors and core banking solutions provide collateral management capabilities for physical collaterals. The technical and functional capabilities are different for both the systems. Financial collaterals management systems require automated and real time valuation capabilities with high turn over in the collateral and collateral valuations. Financial collateral systems should have capabilities for management and valuation of various types of financial assets. Physical collateral require rich functionalities to manage and value various types of physical collaterals and rights on physical collaterals.

Under the Basel Accord, risk is managed at obligor level. Therefore, a centralised collateral management systems is required to manage financial and physical collaterals across the product processors.

Market-related data and models
These models are based on structural and reduced form. Market-related data such as equity price, bond price, spread, rating, rating migration and balance sheet structure are the major inputs to the model. The borrower data is a major component of these systems. Moody's expected default frequency (EDF) based system provides both data and model.

Probability of default and rating migration
Rating agencies provide rating and rating migration data for obligors, industries and countries. For the past five years or so, rating firms are providing web-based private firm models. In the process, they collect private firm financial and qualitative data; risk measures in the form of rating and scores; default and loss history. (Since all this data are shared over and web by the financial institutions using the data to measure the PD. It will take some time before meaningful PD migration data for private firm are available.

LGD modelling
There are two approaches to measure LGD: by building a regression equation around the historical loss history and by measuring the LGD from the bond pricing. Both approaches at present are only suitable for the US. No suitable software is available for modelling LGD for exposures outside the US.

Exposure at default

Every bank broadly offers three types of products: funded products such as loans and overdrafts; non-funded products such as letters of credit; and guarantees and derivative products. Exposure at default (EAD) for funded and non-funded products is estimated through historical customer behaviour and that for derivative products is estimated through current and potential exposures.

Third-party software tools are available to model customer behaviour and EAD for funded, non-funded and derivative products. Generally, the product processors for funded and non funded exposures do not have the capability to model EAD, External, modelling software is used to find an analytical solution.

Various treasury and asset management solutions provide valuation facilities for widely used and vanilla derivative products. To build analytics for interest rate and credit, pre-built tools and libraries in Microsoft Excel, VB and C++, C#, Java are available or can be built. These libraries can be integrated into other systems.

Portfolio management

Portfolio management solution includes, solutions for measurement of economic capital, correlations, unexpected loss. The credit risk or the credit portfolio risk is measured in terms of unexpected loss or economic capital. With the given level of data and estimation techniques for PD, LGD, EAD and correlation available with the banks, it will take some time before the portfolio credit risk in terms of economic capital or unexpected loss can be estimated by way of output from a software.

To the author's knowledge no one as yet is offering a complete or substantially complete solution for portfolio management (see the chapter on *Portfolio Models*). Point solutions on some aspects of credit portfolio management have started appearing. For example, Standard & Poor's (S&P's) offer Default Filter – "a tool kit for developing, validating, and stress testing default probability models based on the institution's own credit factors" – under portfolio management offerings. Some preliminary unexpected loss and economic capital based models have started appearing on the market. These models help in pricing deals.

Regulatory capital

Regulatory capital estimates provided through various equations of Basel II are still the only measures available for credit risk. These measures are also being used for RAROC and pricing. Reveleus, Algorithmics, Sungard, FinArch and Fermat have emerged as leaders in computation of regulatory capital.

In addition to the regulatory capital computation, these solutions provide regulatory reporting, RAROC and pricing.

RAROC and pricing

For RAROC and pricing purposes, banks are using the regulatory capital measures and not the economic capital. As the economic capital estimation will develop, the analysis, reporting and use of RAROC and pricing will be based on economic capital. Moody's is offering RAROC and pricing solution on top of their EDF solutions.

DATA COLLECTION AND ANALYSIS TOOL

Readers may like to refer to the chapters on credit ratings and credit scoring to understand about the information collected. The data collection and analysis tool (DCAT) is used for computing credit scores and ordinal and cardinal credit ratings. In assets acquisition process, DCAT functionality mainly covers the receipt and storage of historical financial and non financial data and using the data collected build and validate credit scoring and credit rating systems.

It may be noted that the functionality explained in the following subsections is an indicative superset of functionality. No single commercially available system may have the length or breadth explained here. Furthermore, loan approval systems are generally separate systems and not extended credit rating systems.

Receive and store historical credit bureau reports, financial and non-financial data

Borrowers can be divided into two types – individuals and business.

Individual borrowers have credit reports from credit bureaus. Credit bureaus provide the credit scores and credit reports for individuals. Credit bureaus have started many value-added services like combining credit reports from all other credit bureaus (3-in-1 reports in the US) and also provide updates on credit report data and credit scoring models.

Broadly there are two types of business borrowers – borrowers which are public companies and others – and three ways to receive financial and non-financial information:

❑ Receipt of data from third party data provider. This may be in bulk for many borrowers, or for an individual borrower through electronic media or a network. Data provider supply data in two modes – pull or push; the technologies for pull and push are different.

❑ Receipt of financial and non-financial data directly from a borrower through electronic media or on paper;

❑ Receipt of incremental information for financial statements and non-financial information from various sources (third party data provider, from borrower, from internal research and market intelligence).

All the three sources have different time efficiency. Also the level of aggregation and information items also differs. Reconciliation of the information from more than one source is always a challenge. Industry has not yet agreed on standards for information relevant for credit risk. Financial institutions have taken reference data initiative to improve the data quality and reconciliation. XML technology helps to solve this problem to some extent, but cross-referencing, benchmarking and comparing non-financial information across borrowers and across industries is a really major challenge. Non-financial information includes various types of files and file sizes are very large (running into tens of megabytes). Standards like 15022 and 20022 have emerged.

The base line financial information varies across the accounting regimes. Templates provided DCAT software are available generally only for the US, the UK, Japan and European Union. Templates are generally available for the Accounting Standards applicable to the regime and also under International Accounting Standards. To standardise the format at a technology level the XML Business Reporting Language (XBRL) is being adopted by various regulatory and other agencies.

Still, there is no standardisation for non-financial data. Furthermore, no single source is of data for non-financial data is sufficient. Generally, different type of value addition is done by the data supplier.

In addition to financial and non-financial information, systems do capture and store risk factors such as credit rating or credit score and facility rating (refer chapters on Credit Scoring and Credit Rating for further details).

Earlier edition of DCAT solutions were called financial statement spreading solutions.

Build and validate credit scoring and rating models

The functionality for building and validating credit scoring or credit rating models varies according to the type of borrower, the non-financial information to be considered, the products to be sold to the borrower (asset types) and the quantum of borrowing. There are three types of borrower – retail, small and medium enterprises (SMEs) and corporates. Corporates are further divided into two types – private and public. Within the non-financial information, the information to be considered, the benchmarks and the relative weights vary. Products include retail financing, leases, trade finance, dealer products, loan products, fund based lines of credit, non-fund based lines of credit etc.

Systems provide capabilities to build separate models for credit scoring, borrower rating and facility rating. Systems generally have the capabilities to build and store models suitable for different products and borrowers. Systems also provide capabilities to build champion challenger models and what-if analysis. Vendors are extending their existing validation capabilities to the validation framework required by the Basel Accord.

Projection of financial statements, especially cash flow statements, is usually a major challenge. Systems provide basic capabilities for building projected financial statements and cash flow; systems also provide the capabilities to interface with third-party financial statement projection systems.

Use credit scoring and rating models

Credit scoring and rating systems need to provide broadly the following capabilities:

❑ Extract and store credit data from bureaus like Equifax, Experian, Trans Union, D&B, FICO, etc. Also receive and store behavioural scores and collection ratings.

❑ Store application data including financial data and non-financial data. Store and use customer and customer group profile.

❑ Define and enable use of credit scoring model based on parameterised business rules over data elements from application data, bureau data, and payment history data.

❑ Define multiple credit scoring models based on different data elements and weightages and store version and history of parameters and models.

❑ Provide automated and manual scoring. Generate and store multiple scores from different scoring models.

❑ Integrate with channels and customer relationship management applications.

❑ Enable separate obligor and facility ratings.

In the US for example, the three credit bureaux have together started offering credit scoring systems which includes data design characteristic levelling methods and segmentation techniques.

Credit scoring or rating systems generally do not have workflow capabilities. Systems with retail focus do have the capabilities to automate data capture, processing and exception management. The credit scoring systems generally aim to:

❑ reduce inconsistency in credit decision making across products and business groups;

❑ measure credit risk through a single number;

❑ implement internal rating systems;

❑ implement multiple scoring models for different industries, customers and products;

❑ implement changing business rules;

❑ design collection strategies.

Loan approval and pricing systems

For loan approval and pricing systems the primary goal is to automate the approval, pricing and release of payments. These systems have strong workflow capabilities, collateral management and payment interfaces.

The systems generally provide the following capabilities

- ❑ Stores historical financial and non financial data, credit reports, credit scores, credit ratings.
- ❑ Extract collateral valuation from systems like Blue Book, Taylor Martin, Green guide, Auction Guide, Reuters, and other data supplying systems.
- ❑ Build workflows. This includes:
  - ❑ Process flows based on business group, product type, transaction and transaction amount.
  - ❑ Role based work qeues and linking users to roles and implementation of maker and checker.
- ❑ Build and use the templates. This includes:
  - ❑ Product definition;
  - ❑ Repayment schedule;
  - ❑ Collateral requirements;
  - ❑ Tenure;
  - ❑ Documentation;
  - ❑ Interest, commission, fee and charges;
  - ❑ Limits;
  - ❑ Auto populate;
  - ❑ Business rules for collateral, customer scores and ratings, cost of funds, auto compute quote etc.
- ❑ Integrate with back end systems, product processors, and funding systems:
  - ❑ Extraction of credit scores, behavioural scores and credit rating;
  - ❑ For release of approved limits;
  - ❑ Extract payment history.
- ❑ Auto decisioning.

Technology and examples

Systems are usually based on vendor independent technologies. Since volume and high-availability is not a key requirement, most of the systems are based on Microsoft technologies with SQL/Oracle data base servers. Systems used for individual segments of the retail industry (such as credit cards) where volume and high availability is a key issue are available based on IBM, Java and other high availability technologies.

These systems usually have a n-tier architecture. Microsoft-based systems use the .NET framework with various compatible web servers (generally IIS). To provide flexibility in interfaces and in building business rules and templates, XML, SOAP and WSDL technologies are used. Microsoft Office, Adobe Acrobat and scanned documents are the document types used by the system. Low-end Intel-based servers are generally sufficient for most of the systems. For web-based systems, an access to the Internet is required. HTTP access is provided to the servers.

Important tools in this area are the tools from Moody's KMV – Risk Analyst and Financial Analyst.

Financial Analyst

This is a financial statement spreading solution. Financial Analyst is a solution for systematic credit analysis. The centralised database structure of Moody's Financial Analyst offers a framework for forecasting key variables, the analysis of cash flow and building financial data templates consistent with the accounting and credit policies.

Mood'y KMV Risk Analyst

Risk Analyst collects, analyses, manages and reports analyzes and stores historical and projected financial statements and non-financial data to measure credit ratings or a borrower's creditworthiness. The tool provides a configurable framework to deploy a standard set of analytical facilities to address internal credit rating practices for public and private firms. The tool provides a centralised database, which is useful for development and evaluation of credit rating models.

The system helps in computing EAD, LGD, and expected loss to support internal rating and facilitate computation of regulatory capital. The system also provides capabilities to stress test the scenarios on the obligor's financial conditions. The system also provides capabilities to generate rolling statements. The system also provides financial statement reporting templates for various industries and countries.

CREDIT RATING

Credit rating tools can be analysed in the following four ways:

❑ private versus public firms;
❑ models versus data;
❑ credit scoring versus market information based tools;
❑ models accessible over the web.

Tools for private and public firms

The credit assessment of private firms is very difficult owing to the non-availability of data. The most important differentiation between the public and private firm models is the absence of market-related information for private firms. Rating agencies have started providing web-based tools for assessing private firms. Initial private firm models have been based on public firms and these models are now being improved as banks subscribe to these models and start uploading private firm data.

CreditEdge

CreditEdge provides daily Moody's KMV EDF™ credit measures for public companies and financial analysis data from a variety of forward-looking, timely data from a variety of forward-looking, timely sources to support trading and investment decisions.

Models *versus* data

Rating agencies are now supporting industry- and region- or country-specific credit rating models for the mid-market or private firm segments. Models are available for North America, Europe and Japan for banks, insurance companies, utilities, etc. For Europe, S&P's have 110,000 small and medium private firms in their database. The rating agencies have started bundling the borrowers and the rating database within the model. The supplied model is calibrated or trained on the supplied data. The bundled data is either a corporate credit rating database or the private firm database collected from user institutions that used the model through the Internet. A toolkit is also bundled with the supplied model, and this is useful to extract data for calibrating and validating the internal model of, for example, the bank. So the supplied model works as a tool for benchmarking plus an external database source. The toolkit and the database are different for calibration and validation.

Data to calibrate models

These models are used to calibrate the internal models developed by banks. These models generate default rate, rating migration and default correlation statistics, tailored to the specific requirements of the bank. Data can be exported to Microsoft Excel for further modelling and benchmarking. The data generated can be used for computation of economic capital, unexpected loss and risk pricing. An example is CreditPro from S&P's. This model is useful for generating the following data for an industry sector, country or geographic region, or for a year, quarter or month:

❑ marginal default rates;
❑ cumulative default rates;
❑ transition matrices;
❑ company counts;
❑ default correlations.

Data to validate the model
Default Filter from S&P's is a toolset for default probability measurement with facilities for default probability model development, with ongoing data management, validation and stress testing. The toolset also helps to automate credit data centralisation, data spreading and data scrubbing. The utilities provided assist in reviewing, cleaning and storing historical financial, credit and macroeconomic data. These utilities provide ready-made solutions to common internal implementation hurdles. The model also helps in validating its accuracy with respect to defaulter, non-defaulter and portfolio default rate. The model provides extensive facilities for report generation and because of these capabilities the model is also used for validating portfolio models.

Credit scoring *versus* market-information-based tools
Within credit scoring there are various techniques (please refer to the chapter on credit scoring for a description of credit scoring techniques).

Moody's uses EDF as a credit measure for probability of default for public firms. For a firm the EDF is computed from three drivers: the market value of its assets, its volatility and its current capital structure. Moody's have also developed private firm EDF measures; however, such measures are based on research and benchmarking and not from the market data.

CreditMonitor
CreditMonitor from Moody's is an EDF-based model for public and private firms, which enables financial institutions and regulators to

fundamentally improve their credit risk processes. The model helps to understand why default probabilities have changed during the period by examining the details contributing to these changes. CreditMonitor allows users to measure and manage credit risk with greater accuracy.

Models accessible over the Web

Tools or models sold to the banks can be made available over the Internet. In these cases the vendor (which is in most cases a rating agency) maintains, updates and validates the model. The benefit to every one (the subscriber and the vendor) is that the borrower data used to compute the credit rating is used to further refine the model. This type of sharing is otherwise not possible. Depending upon the contract, the vendor may sanitise and share the data that it has uploaded for use in the model. Web-access models are mostly used for benchmarking, therefore generally, along with the model estimates for default, the historical default rates for similar and rated companies are also provided.

These tools are more like a service than a software tool, as the vendor provides ongoing data management, validation and stress testing and other consultancy services around these offerings. Default Filter is one such example from S&P's.

Using the Web is probably the only way to build and use models for private firms.

CreditModel

S&P's have used the "Proximal Support Vector Models" to develop CreditModel. According to S&P's, support vector models have successfully solved problems that have stumped neural network models. These tools are useful for:

❑ generating a quick indication of credit quality;
❑ credit scoring for loan origination, pricing, syndication and securitisation;
❑ sensitivity analysis;
❑ benchmarking internal ratings;
❑ measuring changes in the credit quality for a smaller period (such as every quarter – usually rating every quarter may be costly);
❑ providing two types of technological advantages

❑ the technology used is open, web and XML-based – this enables interfaces with external systems;

❑ the systems have built-in data cleansing and data centralisation tools, or provide interfaces to external data cleansing or data centralisation tools.

COLLATERAL MANAGEMENT SYSTEM

Collaterals reduce the risk either by reducing the exposure or by reducing the LGD. Since Basel 1996 amendment, financial institutions are managing pre-settlement risk on traded assets through margins (collaterals). Therefore, collateral management systems exists in banks since a decade or so. The new generation treasury systems and core banking product processors both provide built-in collateral management functionalities. However, the facilities provided therein may not be sufficient for the following reasons.

❑ Collateral management is very difficult if collaterals are managed in different product processors (treasury, core banking, trade finance, etc). Cross-collateralisation is extremely difficult to manage across the product processors.

❑ Each product processor provides collateral management facilities that are most suitable for the products and collateral that it is supporting. For example, core banking products generally do not provide automatic daily valuation (or on-line valuation) of financial assets.

Therefore, to consider collaterals in day-to-day business operations and decisions, banks have started implementing centralised collateral management solutions for trading and banking assets.

A good centralised collateral system

❑ supports collaterals for traded and banking assets;

❑ captures corporate, legal and reference data for a wide variety of credit support, collateral and margin agreements;

❑ supports collateral agreements;

❑ supports a wide variety of collateralisation terms agreed between multiple parties and multiple counterparties;

❑ captures the static and non-static trade, asset and market data;

❑ monitors exposures and collateral values between parties and counterparties;

❑ provides inter-party reconciliation of trades and collateral;
❑ monitors and tracks all collateral held, pledged and in-transit, and the related cash flows;
❑ supports re-hypothecation and collateral utilisation reporting, which is useful for cross-collateralisation;
❑ provides output to a central limit management system;
❑ supports reporting for various stakeholders;
❑ integrates with systems computing regulatory capital for credit risk.

LOSS GIVEN DEFAULT

Core LGD models are also not yet available commercially; however, rating agencies have attempted to provide some models around the data they have collected over the past two decades or so.

LGD data are very scarce. No bank has data sufficient for modelling. Every bank needs external data. S&P's and Moody's provide LGD models and LGD data offerings for the US. Both have compiled loss data in the US for various instruments and industries for the losses over the last two decades.

S&P's have compiled 3,195 defaulted bank loans and high-yield bonds, as well as other debt instruments, and Moody's have compiled data for 1,800 losses. Moodys cover 900 defaulted firms and S&P covers 700 defaulted firms. While Moody's offers a loss database bundled with a LGD model, S&P's offers a loss database bundled with a model or as a separate product.

These are generally regression models linking systemic, industry, firm-specific and collateral-related characteristics to the historical loss. These models bring value for the banks in the form of a calibrated model with LGD data. Therefore, in the absence of historical data for most of the markets, LGD models are not available outside the US.

Stochastic and reduced form models for recovery rate are yet to be available commercially.

REGULATORY CAPITAL COMPUTATION

The Basel Committee has undertaken five Quantitative Information Studies (QISs). Each of the studies have measured the regulatory capital according to the computation method as proposed by the respective consultative paper or the Accord at the

Table 1 Computation of capital

Computation of capital	Computation of provisions	Supervisory deduction
Tier I Capital	Analysis of general	Standard approach
Tier II Capital	provision	Foundation approach
Tier II Capital after general	❏ eligible for inclusion	Advanced internal-
provision	in the capital	ratings-based approach
Total Tier I and II Capital	❏ not eligible for	
Tier III Capital	inclusion in the capital	
Tier I Capital eligible for	Analysis of specific	
market risk charge	provision	
Eligible Tier III Capital		

time the studies were conducted. The committee has provided Microsoft Excel files to compute the regulatory capital for each of the asset classes. These excel sheets can be used, and are being used, by banks to compute regulatory capital.

Microsoft Excel-based software

Analysis of QIS3, QIS4 and QIS5 brings out the steps in the regulatory capital computation, as shown in Tables 1–2.

Exposure and RWA computations differ for each asset class while the broad approach is broadly the same for standardised, IRB foundation and IRB advanced approach. A generic approach is explained in Figure 3.

The following table explains the computation of regulatory capital for corporate assets under the foundation internal-ratings-based approach. For others refer to the Microsoft Excel file provided on the BIS Website.

To a large .extent, the information and capital treatments requested in QIS4 and QIS5 reflect provisions of Basel II on capital computation.

Analytical software for regulatory capital
Case study: Reveleus Basel II Analytics (www.reveleus.com). Reveleus™ is a suite of analytical applications for the financial services industry focused on the areas of risk management, customer insight and enterprise-wide financial performance. For regulatory capital computation, the exposure module integrates data from multiple product lines

Table 2 Computation of exposure and RWA for (a) banking book and (b) trading book

Drawn exposures	Undrawn lines	Repo style	OTC derivative exposure	Other off-balance sheet exposure
(a)				
Total Corporate Sovereign Retail SME Equity Purchased Receivable	Total corporate Sovereign Retail SME Equity	Total corporate Sovereign Retail SME Equity	Total corporate Sovereign Retail SME Equity	Total Corporate Sovereign Retail SME Equity
RWA for counterparty exposures	**RWA for specific charge**	**General market risk charge**	**RWA for the risk described in Part 4 of the Trading Book Paper**	**Settlement risk**
(b)				
Repo style OTC derivative	Standard method Internal model method	Market risk charge	Risk-weighted assets	Risk-weighted assets

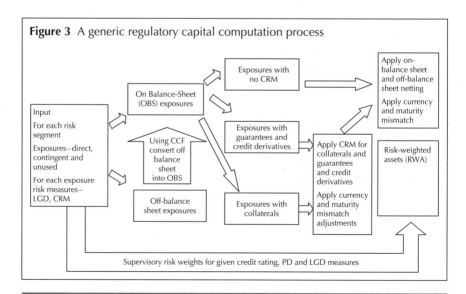

Figure 3 A generic regulatory capital computation process

within a bank to compute exposure at the obligor level or pool level within a risk segment class. Exposure for different products like loans, investment and loan equivalent exposures for off balance sheet contingent products and financial derivatives are computed. Regulatory capital computation under standard and internal foundation and internal advanced options are pre-built and delivered on an advanced data warehouse based framework. The collateral module provides mechanisms to apply various risk mitigation techniques including pooling of collateral to the pools of facilities. Maturity and currency mismatches are applied for assets with collaterals, guarantees and credit derivatives and for off balance sheet and on balance sheet netting arrangements. The risk-weighted assets for every exposure or group/pool of exposure before and after the application of adjustments for credit mitigation, credit protection and RWA are computed for on balance sheet assets, off balance sheet assets including derivative products, and foreign exchange products.

PORTFOLIO MANAGEMENT
Case study: a commercially available solution –
Portfolio Manager
Portfolio Manager by Moody's KMV is for measuring and managing portfolios. It enables users to measure portfolio-level credit risk, to compare that risk to the economic returns, to benchmark portfolio risk and to identify specific actions to improve portfolio performance. It allows users to assess the amount of risk that loans, bonds, credit derivatives and other instruments add to a portfolio. In addition, Portfolio Manager quantifies potential performance improvement and the diversification benefits of originations, hedges and trades. Portfolio Manager leverages the empirical credit migration data in Moody's KMV EDF™ credit measures and detailed data on corporate asset correlations. Portfolio Manager is a transparent solution that enables senior managers to examine the drivers of portfolio risk and capital and gain an intuitive understanding of the impact of rating, recovery, term, industry, size and other drivers of portfolio risk. The Portfolio Manager:

❏ generates distributions of portfolio loss and value using analytic and Monte Carlo-based risk measures for volatility and tail risk;

❏ measures risk–return characteristics of every credit exposure, including aggregates of relatively homogeneous assets such as credit card lines and mortgage portfolios;

❏ analyses the counterparty credit risk of products such as interest-rate swaps and other trading desk instruments;

❏ assesses each exposure in terms of stand-alone risk, portfolio risk contribution (taking into account its correlation with all other exposures) and risk-adjusted return;

❏ improves credit portfolio performance by
  ❏ identifying outliers and concentrations that impact performance at the exposure, obligor and industry levels,
  ❏ quantifying the diversification benefit of additional exposures,
  ❏ determining appropriate buy–sell quantities with the Trades Optimiser to enhance performance;

❏ provides insight into the contribution of individual positions to portfolio capital and other risk measures;

❏ uses empirical-based parameterisation drawn from the Moody's credit data sources.

DATA SOURCES

In addition to the data supplied by rating agencies, data are collected and supplied by central banks or institutions under the authorisation of central banks and other private institutions.

Credit registers

Credit registers being maintained by supervisors in many countries are another important source of data. Credit registers are maintained either by supervisors or by private organisations and they can play an important role in credit risk measurement. These registers were originally meant to be a key tool for the supervisory authorities to validate the internal models developed by the banks. The extensive use of credit register information aims to improve the identification and control of credit risks. They probably constitute one of the most important mechanisms available to address and eventually resolve certain validation and benchmarking challenges that Basel II has been posing.

There are various forms of credit registers in use. The most important characteristics of credit registers have been:

❑ compulsory reporting on credit granted above a certain limit by credit institutions;
❑ information on defaults, volume and category exposure, with the ability to distinguish between individuals and firms;
❑ the high level of confidentiality and the low cost of use by participating institutions;
❑ the types of instrument (trade credit, financial credit, lease, etc);
❑ the currency denomination and maturity;
❑ the existence of guarantees or collateral;
❑ the types of guarantor (government or credit institution) and the coverage of the guarantee;
❑ the amount drawn and undrawn of a credit commitment;
❑ whether the loan is current or past due (distinguishing between delinquency and default status);
❑ the industry and domicile of the borrower.

All these are important inputs for modelling, validation and benchmarking for PD, LGD, EAD, unexpected loss and portfolio risk.

Central financial statement database

In addition to the credit registers, supervisors or banking associations maintain central financial statement databases of borrowers. These databases focus on corporate entities. Database sizes vary substantially but, in general, data collection is designed to capture the largest 10–20% of counterparties, which account for 80–90% of the total exposure or business activity.

For data collection and dissemination, central banks use their own branch and network, or operate a joint venture with private banks.

Industry and macroeconomic data/information

Every country and industry has vendors supplying information and data on industry and macroeconomic information. Please refer to the BIS working paper "Credit Ratings and Complementary Sources of Credit Quality Information" for identifying various sources of data and information.

PREPARING FOR BASEL II

The data requirements are driven by the risk components that a bank intends to measure, the asset classes and the product features.

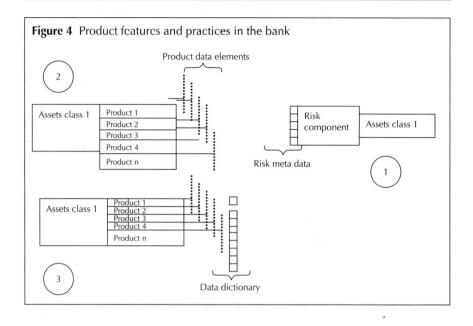

Figure 4 Product features and practices in the bank

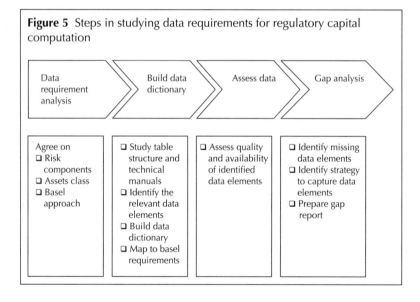

Figure 5 Steps in studying data requirements for regulatory capital computation

While the data requirements to measure risk components for a given asset class can be broadly said to be the same, the data availability will vary according to the product features and practices in the bank, see Figure 4.

This challenge can be addressed in two ways:

❏ building metadata for each of the risk component for each assets class;
❏ mapping the data elements from the products onto the risk metadata.

For a large bank, an intermediate step of building a data dictionary for the product data elements is the best practice to make the data identification exercise reusable. The data dictionary is then mapped onto the risk metadata. See Figure 5.

A typical challenge in the whole exercise would be inadequate documentation about the source systems and local definitions of data elements in the source systems.

Index